Inverse Synthetic Aperture Radar Imaging

Inverse Synthetic Aperture Radar Imaging

Principles, Algorithms and Applications

Victor C. Chen
Ancortek Inc. USA
US Naval Research Laboratory, retired

Marco Martorella
Department of Information Engineering
University of Pisa

SciTech Publishing
an imprint of the IET

Edison, NJ
theiet.org

SciTech PUBLISHING
an imprint of the IET

Published by SciTech Publishing, an imprint of the IET.
www.scitechpub.com
www.theiet.org

10 9 8 7 6 5 4 3 2 1

ISBN 978-1-61353-013-9 (hardback)
ISBN 978-1-61353-035-1 (PDF)

Typeset in India by MPS Limited
Printed in the UK by CPI Antony Rowe Ltd
Printed in the USA by Integrated Books International

Contents

Preface

Inverse synthetic aperture radar (ISAR) imaging has been the focus of many researchers and operational users in the last few decades. Starting from the 1970s, researchers began to investigate radar imaging of a target on a rotating turntable, and then a concept of inverse SAR (ISAR) imaging was proposed. Shortly after that, ISAR systems were built, and ISAR imaging of ships and aircraft was successfully demonstrated. The main idea behind the concept of ISAR imaging is to exploit Doppler information induced by the rotation of targets, which allows for echoes returned from different parts of the target to be distinguished from each other. This makes ISAR systems different from other imaging systems, including direct synthetic aperture radar (SAR) systems. Although SAR and ISAR share the same underlying concept of forming a synthetic aperture, they are substantially different in the way they process the radar received signal and generate a focused image of a target. The importance of ISAR is its potential to form radar images of targets without knowing their motion parameters. Such targets are generally referred to as *noncooperative targets*. Thus, ISAR is able to handle a class of scenarios that conventional SAR cannot.

Because of the nature of ISAR, most early works were of military interests and classified. Since late 1980s, some books that include ISAR topics have been published [1,2], and more and more publications on ISAR topics have begun to appear in conferences and journals.

This book is based on our 20 years' research work on ISAR imaging of moving targets and noncooperative target recognition. The goal of this book is to provide readers with a basic concept of the principles of ISAR imaging of noncooperative targets and a working knowledge of various algorithms of ISAR imaging formation and autofocus. Therefore, a large part of this book is devoted to the basic concepts and mathematical models of ISAR imaging, basic algorithms for motion compensation, ISAR image formation, and image autofocusing. Chapter 1 is an introduction to the concept of ISAR imaging. Chapter 2 discusses the basic principles of ISAR imaging, including ISAR scattering model, signal waveforms, point spread function, image projection plane, cross-range focus processing, and bistatic ISAR. Chapters 3 through 5 are dedicated to detailed algorithms of ISAR image formation, motion compensation, and autofocus.

The second part of the book discusses more recent research work on ISAR imaging. Details about ISAR signal processing issues are included in Chapter 6, whereas Chapter 7 deals with feature extraction. Chapter 8 describes a new technique of refocusing moving targets in SAR images. Chapter 9 discusses frequency modulated continuous wave (FMCW) ISAR imaging. Chapters 10 and 11 introduce applications of bistatic and polarimetric ISAR imaging, respectively. Finally, Chapter 12 includes five case studies of ISAR imaging applications.

MATLAB source codes are also provided and can be used to simulate and process ISAR data and to test some of the algorithms provided in the book. These source codes are given on an as-is basis, and no warranties are claimed. The contributors of the source codes will not be held liable for any damage caused. To request the supplementary files contact books@theiet.org.

We would like to express our sincere thanks to Dr. Elisa Giusti of the University of Pisa for developing part of the MATLAB source codes provided in this book. We would also like to thank Prof. Gang Li of Tsinghua University, Beijing, China, for his contribution to keystone method in Section 5.5 and related MATLAB source codes. We especially thank Prof. Anton Lazarov of the Burgas Free University, Bulgaria, for allowing us to introduce his interesting work on entropy-based ISAR imaging formation and related figures in Section 5.4. We are also grateful to the Italian Space Agency, the University of Adelaide, the Australian Defence Science and Technology Organisation, particularly Dr. Brett Haywood and Dr. Bevan Bates, the ONERA, French Aerospace Laboratory, particularly Dr. Luc Vignaud, the FHR-Fraunhofer Institute, specifically Stefan Brisken, and the South African Council for Scientific and Industrial Research, with special mention to Willie Nel, for providing the real data used in this book to show examples of ISAR images. We wish to express our thanks to Dr. Mark Davis for his constructive suggestions for improving this book and Ms Petrina Kapper for helping with English usage. We are grateful to the staff of IET/Scitech Publishing for their interest and support in the publication of this book.

References

[1] D.R. Whener, *High Resolution Radar*, 3d ed., Arthech House, 1995.
[2] D.L. Mensa, *High Resolution Radar Imaging*, Artech House, 1981.

List of Abbreviations and Symbols

Abbreviations

2-D - Two-dimensional

B-ISAR - Bistatic inverse synthetic aperture radar

BEM - Bistatically equivalent monostatic

CPI - Coherent processing interval

CSA - Chirp scaling algorithm

CW - Continuous wave

DA - Dual apodization

DFT - Discrete Fourier transform

DPCA - Displaced phase center antenna

DSA - Dominant scatterer autofocus

ECSA - Extend chirp scaling algorithm

EM - Electromagnetic

FFT - Fast Fourier transform

FMCW - Frequency modulated continuous wave

FT - Fourier transform

GMTI - Ground moving target indicator

GWN - Gaussian white noise

HS - Hot spot

I and Q - In-phase and quadrature phase

IC - Image contrast

ICBA - Image contrast-based autofocus

ICS - Inverse chirp scaling

IE - Image entropy

IEBA - Image entropy-based autofocus

IF - Intermediate frequency

IFT - Inverse Fourier transform
IOK - Inverse Ω-k
IP - Image peak
IPP - Image projection plane
IPFA - Inverse polar formatting algorithm
IRD - Inverse range-Doppler
ISAR - Inverse synthetic aperture radar
LFM - Linear frequency modulation
LOS - Line of sight
LPF - Low-pass filter
M-ISAR - Multistatic ISAR
MA - Multiple apodization
MC-ATWS - Maximum contrast based automatic time window selection
MIMO - Multiple-input, multiple-output
MPL - Maximum position locator
PFA - Polar formatting algorithm
PGA - Phase gradient autofocus
PPP - Prominent point processing
PRF - Pulse repetition frequency
PRI - Pulse repetition interval
PSF - Point spread function
RCMC - Range cell migration correction
RDA - Range-Doppler algorithm
RMC - Rotational motion compensation
RPF - Range perturbation function
RT - Radon transform
RVP - Residual video phase
SAR - Synthetic aperture radar
SCR - Signal-to-clutter ratio
SFCW - Stepped-frequency continuous wave
SLL - Sidelobe level
SNR - Signal-to-noise ratio
SP-PPP - Single polarization prominent point processing
SPWVD - Smoothed pseudo Wigner-Ville distribution
STAP - Space-time adaptive processing
STFT - Short-time Fourier transform
SVA - Spatially variant apodization

TFT - Time-frequency transform
TMC - Translational motion compensation
WLE - Window length estimator
WVD - Wigner-Ville distribution
ZDC - Zero-Doppler clutter
ZDC-ES - ZDC estimation and subtraction

Symbols

α - Azimuth angle of the target
β - Elevation angle; bistatic angle
γ - Angular acceleration
θ - Rotation angle
λ - Wavelength of EM wave
μ - Chirp rate
ρ - Reflectivity density
τ - Time delay
ϕ - Phase
ω - Angular frequency
Ω - Angular velocity
Δ_r - Range drift
a - Acceleration
A - Amplitude
$A(s)$ - Mean operation over the variable s
c - Speed of EM propagation
f - Frequency
f_c - Carrier frequency
f_D - Doppler frequency
Δf_D - Doppler resolution
k_x - Spatial wavenumber, corresponding to spatial Fourier transform of x
k_y - Spatial wavenumber, corresponding to spatial Fourier transform of y
l_{LOS} - Projected target size along the radar LOS
L - Antenna length
M - Number of rows in a 2-D data matrix
N - Number of columns in a 2-D data matrix
P - Point scatterer
p - Probability distribution function

R - Target range from radar
r - Scatterer displacement from the center of rotation
Δr_r - Range resolution
Δr_{cr} - Cross-range resolution
S - Transformed space from s-space
s - Signal
$S = [s_1, s_2, \ldots s_N]$ - Discrete signal representation
t - Time
T - Time interval
v - Velocity
X-Y-Z - Space-fix coordinate system
x-y-z - Body-fix coordinate system
x'-y'-z' - Reference coordinate system
$I(x, y)$ - Image intensity at (x, y)
$\mathbf{X} = (X, Y, Z)$ - Vector representation of the space-fix coordinate system
$\mathbf{x} = (x, y, z)$ - Vector representation of a point in a target respect to a target-fixed coordinates

CHAPTER 1

Introduction to ISAR Imaging

Radar is capable of detecting, tracking, and imaging targets with high accuracy at long range, day and night, and in all weather conditions. Mainly because of these reasons, radar has been widely used for military and civilian purposes, such as wide-area surveillance, air defense and weapon control, high-resolution radar imaging, remote sensing of environment, weather observation, air traffic control, vehicle collision avoidance and cruise control, harbor and river traffic surveillance, and industry automation [1–6].

Radar transmits electromagnetic (EM) waves to illuminate the environment and receives reflected echoes by objects. From the received signals, after signal and image processing, radar determines whether one or more targets are present and measures distances and velocities of the detected targets. Coherent radar maintains a high level of phase coherence with the transmitted signal. During a coherent processing interval (CPI), based on the Doppler effect, coherent radar can measure target's radial velocity, which is the velocity component along the radar target line of sight (LOS). Modern coherent radar can also generate high-resolution radar image of targets. Since the modern coherent radar appeared, radar imaging has become an attracting feature of modern radar that aims at providing more detailed geometric information about the target of interest. Conventional two-dimensional (2-D) radar images should be interpreted as projections of the target's EM reflectivity function onto an image plane that represents radar image of the target. Such a plane is usually defined through 2-D coordinates, namely, the range coordinate and the cross-range coordinate. The range direction is along with the radar LOS, whereas the cross-range direction is perpendicular to the LOS.

Like any other type of images, radar image is also characterized by its resolution, that is, the ability to separate two closely spaced scatterer centers along the range or the cross-range direction. The resolution along the range is addressed as range resolution and that along the cross-range as cross-range resolution.

Radar imaging always requires high resolution. As it will be clarified later on, fine range resolution can be achieved by selecting wide bandwidth of the signal waveform. However, fine cross-range resolution radar must have very large antenna aperture [7]. Because it is impossible to build and carry around such a large real antenna, a method to synthesize a large antenna aperture, called the synthetic aperture radar (SAR), was introduced, allowing a small antenna aperture to synthesize a large antenna aperture.

From synthesizing a large antenna aperture point of view, the SAR is based on the hypothesis that a target remains stationary during the data collection time interval while the radar antenna moves from one position to the next to synthesize a large size of antenna aperture. On the other hand, a larger antenna aperture can also be synthesized based on the hypothesis that the target moves while the radar remains stationary called the inverse synthetic aperture radar (ISAR).

In the following sections, we will follow this concept through and introduce the fundamental principle of ISAR and imaging algorithms, which will be the focus of this book.

1.1 SAR and ISAR Concepts at a Glance

SAR techniques are often used to synthesize a large antenna aperture from small apertures. Synthetic aperture processing coherently combines signals obtained from sequences of small aperture at different viewing angles to emulate the result that would be obtained using a large antenna aperture [3,7–14]. Coherent processing maintains the relative phases of successive transmitted signals and thus retains both the amplitude and the phase information about the target. A large aperture is synthesized by mounting the radar on a moving platform, generally an aircraft or a satellite, although other carriers, such as helicopters and ground-based rails, have been employed [15,16].

The most common modes operated in SAR are the strip-map mode and the spotlight mode. In the strip-map mode, the antenna beam bears on a fixed direction relative to the moving direction of the platform as illustrated in Figure 1.1a. When the platform moves, an area strip is swept over. If the antenna direction is off the perpendicular of the flight path, it is referred to as squinted strip-map SAR. The strip-map mode can generate wide-area maps of the terrain. The length of the imaged area is determined by the length of the data collection, and the azimuth resolution in the along-track direction is determined by the antenna length, that is, the dimension along the flight direction. It should be noted that, after correcting for the range migration, the *azimuth* can be identified as the *cross-range* and the azimuth resolution becomes the cross-range resolution.

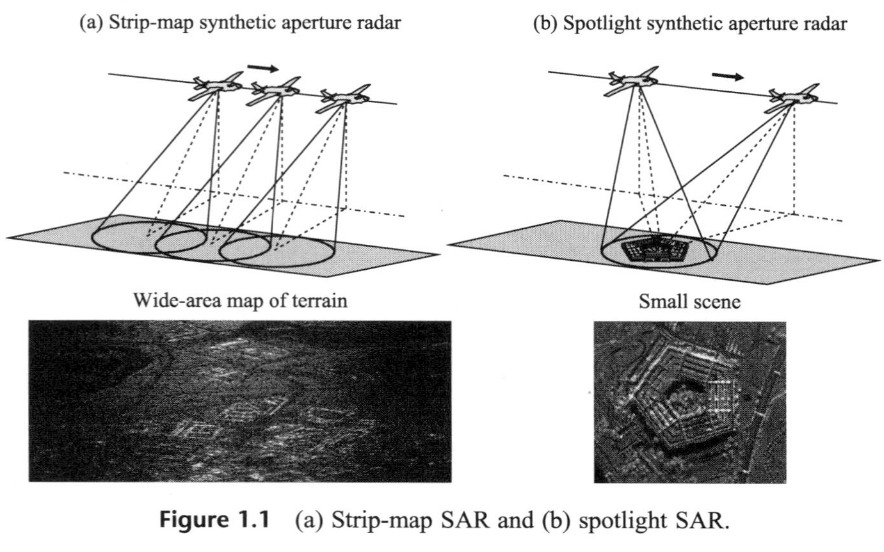

Figure 1.1 (a) Strip-map SAR and (b) spotlight SAR.

In the spotlight mode, the antenna has a narrower beamwidth and points to the same small patch area when the physical aperture moves through the length of the synthetic aperture, as shown in Figure 1.1b. This mode typically generates images of smaller scenes at a finer resolution. The azimuth resolution is determined by angular variation spanned during the formation of the synthetic aperture, and the size of the imaged area is determined by the antenna beamwidth.

To reconstruct the radar image of a target from a sequence of returned signals, it is required that each returned signal must be obtained with a different view of the target. Thus, a relative rotation between the radar and the target is necessary for creating different aspect angles of the target, such that each radar transmitted signal will capture a different view of the target.

Now, we should pay attention to the relative motion between the radar platform and the target. It means the motion is not necessarily produced by a moving platform. If the radar is stationary and the target moves with respect to it, an improvement in cross-range resolution can be obtained [17,18]. To emphasize the concept of *relative motion*, one could argue that whether the configuration called *stationary target and moving platform* or *moving target and stationary platform* really depends on where the reference coordinate system is placed: the former occurs by placing the reference system on the target and the latter by placing the reference system on the radar. According to this view, the differences between SAR and ISAR would depend only on where the reference system is placed. Such a concept is depicted in Figure 1.2, where a spotlight SAR configuration is transformed into an ISAR configuration by moving the reference system from the target to the radar.

Conversely, the same concept may be argued by starting with a controlled ISAR configuration, such as the turntable experiment. In the turntable configuration, the radar antenna is fixed on the ground (typically mounted on a turret), and the target is placed on a rotating turntable, as depicted in Figure 1.3a. By moving the reference system from the radar to the target, a circular SAR geometry can be enabled, as depicted in Figure 1.3b.

Where the reference system is placed determines the type of configuration (i.e., SAR or ISAR configuration), but in practice a subtle yet significant detail exists that substantially defines the difference between SAR and ISAR. This difference depends not on reference

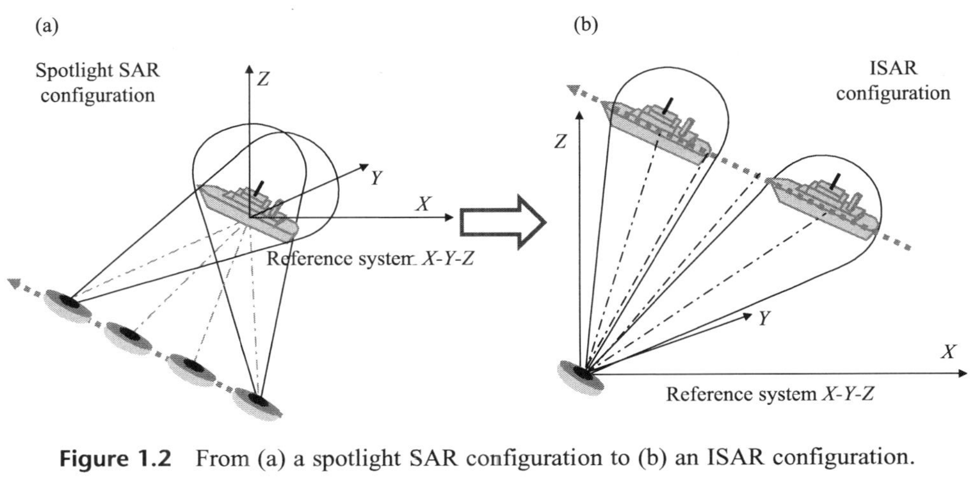

Figure 1.2 From (a) a spotlight SAR configuration to (b) an ISAR configuration.

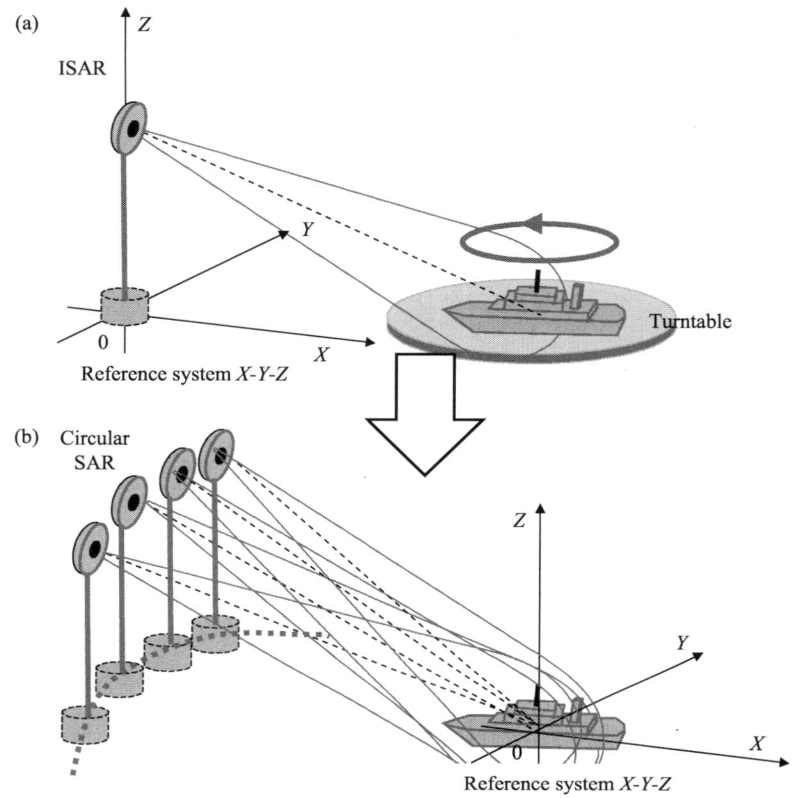

Figure 1.3 From (a) an ISAR configuration to (b) a circular SAR configuration.

system placement (this may be arbitrary to avoid affecting the system physically) but on the target's cooperation. To better explain this concept, one may place the reference system on the target. If such a target moves with respect to the radar and with unknown motion parameters, also called noncooperative target, the synthetic aperture formed during the coherent processing interval differs from that produced by an expected controlled motion of the platform, such as the turntable experiment. Thus, a SAR image formation that follows would be based on an erroneously predicted synthetic aperture and lead to the formation of a defocussed image. A pictorial example of a synthetic aperture formed by a noncooperative target's motion is shown in Figure 1.4a, where the unknown and noncooperative motion of the target generates an unpredictable synthetic aperture as shown in Figure 1.4b.

At this stage, it is necessary to say that the detail of whether the target is cooperative presents a number of issues. They are strictly related to the reason that ISAR is important to determine the relative motion between the radar and the target and, thus, to form radar images of noncooperative targets. In fact, the synthetic aperture formed by an unknown and arbitrary target's motion with respect to the radar is also unknown. This means that the positions of the synthesized aperture elements are not known a priori. Since SAR image processing is based on such knowledge, we should say that any SAR image formation algorithm may not be

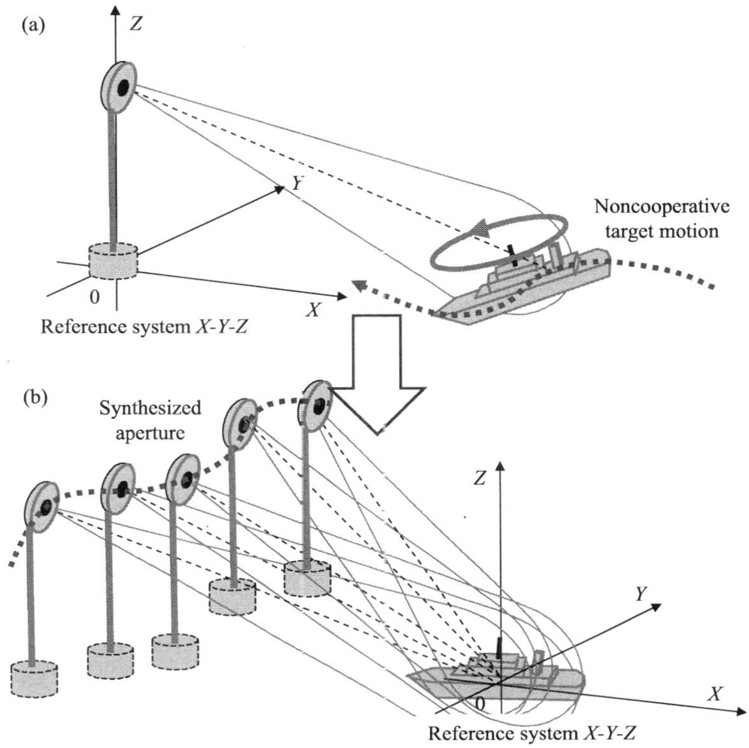

Figure 1.4 Synthetic aperture formed by a noncooperative target motion.

successfully applied to form a focused radar image of a noncooperative target. It is also worth pointing out that in cases where both the radar and the target are moving, if the target's motion is unknown by the radar, the SAR image processing would fail again. Generally speaking, in cases where the radar platform is stationary and a target's motion is noncooperative, ISAR image processing algorithms should be applied instead of using SAR algorithms.

It will be shown in the remainder of this book that ISAR processing is specifically designed to handle radar imaging of noncooperative targets. Because no prior information about the target's motion and, thus, about the synthetic aperture is taken into account, ISAR imaging can be regarded as a blind version of SAR imaging, where the target's motion parameters must be estimated in the process of forming a focused target image.

1.2 Brief Historical Overview of SAR and ISAR

We have to distinguish two starting points when considering the origins of radar imaging: one for SAR and one for ISAR. Although the two approaches to radar imaging have quite a lot in common, some significant differences still mark a line between them. As mentioned in [7], Carl Wiley was one of the first to conceive the SAR concept in 1951 [8], and in 1957 the Willow Run Laboratories of the University of Michigan for the US Department of Defense built the first classified operational system. NASA successfully built the first unclassified SAR systems in the 1960s. The first spaceborne SAR system, SEASAT-A, was launched in 1978. Although this was specifically designed for oceanographic purposes, it also produced important results in other fields, such as in ice and land studies. The results observed by SEASAT-A demonstrated the importance of radar imaging for observing the earth. Since then, several other spaceborne SAR systems now provide improved resolution, wider coverage, and faster revisit times. Several airborne SAR systems have also been developed to overcome limitations of spaceborne SAR systems, such as costs, revisiting time, and resolution. After NASA's first experiments in the 1960s, other important missions have been accomplished, such as SIR-A, SIR-B, and SIR-C, which were completed in 1981, 1984, and 1994, respectively. SAR signal processing algorithms were rapidly developed since the first studies in the early 1950s, and most of the SAR concepts and processing algorithms are collected in a number of books, such as [7,8,10–13,19–21].

ISAR had its beginnings when a few researchers introduced the concept of radar imaging of rotating objects using a stationary radar [9,17,18,22,23]. The main insight in their work was to exploit Doppler information generated by the rotation of an object to distinguish echoes returned from different parts of the object along the cross-range direction that is perpendicular to the radar LOS. The Doppler information along with the range information represented by the time delay of the echoes produces a 2-D radar image mapped onto the image plane that represents projections of the object's reflectivity function.

From an experimental point of view, the US Naval Research Laboratory generated continuous sequence of ISAR images in flight in the late 1970s. From 1984 to 1988, the Naval Research Laboratory demonstrated ISAR imaging using the APS-116 radar with the integration of P-3, and Texas Instruments also integrated the S-3B with the APS-137 radar.

1.3 Fundamentals of ISAR Imaging

In this section, we focus on some ISAR imaging fundamental concepts before introducing all its details. Some of them—such as coherency, Doppler effect, and round-trip delay measurements—are common to other fields of radar but are introduced to provide a review and establish a notation that will be used throughout this book. Other concepts more strictly related to ISAR will also be introduced to provide some initial understanding of ISAR imaging.

1.3.1 Doppler Effect

Radar transmits a signal to a target and receives an echo signal from it. Based on the time delay of the received signal, radar can measure the range of the target. If the target is moving, the frequency of the received signal will be shifted from the frequency of the transmitted signal. This is known as the Doppler effect. The Doppler frequency shift is determined by the radial velocity of the moving target, that is, the velocity component in the direction of the LOS.

The Doppler frequency shift is usually measured in the frequency domain by taking the Fourier transform of the received signal. In the Fourier spectrum, the peak component indicates the Doppler frequency shift induced by the radial velocity of the target's motion. To accurately track the phase information in the radar received signals, the radar transmitter must be driven by a highly stable frequency source to fully maintain phase coherency.

In typical radar applications, the velocity of a target, v, is much slower than the speed of wave propagation c (i.e., $v << c$). In monostatic radar systems, where the same antenna is used to transmit and receive, the round-trip distance traveled by the transmitted signal is twice the distance between the antenna and the target. In this case, the radial velocity of the target with respect to the radar counts for half its value in the generation of the Doppler effect as the EM wave travels from the radar to the target and back. The Doppler frequency generated can be expressed as

$$f_D(t) = -f\left[\frac{2v(t)}{c}\right], \tag{1.1}$$

where f is the transmitted carrier frequency, and $v(t)$ is the radial velocity of the target with respect to the radar. When a target is moving away from the radar, its velocity is defined as positive. As a consequence, the Doppler shift becomes negative.

We will now assume a model for the radar received signal as follows:

$$S_R(t) = A\cos[2\pi(f + f_D)t] = A\cos[2\pi ft + \varphi(t)], \tag{1.2}$$

where A is the amplitude of the received signal, and $\varphi(t) = 2\pi f_D t$ is the phase shift on the received signal due to the target's motion. In the following, we will consider a coherent demodulator with in-phase (I) and quadrature (Q) channels, which allows the full phase information to be retrieved from the signal. By mixing with the transmitted signal

$$S_T(t) = \cos(2\pi ft), \tag{1.3}$$

the output of the I-channel mixer is

$$S_R(t)S_T(t) = \frac{A}{2}\cos[4\pi ft + \varphi(t)] + \frac{A}{2}\cos\varphi(t). \tag{1.4}$$

After low-pass filtering, the I-channel output becomes

$$I(t) = \frac{A}{2}\cos\varphi(t). \tag{1.5}$$

By mixing with the 90° phase shifted transmitted signal

$$S_T^{90^\circ}(t) = \sin(2\pi ft), \tag{1.6}$$

the output of the Q-channel mixer is

$$S_R(t)S_T^{90^\circ}(t) = \frac{A}{2}\sin[4\pi ft + \varphi(t)] - \frac{A}{2}\sin\varphi(t). \tag{1.7}$$

After low-pass filtering, the Q-channel output becomes

$$Q(t) = \frac{A}{2}\sin\varphi(t). \tag{1.8}$$

By combining the I and Q outputs, a complex signal can be formed as follows:

$$S_D(t) = I(t) + jQ(t) = \frac{A}{2}\exp[-j\varphi(t)] = \frac{A}{2}\exp(-j2\pi f_D t) \tag{1.9}$$

Thus, the Doppler frequency shift f_D can be estimated from the complex signal, $s_D(t)$, by using a frequency measurement tool. Because frequency is determined by time derivative of the phase function, the phase difference, $\varphi(t)$, between the received and the transmitted signal is used to estimate the Doppler frequency shift, f_D, of the received signal:

$$f_D(t) = \frac{1}{2\pi} d\varphi(t)/dt. \tag{1.10}$$

From the estimated Doppler frequency, the radial velocity of the target is determined by

$$v = \frac{\lambda}{2} f_D = \frac{c}{2f} f_D, \tag{1.11}$$

where $\lambda = c/f$ is the wavelength of the transmitted frequency. Here, the dependence on the time variable, t, has been omitted for the sake of simplicity.

1.3.2 Motion-induced Radar Imaging

The relative rotation between the radar and a target makes it possible to view the target at different aspect angles. The image generated by the relative rotation is called the motion-induced image and is represented in the joint range and Doppler domain. The concept of range-Doppler imaging was proposed in [9,17,18,20–23] when dealing with radar imaging of a rotating target. The global or space-fixed system (X, Y) and the local or body-fixed system (x, y), which is rigidly fixed in the body, are the two coordinate systems commonly used to describe a target with motion as shown in Figure 1.5. For simplicity, we draw it in only the 2-D X-Y domain.

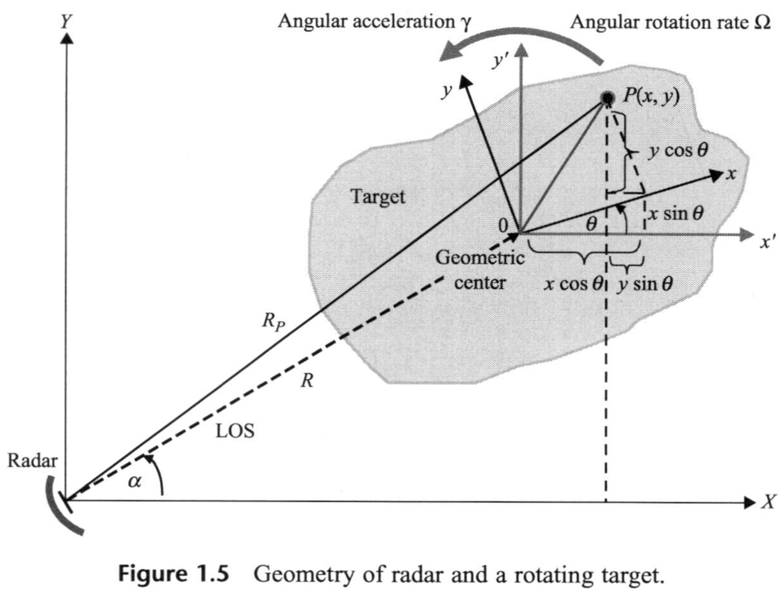

Figure 1.5 Geometry of radar and a rotating target.

We will assume that the radar is at the origin of the global system and the target's origin is located at its geometric center and rotates about the z-axis of the local body-fixed system with an angular rotation rate equal to $\Omega(t) = \Omega + \gamma \cdot t$, where Ω is the angular velocity induced rotation rate, and γ is the angular acceleration. It is worth pointing out that the assumption of constant angular acceleration is a reasonable approximation that holds in the case of short observation time intervals. To describe the rotation of the target, another reference coordinate system (x', y') is introduced, which is parallel to the global system (X, Y) with its origin at the origin of the target body-fixed system. Given the target range, R, by simple geometric calculation, the distance from the radar to a scatterer, $P(x, y)$, on the target becomes approximately

$$R_P(t) \cong R + x\cos[\theta(t) - \alpha] - y\sin[\theta(t) - \alpha], \tag{1.12}$$

and by Taylor expansion

$$\theta(t) = \theta_0 + \Omega t + \frac{1}{2}\gamma t^2, \tag{1.13}$$

where θ_0 is the initial rotation angle in the reference (x', y') system, and α is the azimuth angle of the target.

The baseband version of the radar signal backscattered from the scatterer, P, is a function of R_P as follows:

$$S_P(t) = \rho(x_P, y_P)\exp[-j2\pi f[2R_P(t)/c]], \tag{1.14}$$

where $\rho(x_P, y_P)$ is the reflectivity density function of the point scatterer, P, at (x_P, y_P), and $R_P(t)$ is a function of time.

The Doppler shift induced by the target rotation can be obtained by time derivative of (1.12) with the Taylor expansion of the rotation angle function:

$$\begin{aligned} f_D(t) &= \frac{2}{\lambda}\frac{dR_P(t)}{dt} \\ &= \frac{2}{\lambda}\{-[x\sin(\theta_0 - \alpha) + y\cos(\theta_0 - \alpha)]\Omega - [x\cos(\theta_0 - \alpha) - y\sin(\theta_0 - \alpha)]\Omega^2 t\}. \end{aligned} \tag{1.15}$$

The quadratic term of the Doppler shift is time varying even if the angular rotation rate, Ω, is a constant.

Based on the returned signal from a single point scatterer, the returned signal from the target can be represented as the integration of the returned signals from all scatterers that belong to the target:

$$S_R(t) = \iint_{-\infty}^{\infty} \rho(x, y)\exp\left[-j\frac{4\pi f}{c}R_P(t)\right]dxdy. \tag{1.16}$$

For a target that has both translational and rotational motion and assuming that the azimuth angle α is equal to zero, the range of a point scatterer at (x, y) becomes

$$R_P(t) = R(t) + x \cdot \cos\theta(t) - y \cdot \sin\theta(t). \tag{1.17}$$

The radar returned signal can be rewritten as

$$S_R(t) = \exp\left[-j4\pi f\frac{R(t)}{c}\right]\iint_{-\infty}^{\infty}\rho(x,y)\exp\{-j2\pi[xf_x(t) - yf_y(t)]\}dxdy, \tag{1.18}$$

where $f_x(t)$ and $f_y(t)$ can be seen as two components of the frequency, f, and defined by

$$f_x(t) = 2f\cos\theta(t)/c \tag{1.19}$$

and

$$f_y(t) = 2f\sin\theta(t)/c. \tag{1.20}$$

The radar returned signal (1.18) can also be expressed in terms of the wavenumber $k = 2\pi f/c$:

$$S_R(t) = \exp\left[-j4\pi f\frac{R(t)}{c}\right]\iint_{-\infty}^{\infty}\rho(x,y)\exp\{-j2[xk_x(t) - yk_y(t)]\}dxdy, \tag{1.21}$$

where the two wavenumber components are determined by

$$k_x(t) = 2\pi f\cos\theta(t)/c = k\cdot\cos\theta(t) \tag{1.22}$$

and

$$k_y(t) = 2\pi f\sin\theta(t)/c = k\cdot\sin\theta(t). \tag{1.23}$$

During the entire image processing time interval or the CPI, if the range function, $R(t)$, is known exactly, the phase term due to the target's motion, namely, $\exp[-j4\pi fR(t)/c]$, can be perfectly removed by multiplying the demodulated received signal by the complex conjugate of the same term. Such an operation is typically addressed as *radial motion compensation*, and the signal obtained after such an operation is usually named *motion compensated signal*. Therefore, the reflectivity density function, $\rho(x, y)$, of the target can be obtained by simply taking the inverse Fourier transform of the motion compensated baseband signal:

$$\rho(x,y) = IFT\{S_R(t)\exp[j4\pi fR(t)/c]\}. \tag{1.24}$$

When the radar transmits a sequence of N signals (or pulses) and the received signal of each transmitted signal has M time samples, the radar-collected raw data can be rearranged as a two-dimensional $M \times N$ matrix in delay time (also called fast time), t_m, and pulses (also called slow time), t_n, domain, where $m = 1, 2, \ldots, M$ and $n = 1, 2, \ldots, N$. Because the delay time is related to the range, the raw data are also said to be in the range and pulses domains. Figure 1.6 illustrates how 2-D radar raw data are arranged in the range (fast-time or time-delay) and the pulses (slow-time) domains.

The process of estimating the target's motion and removing the phase term associated with the target's radial motion is called *range alignment* or *range tracking*. This is a fundamental step in the standard motion compensation procedure, also called *coarse motion compensation*. After the phase term is removed, the inverse Fourier transform may be used to reconstruct the reflectivity density function of the target.

To use the Fourier transform effectively, the scatterers must remain in their range cells during the entire CPI, and their Doppler frequency shifts must keep constant. If the scatterers drift out of their range cells or their Doppler frequency shifts are time varying, the image

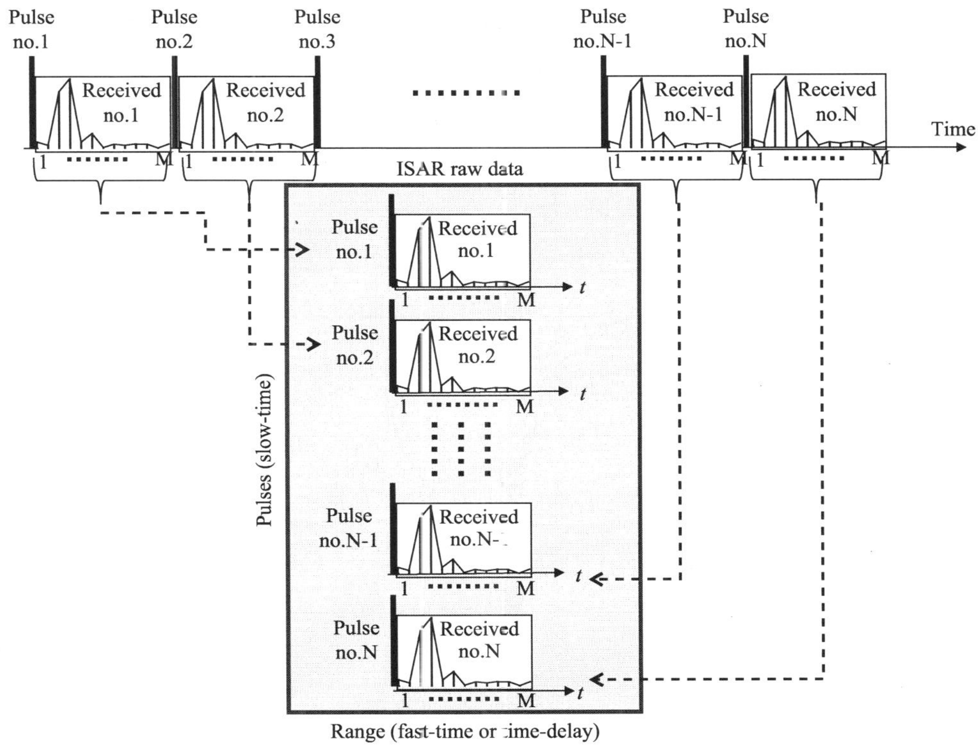

Figure 1.6 ISAR raw data arranged in two dimension to reflect the concept of fast and slow time.

reconstructed using the Fourier transform will be smeared. It should be pointed out even at this stage that range tracking processing is not enough to produce time-invariant Doppler shifts and that a phase correction is needed before applying the Fourier transform. Thus, phase correction, which can be called Doppler tracking, must be applied to obtain a complete phase compensation and hence constant Doppler frequency shifts. Range tracking and Doppler tracking become the standard procedures of motion compensation.

Several algorithms have been developed to deal with the problem of motion compensation, and the aforementioned concepts are meant to provide some insight toward understanding ISAR imaging. Chapters 4 and 5 will provide sufficient details to cover the literature's most commonly used techniques.

Figure 1.7 illustrates the concept of an ISAR imaging system. The radar transmits a sequence of N pulses. For each transmitted pulse, the total number of range cells, M, is determined by the maximum range covered and the size of range cell or bin. Conversely, the total number of pulses, N, during a given CPI determines the Doppler resolution.

After pulse compression and I-Q downconversion in the radar receiver, the baseband complex I and Q data are organized into an $M \times N$ two-dimensional complex matrix, which consists of N range profiles, where each range profile has M range cells. Range profile represents the energy distribution of the received signal as a function of range. Magnitude peaks of the range profile indicate range locations of dominant scatterers. In a range profile,

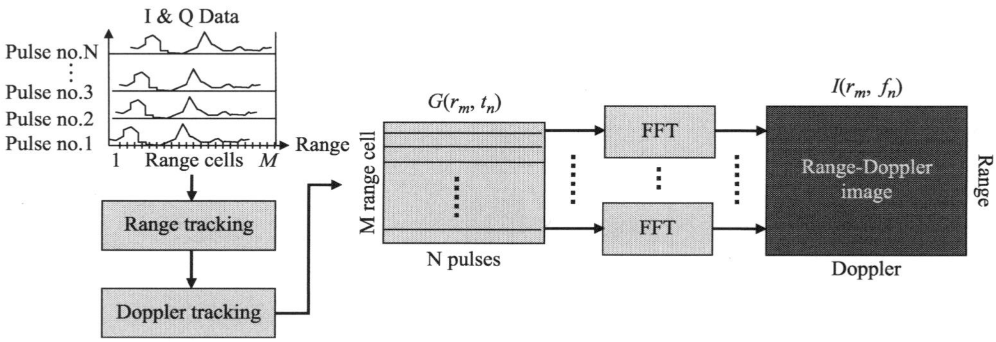

Figure 1.7 The concept of an ISAR imaging system.

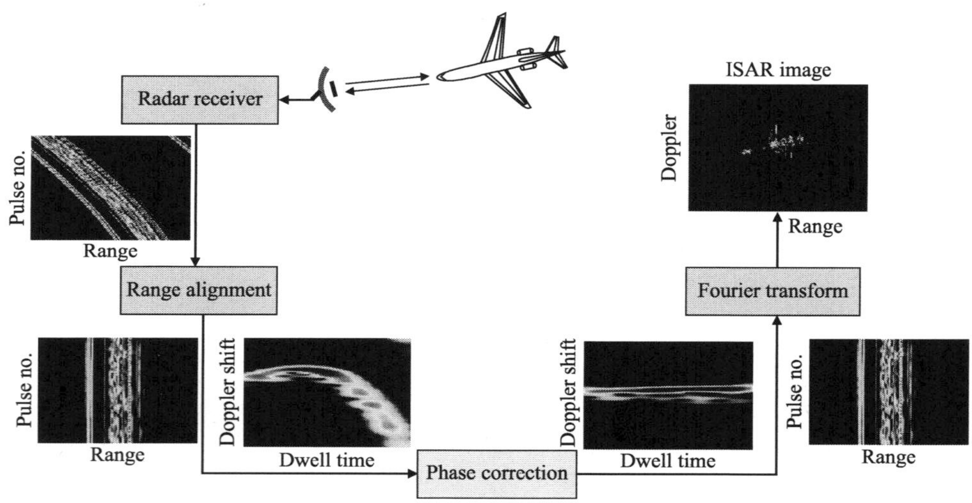

Figure 1.8 Standard motion compensation diagram for generating ISAR image of targets.

if one or more magnitude peaks are detected at certain range cells, it indicates that one or more dominant scatterers have been detected at the corresponding locations. Then, by applying range tracking and Doppler tracking, the aligned-range profiles, $G(r_m, t_n)$, can be obtained, where $m = 1, \ldots, M$, and $n = 1, \ldots, N$, as shown in Figure 1.7.

The Fourier-based image formation takes the Fourier transform at each range cell along N pulses and generates an N-point Doppler spectrum called the Doppler profile. Finally, by adjoining the Doppler spectra for each of the M range cells, the $M \times N$ range-Doppler image is formed:

$$I(r_m, f_n) = FFT_n\{G(r_m, t_n)\}, \tag{1.25}$$

where FFT_n denotes the fast Fourier operation with respect to n. The radar range-Doppler image, $I(r_m, f_n)$, represents the target's reflectivity, $\rho(x, y)$, mapped onto the range-Doppler plane.

Figure 1.8 illustrates the standard motion compensation diagram and shows results of the range tracking and Doppler tracking. The range tracking can keep scatterers in their range

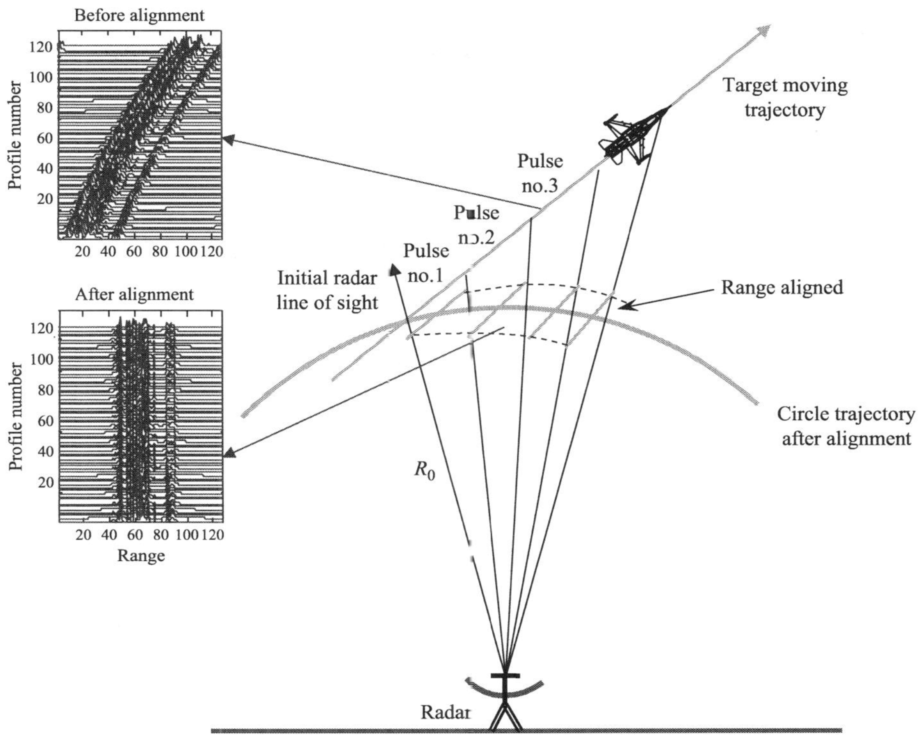

Figure 1.9 Illustration of the perfect circle of a constant range after alignment.

cells and Doppler tracking by phase correction produces constant Doppler frequency shifts as shown in the figure. Therefore, after motion compensation all scatterers on the target appear to be moving with a constant velocity due to its constant Doppler frequency shift and along a perfect circle due to its constant range as illustrated in Figure 1.9. A commonly used method for range tracking is the cross-correlation, which finds misaligned range cells with respect to a reference range profile and then performs the range alignment procedure for all range profiles. The Doppler tracking is performed by a phase compensation method: (1) search for one or several reference range cells by using criteria such as minimum variance; (2) take conjugate phase at the reference range cells; and (3) perform a phase correction for all range cells using the conjugate phase.

If a target is moving smoothly, standard motion compensation is good enough to generate a focused image of the target using the Fourier transform. However, when a target exhibits complex motions (e.g., rotation, acceleration, or jerking), the standard motion compensation algorithm is not sufficient to generate an acceptable image for viewing and analysis. In this case, more sophisticated motion compensation algorithms are needed, such as polar formatting algorithm (PFA).

Polar formatting, which can correct rotational motion for individual scatterers, is required to resample the data such that the sample points on the polar sampling grid are conformed into the desired sample points on a rectangular sampling grid [23,24]. Additionally, to perform polar formatting some initial kinematic parameters of the target are required. (More details about PFA will be provided in Chapter 4 Section 4.2).

If the sophisticated motion compensation is still not sufficient, these individual scatterers may still drift through their range cells, and their Doppler frequency shifts may still be time varying. Thus, if a Fourier transform is simply applied, the resulting image can still be unfocused and smeared. In these cases, the conventional Fourier transform can be replaced by a time-frequency transform [25]. Because of the time-varying behavior of the Doppler frequency shift, a high-resolution time-frequency transform should be applied to perform Doppler processing and efficiently solve the problem of image smearing. Thus, the smearing effect caused by time-varying Doppler frequency shifts can be mitigated in the resulting ISAR image. (Details of time-frequency transform-based ISAR image formation are provided in Chapter 3 Section 3.2).

1.3.3 Range-Doppler Interpretation of ISAR Imaging

To clarify how range and Doppler measurements are related to ISAR imaging, we can use the same signal model introduced in Section 1.3.2. In particular, we can express the model of the received signal after demodulation and radial motion compensation as follows:

$$S_R(t, f) = \iint_{-\infty}^{\infty} \rho(\tau, f_D) \exp\{-j4\pi f/c \cdot [x \cos\theta(t) - y \sin\theta(t)]\} dx dy, \tag{1.26}$$

where $\rho(x, y)$ is replaced by $\rho(\tau, f_D)$ as described next.

On the left-hand side of (1.26), a frequency variable, f, represents the dependence of the received signal on the transmitted frequency. Assuming the angle variation is near zero, that is, $\theta(t) = \Omega t \to 0$, (1.26) can be approximated as follows:

$$S_R(t,f) = \iint_{-\infty}^{\infty} \rho(\tau, f_D) \exp\{-j4\pi f/c \cdot [x - y\Omega t]\} dx dy, \tag{1.27}$$

where $\cos\theta(t) \cong 1$, and $\sin\theta(t) \cong \Omega t$. To represent the time delay, τ, and Doppler shift, f_D, coordinates, (1.27) can be rewritten as follows:

$$S_R(t,f) = \iint_{-\infty}^{\infty} \rho(\tau, f_D) \exp\{-j2\pi[f\tau - t f_D]\} d\tau df_D, \tag{1.28}$$

where

$$\tau = 2x/c \tag{1.29}$$

represents the round-trip delay expressed as a function of the range, x, and

$$f_D = 2f\Omega y/c \tag{1.30}$$

represents the Doppler frequency shift expressed as a function of the cross-range, y. We should point out that $\rho(\tau, f_D)$ in (1.28) represents the reflectivity function projected onto the range-Doppler image plane.

Then, the range-Doppler image can be obtained by simply taking the inverse Fourier transform (IFT) of (1.28) with respect to t and f as follows:

$$\rho(\tau, f_D) = IFT\{S_R(t, f)\}, \tag{1.31}$$

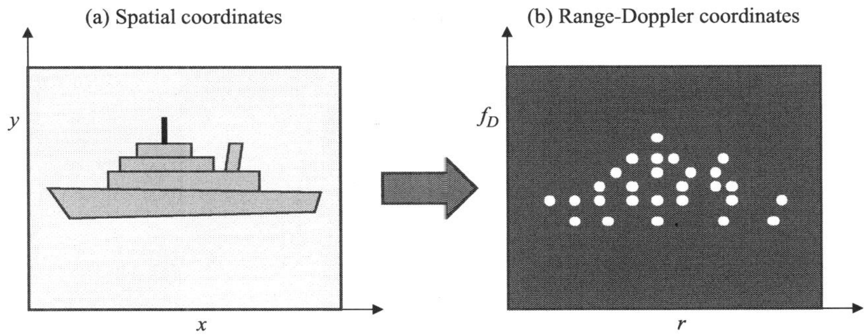

Figure 1.10 Range-Doppler scaling from the spatial coordinates.

where $\rho(\tau, f_D)$ is the target's reflectivity function mapped onto the range-Doppler plane, which is represented by the radar range-Doppler image $I(r_m, f_n)$ in (1.25).

By looking at (1.31), we notice that the range-Doppler image of the target is actually scaled in terms of delay time and Doppler shift. An illustration of the range-Doppler scaling from the spatial coordinates is shown in Figure 1.10.

To scale the ISAR image in range and cross-range coordinates, a scaling operation must be performed by applying the inversion of equations (1.29) and (1.30). It appears clearly that (1.29) can be inverted easily. Conversely, to be able to invert (1.30), we must know the value of Ω, which, as it will become obvious later, is typically unknown because it depends on the target's motion. This problem is known as the cross-range scaling problem. (It will be discussed in detail in Chapter 7 Section 7.1.)

1.4 ISAR Image Resolution

ISAR image resolution is the ability to resolve separated scatterers in the ISAR image. To derive the resolution, we will refer to the range-Doppler interpretation in Section 1.3.3 as it provides simple means to do it. More specifically, we will separate the two dimensions, namely, range and Doppler, to easily calculate the two resolutions. This will allow us to conclude that 2-D image resolution is the Cartesian product of the two resolutions.

1.4.1 Range Resolution

Typically, in modern coherent radar systems a concept of range compression is used to achieve high range resolution without transmitting very short pulses [1,6,14]. The idea behind the concept is that if a signal with wide bandwidth is transmitted, it can be compressed at the receiver by using a suitable signal processing method called the *matched filter*. The result of the compression is that the resolution in the delay time, τ, is inversely proportional to the bandwidth of the transmitted signal, that is, $\Delta\tau \simeq 1/BW$, where BW is the bandwidth of the transmitted signal. Thus, the range resolution is obtained by inverting (1.29) and renaming x with Δr_r in

$$\Delta r_r = c/(2 \cdot BW). \tag{1.32}$$

1.4.2 Doppler and Cross-range Resolution

Doppler resolution refers to the ability to distinguish two sinusoidal components in the Doppler frequency domain. If two sinusoid signals are observed during a limited time interval, T, they can be resolved only when the difference of the two oscillating frequencies is greater than $1/T$. By applying the concept to the Doppler resolution,

$$\Delta f_D = 1/T. \tag{1.33}$$

As a consequence, the cross-range resolution can also be obtained. In fact, by combining (1.30) and (1.33) and renaming y with Δr_{cr} we have

$$\Delta f_D = 2f\Omega \cdot \Delta r_{cr}/c = 1/T. \tag{1.34}$$

By inverting (1.34), we finally obtain an expression for the cross-range resolution:

$$\Delta r_{cr} = c/(2f\Omega T). \tag{1.35}$$

Based on (1.35), if $f = 10$ GHz and $T = 1$ sec, to achieve 0.3 m cross-range resolution the target angular rotation rate must be 2.86°/s. A longer integration time may provide higher cross-range resolution but causes phase-tracking errors and makes Doppler smearing. Because the Doppler resolution, Δf_D, is inversely proportional to the image integration time, T, the cross-range resolution is proportional to the Doppler resolution with a scaling factor: $\Delta r_{cr} = \left[\frac{c}{2\Omega f}\right] \Delta f_D$, where $2\Omega f/c$ is called the scaling factor.

It should also be remarked that the cross-range resolution depends on the target's motion through the parameter Ω; therefore, it is not known a priori. This is typically a problem as ISAR resolution performances are not entirely predictable and depend on the radar-target dynamics, which are not controlled by the radar.

1.5 Main Differences Between ISAR and Optical Images

It is important at this stage to point out differences between ISAR and optical images, such that we will be not confused about the interpretation of ISAR images with respect to optical images. There are, in fact, a number of differences between ISAR and optical images although the following two differences are quite significant to be discussed earlier rather than later. They are the geometrical and radiometric differences between them as follows.

1.5.1 Geometrical Differences

Geometrically, the motion-induced ISAR image is different from the optical image, such as simply those observed by the human eye. This is because the ISAR image not only does not look like the photo image of a target but also is the result of a completely different projection with respect to the optical image. Figure 1.11 illustrates the different view angles between an ISAR image and an optical image observed by a human observer. To have the same view of a target, the view angle from a human observer must be 90° off the view angle of the radar. In ISAR, the image projection plane determines the view of the ISAR imaging system. As shown in Figure 1.11, the profile (or side) view of a ship can be observed from a human eye that looks at the side of the ship. However, the same profile view of the ship can be produced

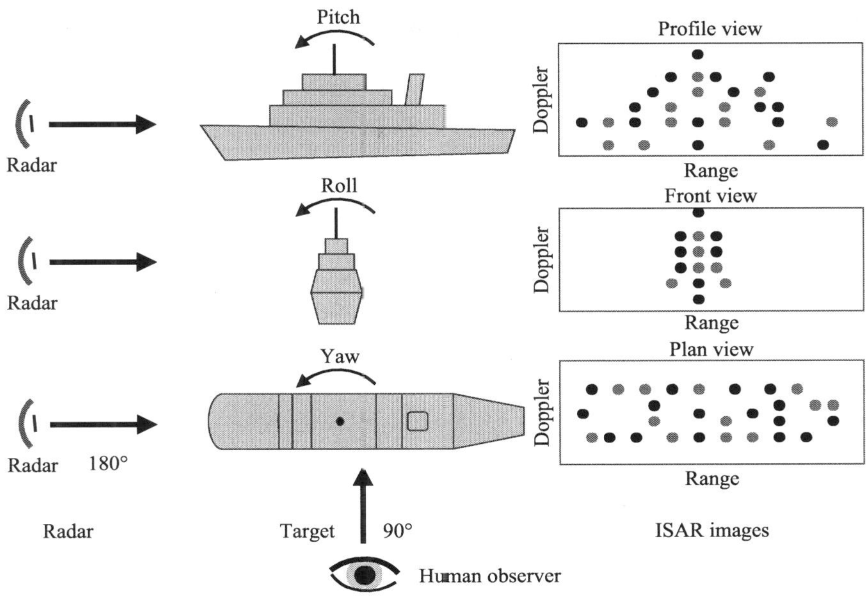

Figure 1.11 Profile view, front view, and plan view of a ship image from ISAR and from a human eye.

only by a pure pitch motion of the ship and imaged by an ISAR looking directly at the front of the ship (LOS direction aligned with the bow–stern ship line) as illustrated in Figure 1.11. The front view of the ship can be observed by a human eye that looks at the front of the ship along the bow–stern ship line. Then, the same front view of the ship can be produced only by a pure roll motion of the ship and imaged by an ISAR looking at the side of the ship (broadside) as illustrated. Finally, the plan (or top) view of the ship can be observed by a human eye that looks at the ship from the top. The same plan view of the ship can be produced only by a pure yaw motion of the ship and imaged by an ISAR positioned at any point on the sea surface. In general, if the ship has roll, pitch, and yaw motion simultaneously, the image becomes a perspective view of the ship, which does not necessarily correspond to any among top, side, or front views.

1.5.2 Radiometric Differences

Another important difference between ISAR and optical images is related to the different types of EM waves, that is, microwaves captured by the radar and light waves observed by an optical sensor or the human eye. Although in both cases the carrier waves are EM waves, the frequency of such carrier is drastically different. For the ISAR case, EM waves are in the region of the microwave (10^9 to 10^{10} Hz). For the optical case, light waves are in the visible region (10^{14} to 10^{15} Hz). The way that objects appear in different parts of the spectrum depends on the frequency and the object characteristics. Thus, if the wavelength of the EM waves is smaller than the feature of an object, the object will be observable. For this reason, by using an EM wave at a high enough frequency an object with a certain size should be

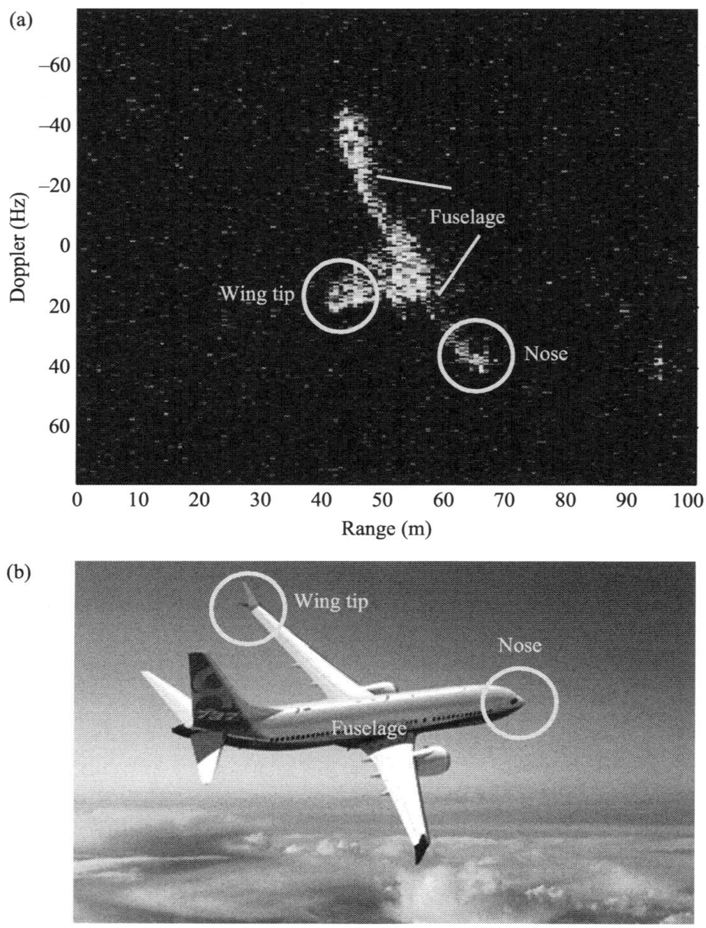

Figure 1.12 Boeing 737 (a) ISAR image and (b) optical image.

observable up to some level of detail. For the same reason, an optical microscope cannot observe an object with a size smaller than the optical wavelength.

To illustrate the difference between ISAR images and optical images, Figure 1.12 shows the ISAR and optical images of a Boeing 737. Because of the aforementioned radiometric differences, the ISAR image is quite different from the optical image. However, some important features in the object, such as nose, wing tips, and fuselage of an airplane, can still be recognized from the ISAR image shown in Figure 1.12a.

1.6 Conclusions

Introducing a book is never easy. The aim of this introductory chapter is to cover some fundamental concepts on ISAR and to highlight some characteristics that make these systems unique, especially related to SAR. Although SAR and ISAR share the same underlining concept of forming a synthetic aperture, they are substantially different in the way they process the radar received signal and generate a focused image of a target.

Moreover, some basic concepts such as Doppler effect and image resolution have been introduced along with that of motion-induced radar imaging. Detailed signal modeling and image formation algorithms have been left out since they deserve focused attention, and they will be treated in the next few chapters.

References

[1] Skolnik, M.I., *Radar Handbook*, 3rd ed., McGraw-Hill, 2008.

[2] Ulaby, F.T., R.K. Moore, A.K. Fung, *Microwave Remote Sensing: Active and Passive*, Vol. 3, Artech House, 1986.

[3] Oliver, C., and S. Quegan, *Understanding SAR Images*, SciTech Publishing, 2004.

[4] Hanssen, R.F., *Radar Interferometry: Data Interpretation and Error Analysis*, Springer, 2001.

[5] Van Zyl, J.J., *Synthetic Aperture Radar Polarimetry*, Wiley, 2011.

[6] Richards, M.A., J.A. Scheer, and W.A. Holm, *Principles of Modern Radar*, SciTech Publishing, 2010.

[7] Franceschetti, G., and R. Lanari, *Synthetic Aperture Radar Processing*, CRC Press, 1999.

[8] Wiley, C.A., "Synthetic Aperture Radars," *IEEE Trans. on Aerospace and Electronic Systems*, vol. 21, pp. 440–443, 1985.

[9] Mensa, D.L., *High Resolution Radar Imaging*, Artech House, 1981.

[10] Carrara, W.G., R.S. Goodman, and R.M. Majewski, *Spotlight Synthetic Aperture Radar: Signal Processing Algorithms*, Artech House, 1995.

[11] Soumekh, M., *Synthetic Aperture Radar Signal Processing with MATLAB Algorithms*, Wiley, 1999.

[12] Jakowatz, C.V., D.E. Wahl, P.H. Eichel, and D.C. Ghiglia, *Spotlight-Mode Synthetic Aperture Radar: A Signal Processing Approach*, Springer, 1996.

[13] Curlander, J.C., and R.N. McDonough, *Synthetic Aperture Radar: Systems and Signal Processing*, Wiley, 1991.

[14] Whener, D.R., *High Resolution Radar*, 2d ed., Arthech House, 1995.

[15] Klausing, H., N. Bartsch, and C. Boesswetter, "A MM-Wave SAR-Design for Helicopter Application (ROSAR)," in *16th European Microwave Conference*, September 8–12, pp. 317–328, 1986.

[16] Pieraccini, M., G. Luzi, and C. Atzeni, "Terrain Mapping by Ground-Based Interferometric Radar," *IEEE Trans. on Geosci. Remote Sensing*, vol. 39, pp. 2176–2181, 2001.

[17] Chen, C., and H. Andrews, "Target-Motion-Induced Radar Imaging," *IEEE Trans. on Aerospace and Electronic Systems*, vol. 16, no. 1, pp. 2–14, 1980.

[18] Chen, C., and H. Andrews, "Multifrequency Imaging of Radar Turntable Data," *IEEE Trans. on Aerospace and Electronic Systems*, vol. 16, no. 1, pp. 15–22, 1980.

[19] Sherwin, C.W., J.P. Ruina, and R.D. Rawcliffe, "Some Early Developments in Synthetic Aperture Radar Systems," *IRE Trans. on Military Electronics*, vol. 6, pp. 111–115, 1962.

[20] Walker, J.L., "Range-Doppler Imaging of Rotating Objects," *IEEE Trans. on Aerospace Electronic Systems*, vol. 16, no. 1, pp. 23–52, 1980.

[21] Ausherman, D.A., A. Kozma, J.L. Walker, H.M. Jones, and E.C. Poggio, "Developments in Radar Imaging," *IEEE Trans. on Aerospace and Electronic Systems*, vol. 20, no. 4, pp. 363–400, 1984.

[22] Brown, W.M., and R.J. Fredricks, "Range-Doppler Imaging with Motion through Resolution Cells," *IEEE Trans. on Aerospace and Electric Systems*, vol. 5, no. 1, pp. 98–102, 1969.

[23] Walker, J.L., W. Carrara, and I. Cindrich, "Optical Processing of Rotating-Object Radar Data Using a Polar Recording Format," Technical Report RADC-TR-73-136, AD 526 738, Rome Air Development Center, Rome, NY, May 1973.

[24] Lipps, R., and D. Kerr, "Polar Reformatting for ISAR Imaging," in *Proceedings of the 1998 IEEE Radar Conference*, pp. 275–280, 1998.

[25] Chen, V.C., and H. Ling, *Time-Frequency Transforms for Radar Imaging and Signal Analysis*, Artech House, 2002.

CHAPTER 2

Basic Principles of ISAR Imaging

As indicated in Chapter 1, inverse synthetic aperture radar (ISAR) is different from synthetic aperture radar (SAR) because the former has a larger antenna aperture based on target motion while the radar remains stationary and generates a high-resolution image of moving targets. In this chapter, we will introduce principles related to how ISAR is capable of generating such a high-resolution image of a moving target: ISAR scattering model, ISAR signal waveforms, ISAR image projection plane, point spread function, ISAR image processing, and bistatic ISAR.

2.1 ISAR Scattering Model

Electromagnetic (EM) scattering occurs when a target is illuminated by radar-transmitted EM waves. The incident EM waves induce electric and magnetic currents on the surface and within the volume of the target that will generate a scattered EM field. The scattered EM waves are reflected in all possible directions. If the target is located far enough from the radar, the incident wavefront can be treated as a plane wave. The power of EM waves scattered to all possible directions is measured using a bistatic scattering cross section of the target. For the direction that is back toward the radar, the bistatic scattering becomes backscattering, and the cross section is a backscattering cross section called the radar cross section (RCS).

2.1.1 Monostatic RCS

The RCS is a measure of the reflective strength of a target, which is defined as 4π times the ratio of the power-per-unit solid angle scattered in a specified direction to the power-per-unit area in a plane wave incident on the scatterer from the specified direction. It is formulated by $\sigma = \lim_{R\to\infty} 4\pi R^2 |E_s|^2/|E_i|^2$, where E_s is the intensity of the far-field scattered electric field, E_i is the intensity of the far-field incident electric field, and R is the distance from the radar to the target. High-frequency RCS prediction methods and exact RCS prediction formulas can be found in [1].

RCS of a target is used to characterize the target. It depends on the target's size, geometry, material, transmitted frequency, polarization of the transmitter and receiver, and aspect angles of the target relative to the radar transmitter and the receiver, respectively [1–4]. The RCS

Table 2.1 Typical RCS Values

Targets	RCS Values at X-band (8-12 GHz)
Bird	−20 dB
Human	−3–0 dB
Jet–Plane	7–20 dB
Automobile	20 dB
Truck	23 dB
Bomber	10–30 dB
Ship	35–50 dB

is usually measured using square meters (m^2) or log-scale (dB). A typical RCS is 0.0005 m^2 for insects, 0.01 m^2 for small birds, 0.5 m^2 for humans, and 100,000 m^2 for a large ship. According to the radar range equation [4], the maximum detectable range of a target is proportional to the fourth root of its RCS. Table 2.1 lists some typical RCS values for some targets.

When a target surface is illuminated by EM waves, each subsurface produces a voltage. The vector sum of the subsurface voltages determines the total RCS of the target. The RCS is defined by the square root of the magnitude of the vector sum. The accuracy of the RCS calculation depends on the modeling accuracy of these subsurfaces and the interactions between them. If the modeling is not accurate, the calculated RCS may not agree with the measured RCS.

2.1.2 Bistatic RCS

The bistatic RCS of a target is a measure of the energy scattered from a target in the direction of the receiver. It is more complex than the monostatic one as it depends on both the target's aspect angle and the bistatic angle and therefore adds another degree of freedom to the problem of RCS modeling and prediction.

Bistatic radar configurations should be distinguished in three regions: pseudo-monostatic, bistatic, and forward scattering. The first region concerns very small bistatic angles (typically less than 5 degrees for complex targets) and, under some conditions involving target's smoothness and very small wavelength, can be considered equivalent to the monostatic RCS viewing from the bisector between the transmitter, target, and receiver [5]. For the second region, the bistatic RCS cannot usually be derived from the monostatic one. Moreover, as already mentioned, it varies depending on two angles rather than only one as in the case of monostatic RCS [6]. The third bistatic region concerns the forward scattering region, where the bistatic angle is close to 180°. Here the bistatic RCS can be predicted with a simple formula that relates the latter with the target's silhouette, $\sigma = 4\pi A^2/\lambda^2$, where A is the target's silhouette area.

2.1.3 EM Scattering Mechanisms and EM Prediction

The mechanism of the EM scattering from a target is a complicated process that includes reflections, diffractions, surface waves, ducting, and interactions between them. Reflection is from surfaces and has the highest RCS peak among other scattering mechanisms. Diffraction is from discontinuities (e.g., edges, corners, vertices) and is less intense than the reflection. The surface wave is associated with the current traveling along the surface of the target body. Ducting occurs when a wave enters into a waveguide-like structure (e.g., the inlet cavity of an aircraft). Spiky features and lobes in the RCS may also be associated with multiple reflections, diffraction, and other scattering mechanisms.

The RCS prediction method is an analytical method of calculating the RCS. The incident wave induces a current on the target and thus radiates an EM field. If the distribution of induced currents is known, it can be used in the radiation integral to calculate the scattered field and therefore the RCS. The commonly used methods for RCS prediction are physical optics (PO), ray tracing, method of moments (MoM), and finite-difference (FD) methods. The PO method is a high-frequency approximation used to estimate the surface current induced on a body. It is accurate in the specular direction but inaccurate when computing at angles far from the specular directions or in the shadow regions. However, surface waves are not included in the PO method. To improve the accuracy for the current distribution near the edges, the physical theory of diffraction (PTD) may be used.

The ray-tracing method is used to analyze large objects with arbitrary shape. Geometric optics (GO) is the classical theory of ray tracing and provides a formula for computing the reflected and refracted fields. It can also be supplemented by the geometric theory of diffraction (GTD). Computer codes based on GTD, PTD, and their hybridizations have been developed for the prediction of high-frequency scattering from complex perfectly conducting objects.

The XPATCH code, which is based on the shooting-and-bouncing ray technique and PTD, has been widely used to generate the target's RCS signatures for noncooperative target recognition [7]. It allows the backscattering to be calculated from complex geometries. Other computer codes for RCS prediction include RAPPORT code developed by Netherlands, numerical electromagnetic code (NEC), electric field integral equation (EFIE), and finite difference time domain (FDTD) [3,8].

A simple RCS prediction code based on the PO method called the POFacet is available [9,10], which can calculate monostatic RCS as well as bistatic RCS of a static object. The bistatic RCS determines the EM power flux density scattered by an object in an arbitrary direction. In using the PO facet RCS prediction method, an object is usually approximated by a large number of subdivision surfaces (triangular meshes) called facets, which produce a continuous surface representing the object. The total RCS of the object is the superposition of the square root of the magnitude of each individual facet's RCS. The scattered field of each triangle is computed by assuming that the triangle is isolated and other triangles are not present. Besides, multiple reflections, edge diffraction, and surface waves are not considered. Shadowing is only approximately included by considering a facet to be completely illuminated or completely shadowed by the incident wave.

Another simple computer simulation model is based on approximate and simplified complex scattering solutions [11]. A complicated target is decomposed into basic geometric shapes, such as spheres, ellipsoids, cylinders, and plates. Thus, the formulas of the simple components are available for applying to these decomposed shapes. But this model is not an exact solution. An example RCS of a ground launching cruise missile that was decomposed into basic geometric shapes is calculated in [11] and shown in Figure 2.1.

Nevertheless, the simplest model of the EM scattering mechanism is still the point scatterer model. A target can be defined in terms of a three-dimensional (3-D) reflectivity

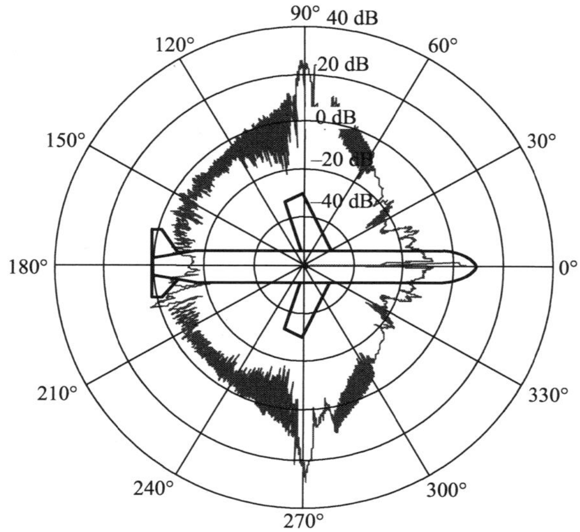

Figure 2.1 RCS versus view angles of a ground launching cruise missile model [11].

density function characterized by point scatterers. Compared with other EM scattering models, the point scatterer model can easily incorporate the target's motion in EM scattering and isolate the EM scattering from each individual motion component.

2.1.4 Radar Backscattering from Point-like Models of Targets

Radar backscattering from a target can be simulated if we define a scattering model of the target. Given the target initial location and its motion parameters, at each time instant the location of the target can be determined. Based on the geometry of radar and the target, the radar-received signal can be easily formulated.

Suppose the radar transmits a sequence of narrow rectangular pulses at a carrier frequency, f_c, and with the pulse width, ΔT, and the pulse repetition interval (PRI), T_{PRI}, the radar baseband signal reflected from a point target can be expressed by

$$s_B(t) = \sum_{n=1}^{N_P} \sigma^{1/2} \operatorname{rect}[t - n \cdot T_{\text{PRI}} - 2R(t)/c, \Delta T] \cdot \exp\{-j4\pi f_c R(t)/c\}, \tag{2.1}$$

where σ is the RCS of the target, N_p is the total number of pulses received, $R(t)$ is the distance from the radar to the point target at time t, and the rectangular function rect is defined by

$$\operatorname{rect}(t, \Delta T) = \begin{cases} 1 & |t| \leq \Delta T/2 \\ 0 & |t| > \Delta T/2 \end{cases}. \tag{2.2}$$

Figure 2.2a shows a range profile after transmitting one pulse signal to a target. After transmitting N_p pulses and rearranging the N_p received signals, a two-dimensional (2-D) pulse-range matrix, or called the range profiles, can be obtained as shown in Figure 2.2b.

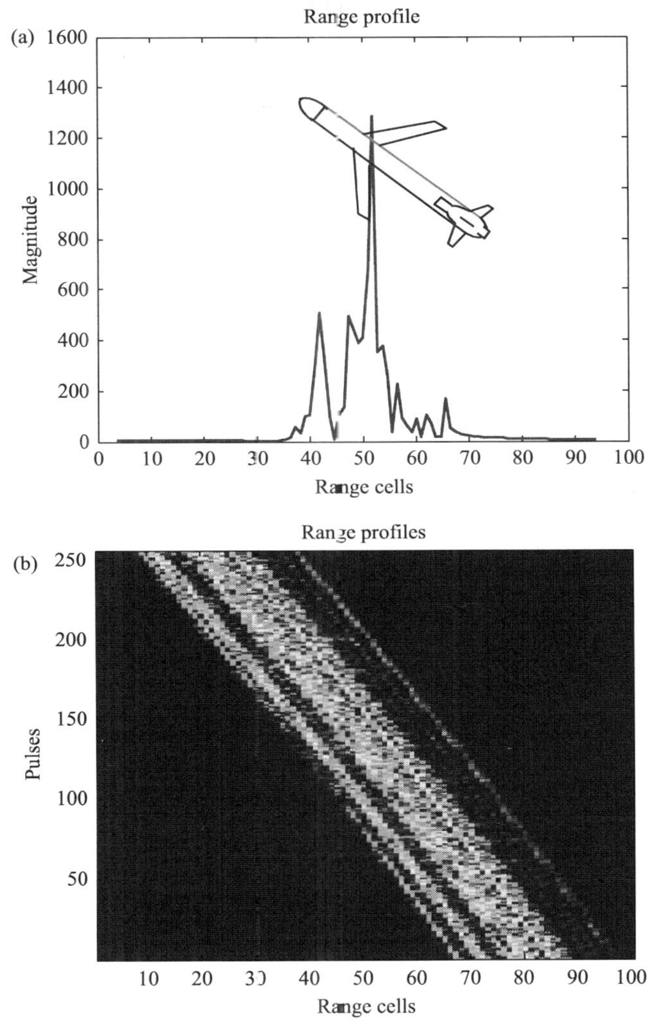

Figure 2.2 (a) A range profile after transmitting one pulse signal to a target. (b) after transmitting N_p pulses and rearranging the N_p received signals, the 2-D range profiles of the target.

2.2 ISAR Signal Waveforms

ISAR generates high-resolution range-Doppler image of a moving target. For high-resolution and unambiguous range and Doppler measurements, ISAR signals must have wide bandwidth and satisfy certain requirement for the selection of its pulse repetition frequency (PRF). To have wide bandwidth, a multifrequency signal or a modulated continuous wave (CW) signal should be used, such as the stepped-frequency signal, the linear frequency modulation (LFM) signal, and a train of modulated pulses. To have high Doppler frequency measurement, the PRF must be sufficient high to avoid Doppler ambiguities. Similarly, to measure a target at a long distance, the PRF must be lower enough to avoid range ambiguities. In general, multiple PRFs can be used to resolve the range and Doppler ambiguities.

When a radar transmits a signal, $s_T(t)$, to a target, the received signal, $s_R(t)$, returned from the target is proportional to the transmitted signal with a round-trip delay $s_T(t-\tau)$ and scaled by the reflectivity function, ρ, of the target,

$$s_R(t) \propto \rho s_T(t-\tau) = \rho \exp\{j2\pi f(t-\tau)\} \quad (0 \le t \le T), \tag{2.3}$$

where T is the total signal time duration, ρ is the target reflectivity, and f is the transmitted frequency of the signal.

The round-trip travel time, τ, is determined by the distance of the target, R, and the speed of electromagnetic wave propagation, c:

$$\tau = 2R/c. \tag{2.4}$$

When the target is moving with a radial velocity, V_R, relative to the radar, the radar signal must travel an extra longer or shorter distance, depending on the projection of target velocity along the line of sight (LOS) to reach the target. The signal received at time, t, is reflected from the target at time $t - \tau(t)/2$, and the round-trip travel time is a time-varying delay, $\tau(t)$.

The radar signal must have a waveform suitable to the required function of the radar. CW radar transmits an electromagnetic wave with constant amplitude and frequency in an infinite time duration and can measure only the Doppler frequency shift of a moving target without range resolution. To achieve a desired range resolution, the transmitted signal must have enough bandwidth, which can be obtained through the modulation of the transmitted CW waveform. Frequency modulated CW (FMCW) signal waveform is a typical modulation scheme. Other commonly used modulation schemes include the stepped-frequency waveform, pulse Doppler waveform, and coherent pulse (amplitude, phase, and frequency) modulation waveforms.

The simplest signal waveform is the CW waveform with a CW carrier frequency, f_c:

$$s_T(t) = A \cdot \exp\{j2\pi f_c t\}. \tag{2.5}$$

If the CW signal is transmitted, the received signal is

$$s_R(t) = A_R \cdot \exp\{j2\pi f_c(t-\tau)\}, \tag{2.6}$$

where τ is the round-trip time delay between the radar and a target. If the target is moving, the received signal reflected from the target becomes

$$s_R(t) = A_R \cdot \exp\{j2\pi(f_c + f_D)(t-\tau)\}, \tag{2.7}$$

where f_D is the Doppler shift of the carrier frequency, f_c.

By mixing the transmitted signal, $s_T(t)$, and the received signal, $s_R(t)$, the Doppler shift frequency can be extracted through a low-pass filter (LPF) as depicted in Figure 2.3. The output of the mixer can also be applied to an intermediate frequency (IF) band filtering to eliminate low frequency components of interference and noise. The minimum detected target's velocity is determined by the lower limit of the IF band filter.

By using an additional mixer with a 90° phase shifted CW oscillator signal as shown in Figure 2.4, the sign of the Doppler frequency shift can be determined. The two mixers'

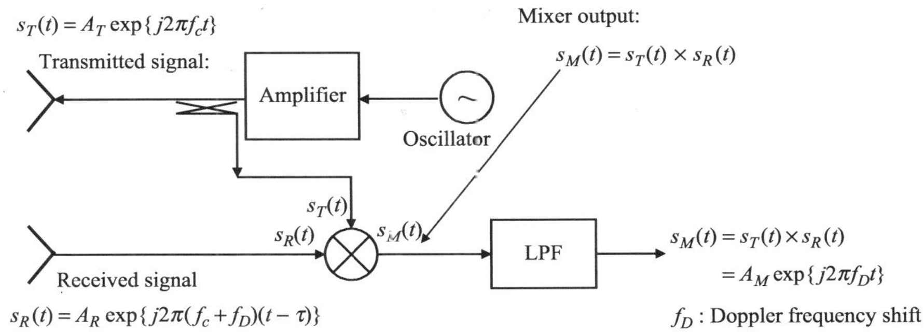

Figure 2.3 A block diagram of a CW Doppler radar.

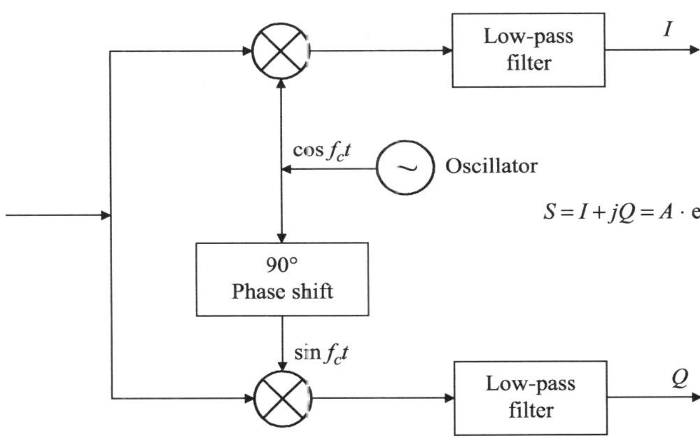

Figure 2.4 Quadrature demodulator.

scheme is called the quadrature demodulator with an in-phase (I) and a quadrature (Q) phase demodulator. The output of the quadrature demodulator is a complex signal, S.

The I component is the real part ($I = A\cos\varphi$), and the Q component ($Q = A\sin\varphi$) is the imaginary part of a complex signal:

$$S = I + jQ = A \cdot \exp\{j\varphi\}, \tag{2.8}$$

where $A = \sqrt{I^2 + Q^2}$, and $\varphi = \tan^{-1}(Q/I)$.

To measure range information with the CW signal, the basic approach is to measure the phase shift of the received CW signal. Given the CW wavelength, $\lambda = c/f_c$, the target's round-trip range is $2R = (N + \delta\lambda)\lambda$, where N is an integer number of the wavelength, and $\delta\lambda$ is the fraction of the wavelength. If we do not know N and $\delta\lambda$, we are then unable to measure the target range.

However, one approach can measure the target range using a CW waveform. The first step is to tune the transmitted frequency, f_c, such that the target range reaches an integer number, N_1, of the tuned frequency wavelength λ_1, that is, $2R = N_1\lambda_1$. Then the second step is to keep tuning the transmitted frequency until the target range reaches another integer number, N_2, of the tuned frequency wavelength λ_2, that is, $2R = (N_1 + N_2)\lambda_2$. Thus, by solving the previous two equations, the unknown variable, R, can be found as

$$R = (c/2)N_2/(f_2 - f_1), \tag{2.9}$$

where $f_1 = c/\lambda_1$ and $f_2 = c/\lambda_2$.

In practice, however, we do not use the pure CW form in ISAR. The commonly used ISAR waveforms are the linear frequency modulation (LFM) chirp, FMCW waveform, and stepped frequency. We will discuss more details on these waveforms.

2.2.1 LFM Pulse Waveform

In high range-resolution radar systems, to achieve high range resolution, the transmitted signal must have a wide bandwidth. LFM (or chirp) waveform is commonly used in wideband radars. In the LFM pulse waveform, its carrier frequency linearly changes within a single pulse as shown in Figure 2.5a. The pulse shape is determined by the window function. A commonly used window function is the Gaussian window function, which is sometimes called the Gabor transform and achieves the best time-frequency product among all the possible window functions [12].

The LFM pulse with a Gaussian envelope can be expressed as

$$s_{\text{LFM}}(t) = \left(\frac{a}{\pi}\right)^{1/4} \exp\left\{-\frac{at^2}{2}\right\} \cdot \exp\left\{j2\pi\left(f_c + \frac{\mu}{2}t\right)t\right\}, \tag{2.10}$$

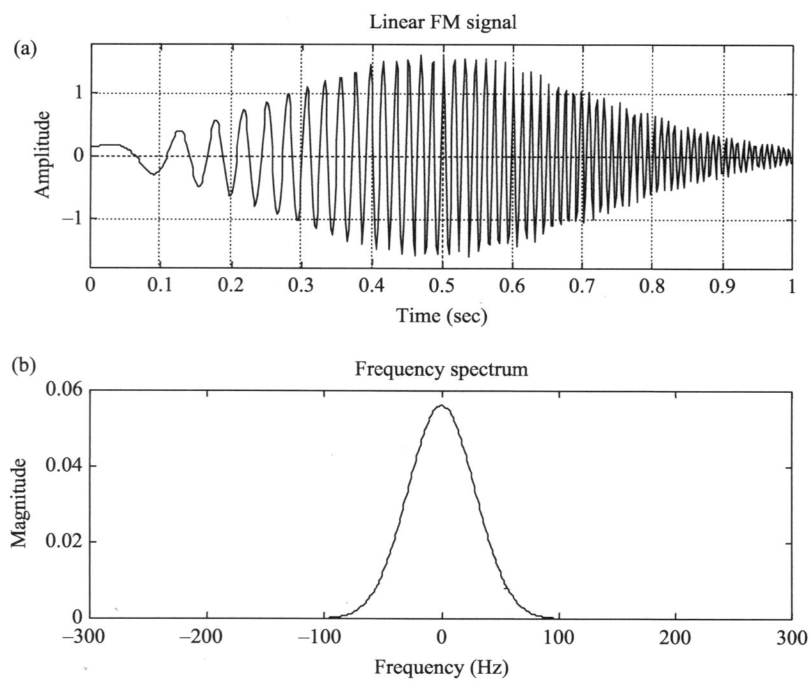

Figure 2.5 An LFM's (a) waveform and (b) frequency spectrum.

where f_c is the carrier frequency, μ is the frequency changing rate or chirp rate, and a determines the width of the Gaussian envelope.

The frequency spectrum of the LFM with a Gaussian envelope is shown in Figure 2.5b and can be derived as

$$S_{\mathrm{LFM}}(f) = \frac{(a/\pi)^{1/4}}{(a - j\mu)^{1/2}} \exp\left\{ -\frac{2\pi^2 a (f - f_c)^2}{(a^2 + \mu^2)} \right\} \cdot \exp\left\{ -j\frac{2\pi^2 \mu (f - f_c)^2}{(a^2 + \mu^2)} \right\}. \tag{2.11}$$

2.2.2 FMCW Waveform

To have a range resolution, radar signals must have enough bandwidth through the modulation of the transmitted CW waveform. The FMCW waveform is a popular solution. Commonly used FMCW waveforms include sawtooth modulation and triangular modulation as shown in Figure 2.6.

For the sawtooth modulation, the transmitted frequency, f_T, is a time-dependent function:

$$f_T(t) = f_0 + B\frac{t}{T}; \quad (0 < t < T). \tag{2.12}$$

For the symmetric triangular modulation, the transmitted frequency becomes

$$f_T(t) = \begin{cases} f_0 + B\dfrac{t}{T/2}; & (0 < t < T/2) \\ f_0 + B - B\dfrac{t - T/2}{T/2}; & (T/2 < t < T) \end{cases}, \tag{2.13}$$

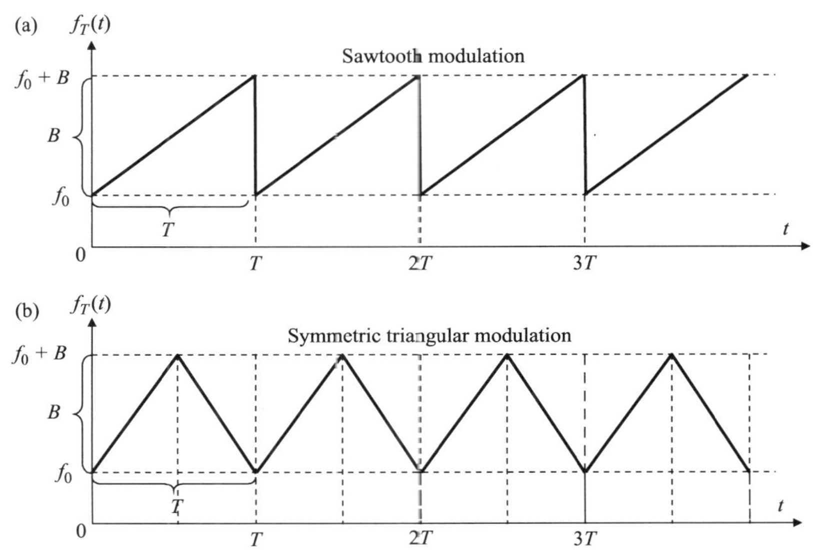

Figure 2.6 (a) Sawtooth modulation and (b) symmetric triangular modulation in the FMCW waveform.

where f_0 is the initial frequency, B is the bandwidth of frequency sweep, and T is the sweep period. At the end of the sweep $t = T$, the frequency becomes $f_1 = f_0 + B$, as shown in Figure 2.6a.

In general, an asymmetric triangular modulation FMCW signal (Figure 2.7) can be expressed by

$$s_{\mathrm{FMCW}}(t) = A \cdot \exp\left\{ j2\pi \left[f_0 + B \int_{-\infty}^{t} m(\lambda) d\lambda \right] \right\}$$

$$= \begin{cases} A \cdot \exp\left\{ j2\pi \left[f_0 + B + B\dfrac{t}{T_0} \right] t \right\} & (-T_0 \le t < 0) \\ A \cdot \exp\left\{ j2\pi \left[f_0 + B - B\dfrac{t}{T_{fb}} \right] t \right\} & (0 \le t < T_{fb}) \end{cases}, \tag{2.14}$$

where T_0 is the sweep time, T_{fb} is the flyback time of the asymmetrical triangular function, and $m(t)$ is convolution of the asymmetrical triangular function $\Lambda_{Asym}(t)$ with a Dirac-function train $\Delta(t) = \sum_k \delta(t - kT)$; $(T = T_0 + T_{fb})$ (Figure 2.8):

$$m(t) = \Lambda_{Asym}(t) \otimes \sum_k \delta(t - kT). \tag{2.15}$$

The instantaneous frequency can be obtained by applying a time derivative to the phase function:

$$f(t) = \begin{cases} f_0 + B + B\dfrac{t}{T_0}; & (-T_0 \le t < 0) \\ f_0 + B - B\dfrac{t}{T_{fb}}; & (0 \le t < T_{fb}) \end{cases}. \tag{2.16}$$

The round-trip time delay, τ, of the transmitted FMCW signal from a target at a distance R is given by $\tau = 2R/c$, where c is the wave propagation velocity. Thus, assume that the

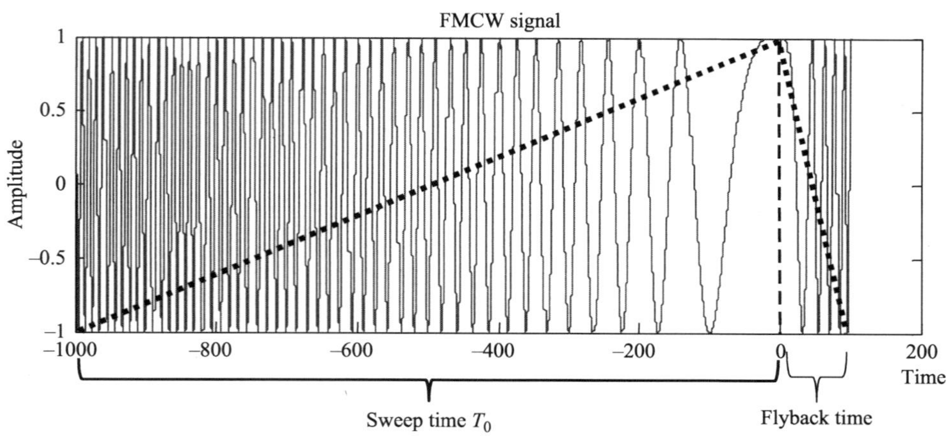

Figure 2.7 FMCW signal waveform.

FMCW is a sawtooth waveform (i.e., the flyback time T_{fb} is zero and thus $T = T_0$) and that the transmitted frequency is $f_T(t) = f_0 + B\frac{t}{T}$; $(0 < t < T)$. The received frequency becomes

$$f_R(t) = f_0 + B\frac{(t-\tau)}{T}. \tag{2.17}$$

By mixing the received signal with the transmitted signal, a sum and a difference of the frequency terms ($f_T + f_R$ and $f_T - f_R$) will be generated at the output of the mixer. The resulting signal is then processed by a low-pass filter to remove the sum term and to keep the difference term (i.e., the beat frequency, f_B). This process is shown in Figure 2.9. The spectrum of the FMCW waveform and the beat signal is shown in Figure 2.10.

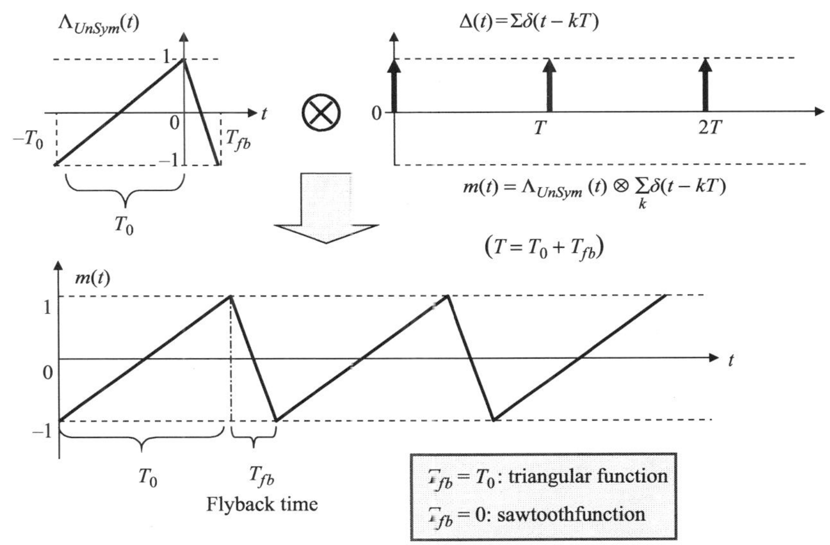

Figure 2.8 Generation of a sequence of asymmetrical triangular function by convolution of the function with a Dirac-function train.

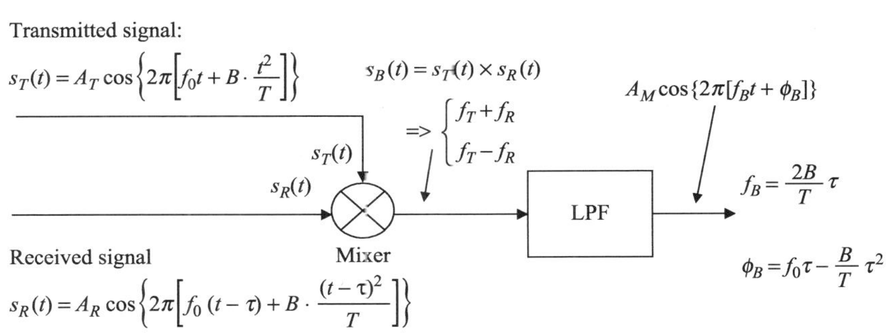

Figure 2.9 Homodyne FMCW radar block diagram.

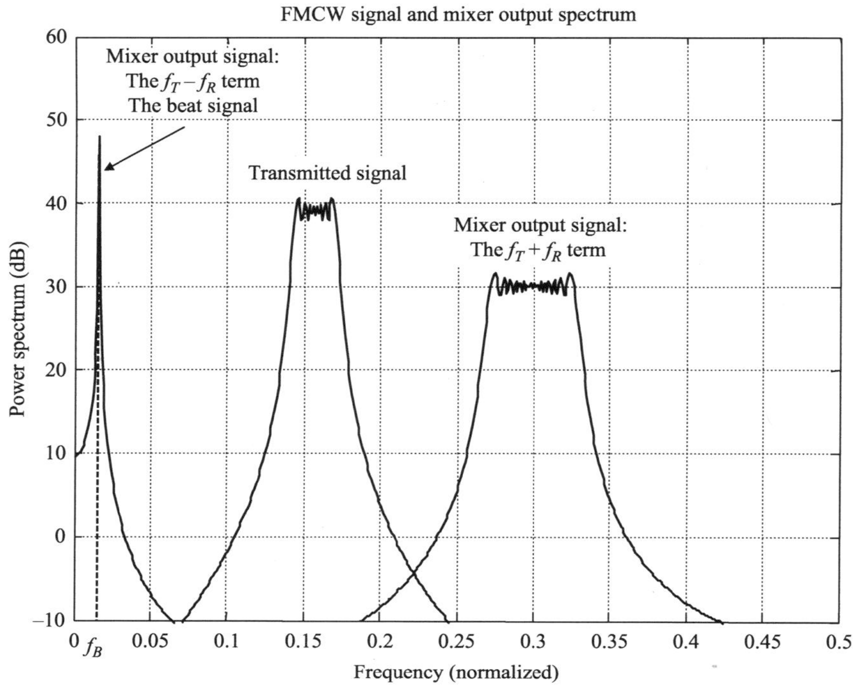

Figure 2.10 Spectrum of the FMCW signal and its beat signal.

The beat frequency is independent of time:

$$f_B = f_T(t) - f_R(t) = 2B\tau/T = (2B/T)\ (2R/c). \tag{2.18}$$

The process of generating the beat frequency from the return signal is shown in Figure 2.11.

Given the beat frequency, f_B, and the radar parameters of B and T, we can measure the distance of the point target:

$$R = cTf_B/(4B). \tag{2.19}$$

If there are multiple point targets, the distance of each point target corresponds to a different beat frequency, which can be measured by the peaks in the frequency.

For a FMCW radar, if the bandwidth $B = 500$ MHz, the sweep period $T = 1.0 \times 10^{-3}$ sec, and the target range $R = 150$ m, then the beat frequency $f_B = 1.0$ MHz and the time delay $\tau = 1.0 \times 10^{-6}$ sec. Thus, the range resolution, ΔR, is determined by the beat frequency resolution, Δf_B: $\Delta R = [cT/(4B)]\ \Delta f_B = 1.5 \times 10^{-4}\ \Delta f_B$.

In cases when a target is moving, a Doppler shift is superimposed onto the beat frequency. The transmitted and received frequency-time signals for moving targets are shown in Figure 2.12a. The beat frequency is no longer a constant value. It is increased by the Doppler shift in one cycle and is decreased by the same amount in the other cycle as shown in Figure 2.12b. The beat frequency switches between two values: $f_{B1} = f_R + f_D$ and $f_{B2} = f_R - f_D$, where f_R is the contribution of target range, and f_D is the contribution of target motion.

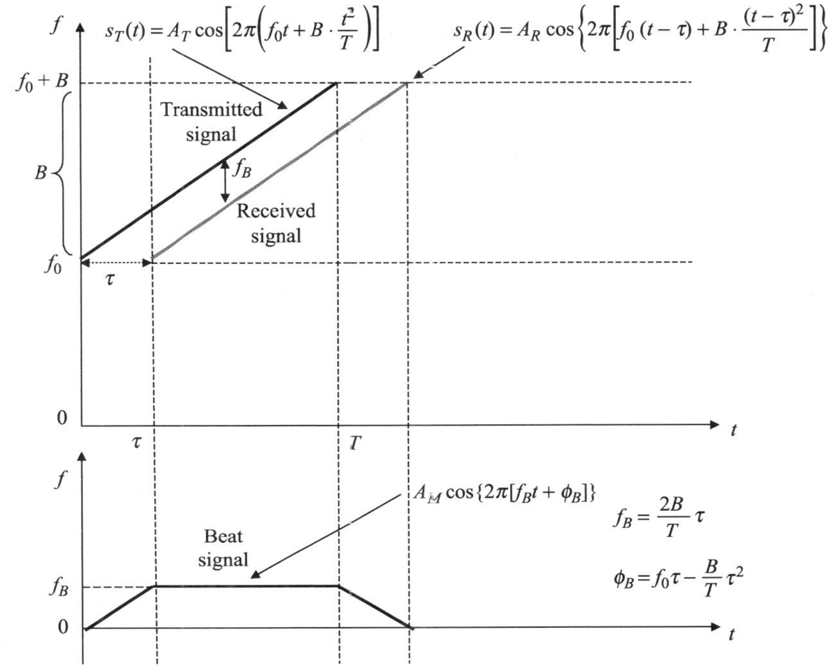

Figure 2.11 The process of generating the beat frequency.

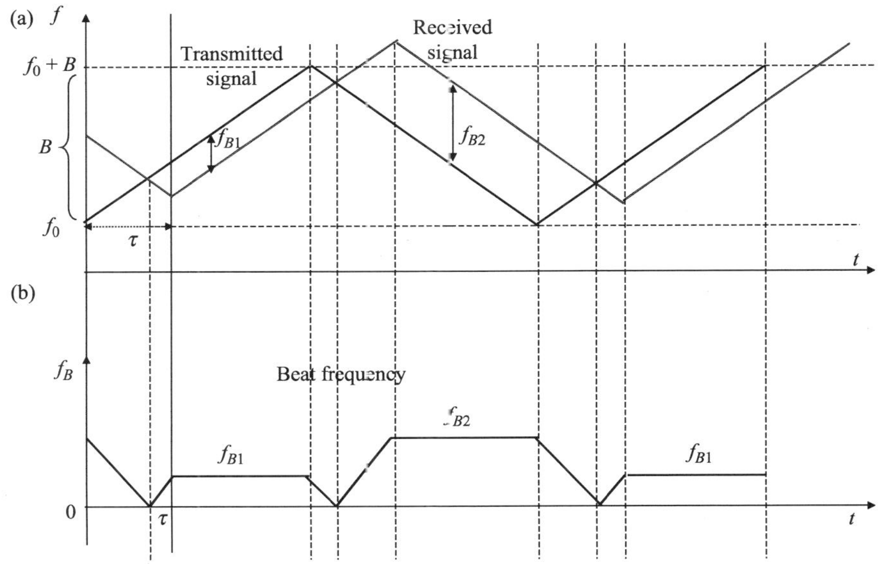

Figure 2.12 (a) The transmitted and received frequency-time signals for moving target and (b) the beat frequencies switched between two values.

Thus, from the two beat frequencies the target range-related beat frequency is $f_R = (f_{B1} + f_{B2})/2$, and the target motion-related beat frequency is $f_D = (f_{B1} - f_{B2})/2$.

2.2.3 Stepped Frequency CW Waveform

To have range resolution, the transmitted signal can be modulated using stepped-frequency signal waveform. This type of radar is called the stepped-frequency CW radar.

The stepped-frequency signal waveform is shown in Figure 2.13a, which can be expressed by a sequence of pulses of coherent continuous wave at increasing frequencies from pulse to pulse by a fixed increment of Δf. Each pulse has a fixed pulse width ΔT, and the pulses are transmitted at a pulse repetition frequency (PRF). The PRI, T_{PRI}, is equal to ΔT, as indicated in Figure 2.13b.

The stepped-frequency signal achieves its wide bandwidth by sequentially changing the carrier frequency step by step over a number of pulses. Thus, it can be described by a sequence of pulses with increased carrier frequencies from one pulse to the next. The stepped carrier frequency is expressed as

$$f_n = f_0 + n\Delta f \quad (n = 0, 1, 2, \ldots, N-1), \tag{2.20}$$

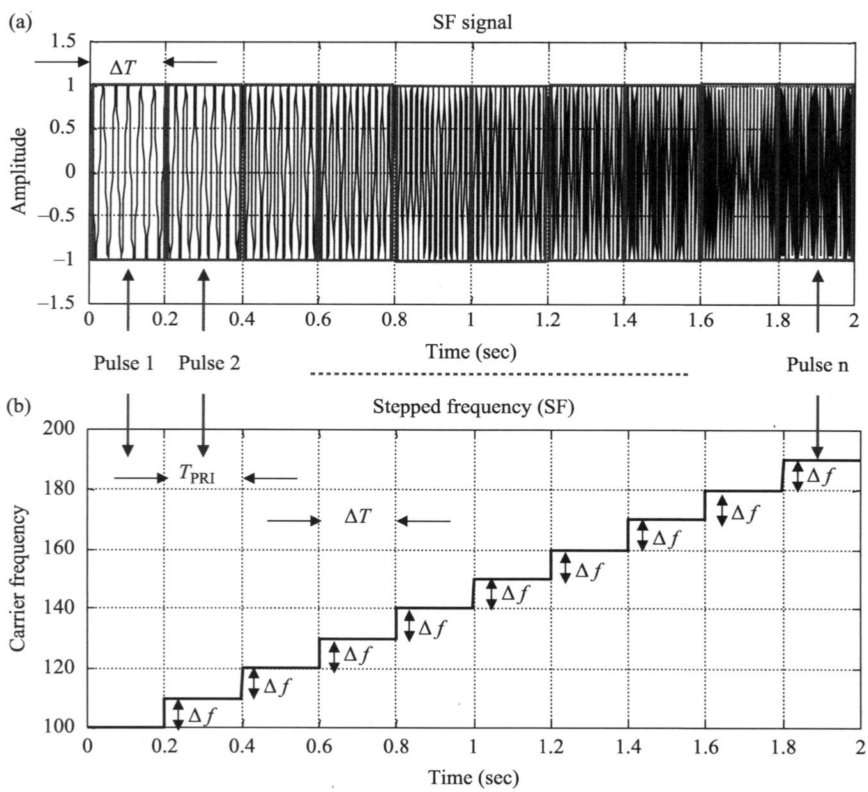

Figure 2.13 (a) Stepped-frequency signal waveform and (b) increasing frequencies from pulse to pulse by a fixed increment.

where Δf is the frequency step. The total bandwidth of the stepped-frequency signal, $N{\cdot}\Delta f$, determines the radar range resolution. Because pulses are transmitted with the pulse width ΔT and T_{PRI}, the stepped-frequency signal is expressed by

$$s_{SF}(t) = \sum_{n=0}^{N-1} \text{rect}(t - nT_{\text{PRI}}, \Delta T) \cdot \exp\{-j2\pi(f_0 + n\Delta f)t\}, \tag{2.21}$$

where the rectangular pulse is defined as

$$\text{rect}(t, \Delta T) = \begin{cases} 1 & |t| \leq \Delta T/2 \\ 0 & |t| > \Delta T/2 \end{cases}. \tag{2.22}$$

The Fourier transform of the rectangular pulse is $FT\{\text{rect}(t, \Delta T)\} = \Delta T\ \text{sinc}(f\Delta T)$, where $\text{sinc}(\bullet)$ is the Sinc function defined by $\text{sinc}(x) = \sin(\pi x)/\pi x$; $(x \neq 0)$ and $\text{sinc}(x) = 1$; $(x = 0)$. Then, the Fourier transform of the time-shifted rectangular pulse $\text{rect}(t - \tau, \Delta T)$ becomes

$$FT\{\text{rect}(t - \tau, \Delta T)\} = \Delta T \text{sinc}(f\Delta T)\exp\{-j2\pi f\tau\}, \tag{2.23}$$

where τ is the time shift. Therefore, the frequency spectrum of the stepped-frequency signal can be derived by taking the Fourier transform of the signal, $s_{SF}(t)$:

$$S_{SF}(f) = \sum_{n=0}^{N-1} \Delta T \text{sinc}[(f - f_n)\Delta T]\exp\{-j2\pi(f - f_n)nT_{\text{PRI}}\}, \tag{2.24}$$

where $f_n = f_0 + n\Delta f$ is the carrier frequency of the n-th pulse.

Assume a radar transmits a stepped-frequency waveform signal $s_T(t) = s_{SF}(t)$ as in (2.21), then the returned signal from a target at a range, R, is

$$s_R(t) = A_R \sum_{n=0}^{N-1} \text{rect}\left(t - \frac{2R}{c} - nT_{\text{PRI}}, \Delta T\right) \exp\left\{-j2\pi(f_0 + n\Delta f)\left(t - \frac{2R}{c}\right)\right\}. \tag{2.25}$$

After quadrature demodulator, the output becomes

$$s_{\text{out}}(t) = A_0 \sum_{n=0}^{N-1} \text{rect}\left(t - \frac{2R}{c} - nT_{\text{PRI}}, \Delta T\right) \cdot \exp\left\{j2\pi(f_0 + n\Delta f)\frac{2R}{c}\right\}. \tag{2.26}$$

If the target is moving with a constant radial velocity, V_R, the target range is a function of discrete time, $t_n = nT_{\text{PRI}}$:

$$R(t_n) = R_0 + V_R t_n. \tag{2.27}$$

Therefore, the phase of the demodulator output becomes

$$\Phi(t_n) = 2\pi(f_0 + n\Delta f)\frac{2(R_0 + V_R t_n)}{c} = 2\pi\left\{f_0\frac{2R_0}{c} + \frac{2R_0}{c}\frac{\Delta f}{T_{\text{PRI}}}t_n + 2\frac{f_0 V_R}{c}t_n + 2\frac{\Delta f V_R}{cT_{\text{PRI}}}t_n^2\right\}, \tag{2.28}$$

where we can see that the target's motion induces a linear phase term $4\pi\frac{f_0 V_R}{c}t_n$, and frequency stepping induces another linear phase term, $4\pi\frac{R_0}{c}\frac{\Delta f}{T_{\text{PRI}}}t_n$. The quadratic phase term, $4\pi\frac{\Delta f V_R}{cT_{\text{PRI}}}t_n^2$, is generated by joint contributions of the target motion and the frequency stepping, called the coupling between the motion and frequency stepping.

Because the time derivative of a phase function is the instantaneous frequency, the instantaneous frequency of the demodulated stepped-frequency signal becomes

$$f(t_n) = \mathrm{d}\Phi(t_n)/dt_n = 2\pi \frac{2R_0}{c}\frac{\Delta f}{T_{\mathrm{PRI}}} + 4\pi \frac{f_0 V_R}{c} + 4\pi \frac{\Delta f V_R}{c T_{\mathrm{PRI}}} t_n, \tag{2.29}$$

where the first constant frequency term is determined by the distance of the target, R_0, and the frequency stepping rate, $\Delta f / T_{\mathrm{PRI}}$, the second constant frequency term is the Doppler frequency shift due to the target's motion, $f_0 V_R / c$, and the third term is a linear frequency term due to the coupling between the target and the frequency stepping.

The instantaneous bandwidth of the waveform is approximately equal to the inverse of the pulse width, which is much less than the effective bandwidth of the stepped-frequency waveform. The effective bandwidth, B_{eff} of the waveform is determined by the product of the number of steps, N, and the stepped-frequency size, Δf:

$$B_{\mathrm{eff}} = N\Delta f. \tag{2.30}$$

The range resolution for a waveform is dependent on the effective bandwidth of the waveform and is given by

$$\Delta r_r = c/(2N\Delta f). \tag{2.31}$$

In practice, instead of transmitting a single stepped-frequency waveform, the stepped-frequency radar usually transmits a sequence of stepped-frequency waveform signals. Figure 2.14 shows the transmitted stepped-frequency waveform signals during time [0, T], which consist of a total of M stepped-frequency signals, called bursts. Each burst consists of N frequency-stepped pulses, where each pulse width is ΔT. In this case, the PRF of the

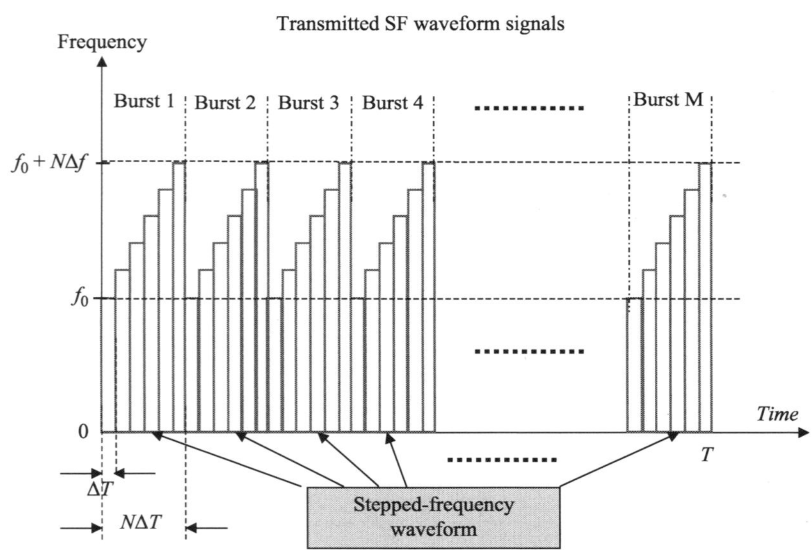

Figure 2.14 Transmitted M bursts of stepped-frequency waveform signals. Each burst has N frequency steps.

stepped-frequency pulses is equivalent to the range sampling rate, and N pulses are equivalent to N range cells. Similarly, each burst in the stepped-frequency radar is equivalent to each transmitted pulse in conventional pulse radar, which determines the CPI, and the total number of bursts, M, determines the Doppler resolution of the stepped-frequency radar. In Section 4.11, we will demonstrate how to use stepped-frequency radar for ISAR imaging.

2.3 Radar Ambiguity Function

The radar ambiguity function is a basic mathematical tool used for characterizing radar performance in terms of target resolution and clutter rejection [13]. The ambiguity function of a signal $s(t)$ is a two-dimensional function in Doppler frequency shift f_D and time-delay τ defined as

$$\chi_S(\tau, f_D) = \int_{-\infty}^{\infty} s(t)s^*(t-\tau)\exp\{j2\pi f_D t\}dt = \int_{-\infty}^{\infty} S^*(f)S(f-f_D)\exp\{j2\pi f\tau\}df, \quad (2.32)$$

where the asterisk refers to the conjugate, and $S(f)$ is the signal frequency spectrum. A high value of the ambiguity function indicates that it is difficult to resolve two nearby targets whose differences in the time delay and in the Doppler frequency shift are τ and f_D, respectively.

The ambiguity function can also be defined in a symmetrical form

$$\begin{aligned}\chi_S(\tau, f_D) &= \int_{-\infty}^{\infty} s\left(t+\frac{\tau}{2}\right)s^*\left(t-\frac{\tau}{2}\right)\exp\{j2\pi f_D t\}dt \\ &= \int_{-\infty}^{\infty} S^*(f+f_D/2)S(f-f_D/2)\exp\{j2\pi f\tau\}df \quad (2.33)\end{aligned}$$

and

$$\chi_S(\tau, f_D) = \chi_S^*(-\tau, -f_D). \quad (2.34)$$

The peak value of the ambiguity function is always at the center of the origin:

$$\chi_S(\tau, f_D) \le \chi_S(0, 0). \quad (2.35)$$

The peak of the ambiguity function $\chi_s(0, 0)$ means that it is impossible to resolve two nearby targets if their differences in the time delay and the Doppler frequency shift are all zeros. An ideal ambiguity function is a thumbtack-type function which has a peak value at ($\tau = 0$, $f_D = 0$) and near zero elsewhere. This means with the thumbtack-type ambiguity function, two nearby targets can be perfectly resolved if their differences in the time delay and the Doppler frequency shift are not zeros. Of course, if $\tau = 0$ and $f_D = 0$, the ambiguity function has an infinite peak that makes two targets ambiguous.

Other properties of the ambiguity function include the following:

1. Ambiguity function of a scaled signal $s(at)$ is

$$s'(t) = s(at) \Rightarrow \chi_{S'}(\tau, f_D) = \frac{1}{|a|}\chi_S(a\tau, f_D/a) \quad (2.36)$$

2. Ambiguity function of a time-shifted signal $s(t-\Delta t)$ is

$$s'(t) = s(t-\Delta t) \Rightarrow \chi_{S'}(\tau, f_D) = \chi_S(\tau, f_D)\exp\{-j2\pi f_D \Delta t\} \tag{2.37}$$

3. Ambiguity function of a frequency-modulated signal $s(t)\exp\{j2\pi ft\}$is

$$s'(t) = s(t)\exp\{j2\pi ft\} \Rightarrow \chi_{S'}(\tau, f_D) = \chi_S(\tau, f_D)\exp\{-j2\pi ft\} \tag{2.38}$$

If we set the Doppler shift to zero, the ambiguity function becomes the auto-correlation function of the signal $s(t)$

$$\chi_S(\tau, 0) = \int_{-\infty}^{\infty} s(t)s^*(t-\tau)dt. \tag{2.39}$$

In the following, we list ambiguity functions of several typical signal waveforms:

1. Rectangular pulse:
   The ambiguity function of a rectangual pulse rect(t, T) is

$$\begin{aligned}\chi_{\text{rect}}(\tau, f_D) &= \int_{-\infty}^{\infty} \text{rect}\left(t+\frac{\tau}{2}\right)\text{rect}^*\left(t-\frac{\tau}{2}\right)\exp\{j2\pi f_D t\}dt \\ &= \int_{-(T-\tau)/2}^{(T-\tau)/2} \exp\{j2\pi f_D t\}dt = \begin{cases}(T-|\tau|)\text{sinc}[f_D(T-|\tau|)] & \text{for } |\tau| \le T \\ 0 & \text{for } |\tau| > T\end{cases}\end{aligned} \tag{2.40}$$

2. Gaussian pulse:
   The Gaussain pulse is represented by

$$\text{Gauss}(t) = \exp\{-\alpha t^2\}, \tag{2.41}$$

and its ambiguity function is also a Gaussian type:

$$\chi_{\text{Gauss}}(\tau, f_D) = \frac{1}{\sqrt{2}}\exp\{-\alpha(\tau^2 + f_D^2)/2\}. \tag{2.42}$$

3. Linear frequency-modulated pulse:
   The LFM pulse is represented by

$$\text{LFM}(t) = \frac{1}{\sqrt{T}}\text{rect}\left(\frac{t-T/2}{T}\right)\exp\{j(2\pi f_0 t + \pi\gamma t^2)\}, \tag{2.43}$$

and its ambiguity function is

$$\chi_{\text{LFM}}(\tau, f_D) = \begin{cases}(T-|\tau|)\text{sinc}[(f_D-\gamma\tau)(T-|\tau|)] & \text{for } |\tau| \le T \\ 0 & \text{otherwise}\end{cases}. \tag{2.44}$$

4. Stepped-frequency signal:
   The stepped-frequency signal can be expressed in (2.21) as

$$s_{SF}(t) = \sum_{n=0}^{N-1} \text{rect}(t - nT_{\text{PRI}}, \Delta T)\cdot\exp\{-j2\pi(f_0 + n\Delta f)t\}, \tag{2.45}$$

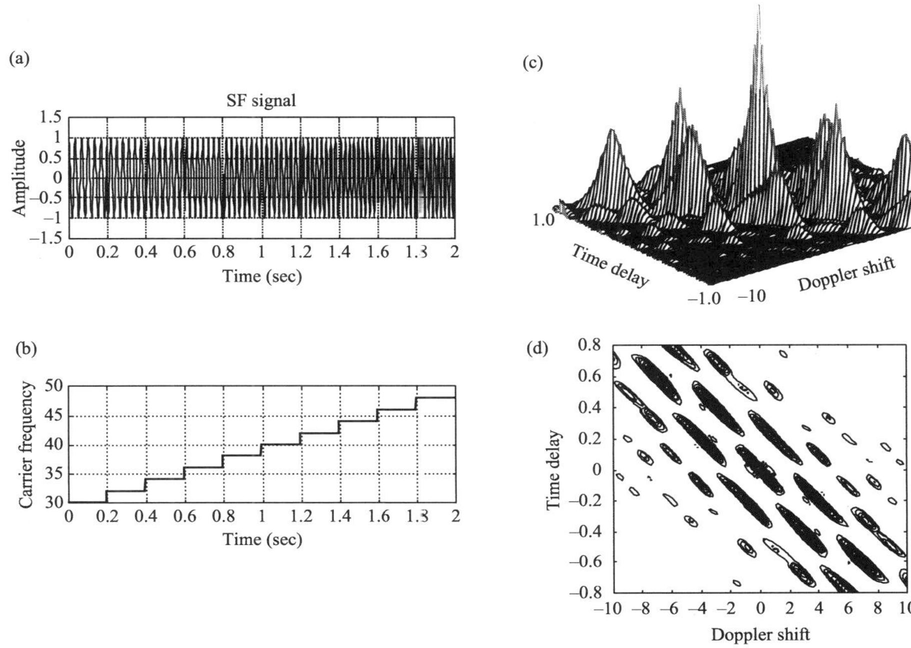

Figure 2.15 Ambiguity function of a stepped-frequency signal: (a) Stepped-frequency signal; (b) stepped carrier frequency; (c) surface; and (d) contour plot of the stepped-frequency ambiguity function.

and the ambiguity function of a stepped-frequency signal waveform is shown in Figure 2.15, where $N = 10$. The stepped-frequency signal in the time domain and its time-frequency distribution are shown in Figure 2.15a and Figure 2.15b, respectively. Figure 2.15c and Figure 2.15d are a surface plot and a contour plot of the ambiguity function of the stepped-frequency signal, respectively [14].

2.4 Matched Filter

Given a signal, $s(t)$, a filter matched to the signal is defined by its impulse response function $h(t) = k \cdot s(\Delta - t)$, where Δ is an arbitrary constant [15,16]. The matched filter provides the maximum output signal-to-noise ratio in white Gaussian noise environment. White noise means the power spectrum of the noise, $P_n(f)$, is uniformly distributed over the entire frequency domain ($-\infty < f < \infty$), that is, $P_n(f) = N_0/2$; and Gaussian noise indicates that the probability density function of the amplitude of the noise is a Gaussian distribution. Given a signal, $s(t)$, in white Gaussian noise, the transfer function of the matched filter is the complex conjugate spectrum of the time-shifted signal $s(t + t_0)$ [15]:

$$H(f) = kS^*(f)\exp\{-j2\pi f t_0\}, \tag{2.46}$$

and its impulse response is a mirror function of the input signal:

$$h(t) = \begin{cases} ks^*(t_0 - t) & (t \geq 0) \\ 0 & (t < 0) \end{cases}, \tag{2.47}$$

where t_0 is a predetermined observation time, and k is a scaling constant.

The matched filter has many important properties:

1. Among all of linear filters, the matched filter is the only one that can provide maximum output SNR of $2E/N_0$, where $E = \int_{-\infty}^{\infty} s^2(t)dt$ is the energy of the signal $s(t)$, and $N_0/2$ is the power spectrum density of the white noise.
2. An optimal filter matched to the signal, $s(t)$, is also optimum to those signals that share the same waveform but different magnitude and time delay: $s'(t) = as(t - \tau)$. This is because of

$$\begin{aligned} H'(f) &= kS'*(f)\exp\{-j2\pi f t_0'\} = akS^*(f)\exp\{-j2\pi f(t_0' - \tau)\} \\ &= aH(f)\exp\{-j2\pi f[t_0' - (t_0 + \tau)]\} = aH(f) \end{aligned}, \tag{2.48}$$

   where $t_0' = t_0 + \tau$. However, a matched filter to the signal $s(t)$ is not optimum to those signals that have the same waveform but different frequency shift, $S'(f) = S(f + \nu)$. This is because the transfer function of the matched filter to the frequency-shift signal is

$$H'(f) = kS^*(f + \nu)\exp\{-j2\pi f t_0\}. \tag{2.49}$$

3. The matched filter is equivalent to a correlator. Because the impulse response of the matched filter is a mirror of the input signal, the output of the matched filter can be expressed as an autocorrelation function of the signal:

$$s_{\text{out}}(t) = \int_{-\infty}^{\infty} ks^*(t_0 - t)s(t - \tau)d\tau = kR_s(t - t_0). \tag{2.50}$$

4. The output of the matched filter to a signal is the Fourier transform of the power spectrum $|S(f)|^2$ of that signal:

$$\begin{aligned} s_{\text{out}}(t, t_0) &= \int S(f)[kS^*(f)\exp\{-j2\pi f t_0\}]\exp\{j2\pi f t\}df \\ &= \int kS(f)S^*(f)\exp\{j2\pi f(t - t_0)\}df = k\int |S(f)|^2 \exp\{j2\pi f(t - t_0)\}df. \end{aligned} \tag{2.51}$$

At an observation time, t_0, which must be equal to or greater than the signal time duration, the output of the matched filter can achieve its maximum value.

For simplicity, the observation time is selected at the origin $t_0 = 0$. Then, the response of the matched filter, $s_{\text{out}}(t, t_0)$, and the ambiguity function, $\chi(\tau, f_D)$, are related by

$$s_{\text{out}}(t, 0) = \chi(t, 0). \tag{2.52}$$

This means that the response of a matched filter can be derived from the ambiguity function by a cut along the line at $f_D = 0$.

2.5 Point Spread Function in ISAR Imaging

In general, the point spread function (PSF) of an imaging system is the system response to a point source. For optical imaging systems, the PSF defines the spread of a point source. It is the impulse response of an imaging system to a point object. The image of an object is a convolution of the object and the PSF given by

$$I(x,y) = \iint O(u,v) \cdot h(x-u, y-v)dudv, \tag{2.53}$$

where $O(u, v)$ is the object function, $I(x, y)$ represents the resulting reconstructed image of the object, and $h(x, y)$ is the PSF of the imaging system representing the linear mapping of the object to the image, as shown in Figure 2.16. The degree of spreading of the point object is a measurement of the quality of an imaging system.

Compared with a passive optics-based imaging system, an active radar imaging system is much more complicated. A radar system response to a point source is related to the transmitted signal waveform, transmitting and receiving antenna patterns, object properties, and image formation processing. The ambiguity function is determined only by signal waveforms without accounting for other factors. Therefore, in radar imaging systems the ambiguity function of the radar signal waveform is not exactly the PSF of the system but is a weighted sum of ambiguity functions.

To derive an appropriate parametric model of the PSF for active radar imaging systems, we should look at the major difference between a radar imaging system and an optical imaging system.

2.5.1 Radar Imaging System and its Mathematical Model

Radar imaging systems can be ground based, airborne, or spaceborne with uniform speed and constant altitude operating at a frequency band. Antenna emits EM waves with a desired beam pattern. The reflected signals are used to form an image of an object of interest.

An important parameter of any imaging system is its spatial resolution, which is defined as the minimum distance at which two different objects are detected by the system as two separated images of objects. Generally speaking, the spatial resolution of a pulse radar imaging system is determined by the length of the transmitted pulse and the antenna beamwidth. The pulse length determines range resolution in the direction of the LOS. The antenna beamwidth determines the azimuth (or cross-range) resolution in the direction perpendicular to the radar LOS. Thus, the produced images are composed of rectangular

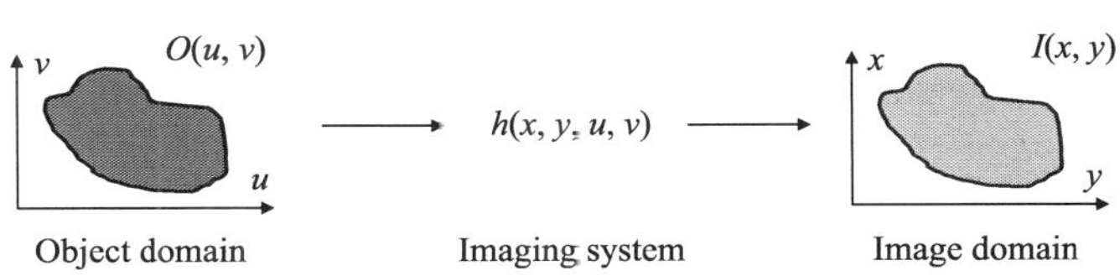

Figure 2.16 Model of an imaging system.

resolution cells whose sides are given by a range cell and a cross-range (or Doppler) cell. To achieve high range resolution, the transmitted pulse should have a very short duration and high-power peak. Pulse compression techniques by frequency modulation are usually used for this purpose. The received signal is processed through a matched filter to compress the pulse width.

The mathematical model of a radar imaging system is composed of a number of system properties. Let $\rho(u, v)$ be the reflectivity density function of an input object to the radar system. The expected output of the system, $I(x, y)$, is obtained by performing the 2-D convolution of the reflectivity density function and the impulse response function as follows:

$$I(x,y) = \iint \rho(u,v) \cdot h(x-u,\, y-v)dudv. \tag{2.54}$$

If the radar system impulse response function becomes a two-dimensional delta Dirac impulse function, namely, $h(x, y) = \delta\,(x, y)$, then the image $I(x, y)$ is equal to the reflectivity density function of the observed object. However, in practice, the impulse response function introduces degradations and blur that affect the appearance of the output image. For this reason, it must be processed to retrieve the object reflectivity.

2.5.2 PSF in ISAR Imaging

In ISAR imaging systems, the 2-D image is represented in the range and cross-range (Doppler) domain. The cross-range is the direction perpendicular to the range direction and contains the relative motion between the radar and the target. In analyzing the PSF, the coupling between the range and cross-range dimensions must be taken into account [17,18].

We will now demonstrate that, also in the case of ISAR imaging, we can express the image as the convolution of the target's EM characteristics (via its reflectivity function) and a system function, namely, the PSF, as shown in (2.54). Such demonstration starts from (1.31): $\rho(\tau, f_D) = IFT[S_R(t, f)]$.

As a first comment to the result in (1.31), we should mention that such result is only ideal. In fact, the reflectivity function $\rho(\tau, f_D)$ would be estimated perfectly if we knew its Fourier transform for all values of the slow time, t, and the frequency, f. Unfortunately, such a function is known only in a limited region of such a bidimensional domain (t, f). If we assume that the Fourier transform of the reflectivity function is known only in a limited domain defined by a new function, $W(t, f)$, we would modify (1.31) as follows:

$$\hat{\rho}(\tau, f_D) = IFT[S_R(t, f) \cdot W(t, f)]. \tag{2.55}$$

It should be clarified that such limitation is already embedded in $S_R(t, f)$ because it is both band and time limited. The scope of the function $W(t, f)$ is to bring forward such limitations. It must also be remarked that $W(t, f)$ also represents the time-frequency characterization of the received signal, $S_R(t, f)$.

If we interpret $\hat{\rho}(\tau, f_D)$ in (2.55) as the delay-time and Doppler ISAR image, we can then write the expression of the ISAR image using a property of the Fourier transform as follows:

$$I(\tau, f_D) = \iint \rho(u,v) \cdot w(\tau - u,\, f_D - v)dudv, \tag{2.56}$$

where $w(\tau, f_D) = IFT[W(t, f)]$ represents the ISAR imaging system PSF expressed in the delay time-Doppler domain.

To be able to rewrite (2.56) into the spatial coordinates (x, y), we will assume that a relationship exists between the coordinate pairs (τ, f_D) and (x, y). More specifically, we will assume that a relationship exists between the delay-time, τ, and the range coordinate, x, and between the Doppler frequency, f_D, and the cross-range coordinate, y, as follows:

$$x = f(\tau), \quad y = g(f_D). \tag{2.57}$$

We can then rewrite (2.56) by operating the change of variables introduced in (2.57) as follows:

$$I(x, y) = \iint \rho(u, v) \cdot w(x - u, y - v) du dv, \tag{2.58}$$

where $w(x, y)$ can be defined as the PSF expressed in spatial coordinates.

2.6 ISAR Image Projection Plane

ISAR image of a 3-D target is projected onto a 2-D range-Doppler plane. For a given target and known target rotation parameters, what the ISAR image looks like is determined by the ISAR image projection plane (IPP) [19–21]. In this section, we will synthesize an ISAR image of the moving target and derive the corresponding IPP, given the target's rotation vector and the radar LOS vector.

2.6.1 Characteristics of Target Motion

A target can be considered a rigid body with six degrees of freedom: three translations along the X, Y, Z directions and three roll, pitch, and yaw rotations $(\Omega_r, \Omega_p, \Omega_y)$ about the local coordinates (x, y, z), as shown in Figure 2.17.

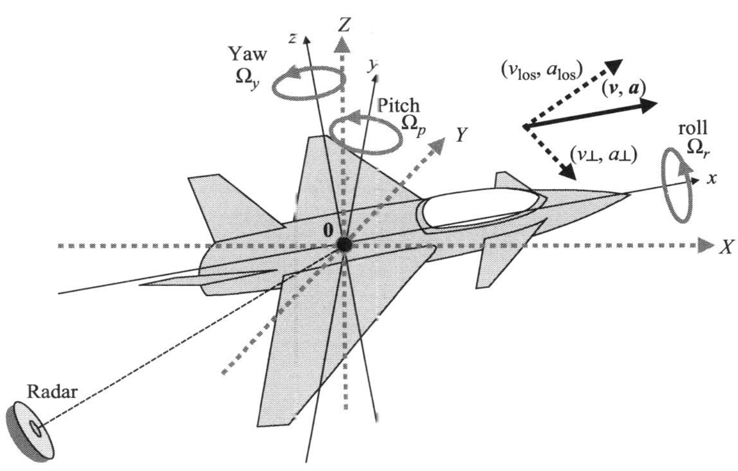

Figure 2.17 The six degrees of freedom of a moving target.

Usually, translational motion of the target is described by its velocity and acceleration ($\boldsymbol{v}$, $\boldsymbol{a}$), which can be decomposed into a component (v_{los}, a_{los}) along the radar LOS and a component ($v_{\perp}$, $a_{\perp}$) perpendicular to the LOS. The component along the LOS generates a Doppler frequency shift. On the other hand, the translation can also change the target aspect angle viewing from the radar. The view angle change is the effect of an apparent rotation of the target, which results in the same effect as target's rotational motion.

Over a short time period, the target rotation can be modeled by a constant rotation rate. Different scatterers in the target produce differential Doppler frequency shifts that form a range-Doppler image of the target. The apparent rotation combined with the self-rotation of the target induces spatially dependent Doppler frequency shifts, which are used to discriminate between scatterers at different spatial positions. However, the combination of these two rotation vectors is not simply a linear vector summation. A higher complexity is expected especially for a target with complex roll, pitch, and yaw motions. In this case, the assumption of constant rotations is valid only for a very short time period, within which the formulation based on lower-order approximation can be used.

ISAR images are formed by coherently processing the radar-received signals. Because of the target's relative motion to the radar, the aspect angle of the target is changed with time. Thus, the location of a scatterer at $\boldsymbol{r}(t)$ becomes a function of time. Assuming that the round-trip time delay from a scatterer at $\boldsymbol{r}(t)$ is $\tau(t)$ and the target is moving with a radial velocity, v_r and acceleration, a_r, the traveled distance of the radar signal before being reflected from the scatterer is

$$c\frac{\tau(t)}{2} = |\boldsymbol{r}(t) - \boldsymbol{R}_0| + v_r\frac{\tau(t)}{2} + \frac{1}{2}a_r\left[\frac{\tau(t)}{2}\right]^2, \tag{2.59}$$

where $\boldsymbol{R}_0$ is the vector range from the center of target to the radar (Figure 2.18).

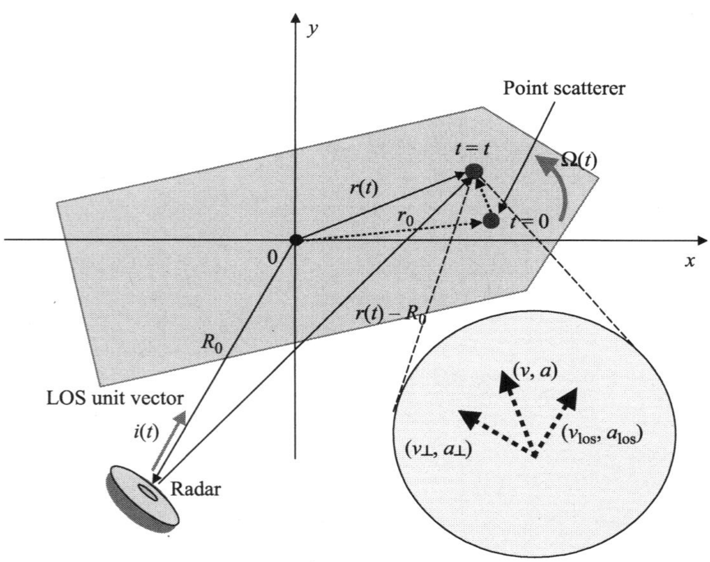

Figure 2.18 A 2-D geometry of a radar and a rotating target.

In most cases, the second-order term is much smaller than the first-order term. Thus, we have

$$c\frac{\tau(t)}{2} = |\boldsymbol{r}(t) - \boldsymbol{R}_0| + v_r\frac{\tau(t)}{2}. \tag{2.60}$$

Thus, the round-trip time delay becomes

$$\tau(t) = 2|\boldsymbol{r}(t) - \boldsymbol{R}_0|/(c - v_r) \approx 2|\boldsymbol{r}(t) - \boldsymbol{R}_0|/c, \tag{2.61}$$

where the position vector $\boldsymbol{r}(t)$ of the scatterer at time t can be derived from its position vector $\boldsymbol{r}(t_0)$ at time t_0 and a rotation matrix $\Re(\theta_r, \theta_p, \vartheta_y)$ [20,21]:

$$\boldsymbol{r}(t) = \Re(\theta_r, \vartheta_p, \theta_y)\boldsymbol{r}(t_0), \tag{2.62}$$

where the roll angle is $\theta_r = \Omega_r t$, pitch angle is $\theta_p = \Omega_p t$, and yaw angle is $\theta_y = \Omega_y t$.

2.6.2 Projection of 3-D Composite Target onto 2-D Image Plane

From (2.1), the phase function of a signal returned from a target is $\varphi(t) = 4\pi f_c R(t)/c$. If target is moving from its initial range with a velocity along the LOS, the phase function can be expressed by

$$\varphi(t) = 4\pi\frac{f_c}{c}\left[R_0 - \int v_{\text{los}}(t)dt\right], \tag{2.63}$$

where c is the speed of wave propagation, f_c is the radar carrier frequency, R_0 is the initial range of the rotation center, and v_{los} is the target's LOS velocity that determines the Doppler frequency shift of the target. The LOS velocity is the projection of the target's velocity vector, $\boldsymbol{v}(t)$, onto the LOS unit vector, $\boldsymbol{i}(t)$:

$$v_{\text{los}}(t) = \boldsymbol{v}(t) \cdot \boldsymbol{i}(t). \tag{2.64}$$

The ISAR image is displayed on a 2-D range and Doppler plane. The Doppler frequency is expressed by

$$f_D(t) = \frac{2f_c}{c}|\boldsymbol{v}(t) \cdot \boldsymbol{i}(t)|. \tag{2.65}$$

If $\boldsymbol{r}$ is the position vector of a scatterer measured from the rotation center, the Doppler frequency shift of the scatterer becomes

$$f_D(t) = \frac{2f_c}{c}[\boldsymbol{\Omega}(t) \times \boldsymbol{r}] \cdot \boldsymbol{i}(t), \tag{2.66}$$

where $\Omega(t)$ is the target's actual rotation vector.

Assuming, at time t, that the angle between the actual rotation vector $\Omega(t)$ and the LOS unit vector $\boldsymbol{i}(t)$ is ζ as shown in Figure 2.19, the Doppler frequency shift can be rewritten as

$$f_D(t) = \frac{2f_c}{c}\Omega(t)r_{cr}\sin\zeta = \frac{2f_c}{c}[\Omega(t)\sin\zeta]r_{cr} = \frac{2f_c}{c}\Omega_{\text{eff}}(t)r_{cr}, \tag{2.67}$$

where the magnitude of the effective rotation vector $\Omega_{\text{eff}}(t) = \Omega(t)\sin\zeta$, and r_{cr} is the actual cross-range displacement of the scatterer.

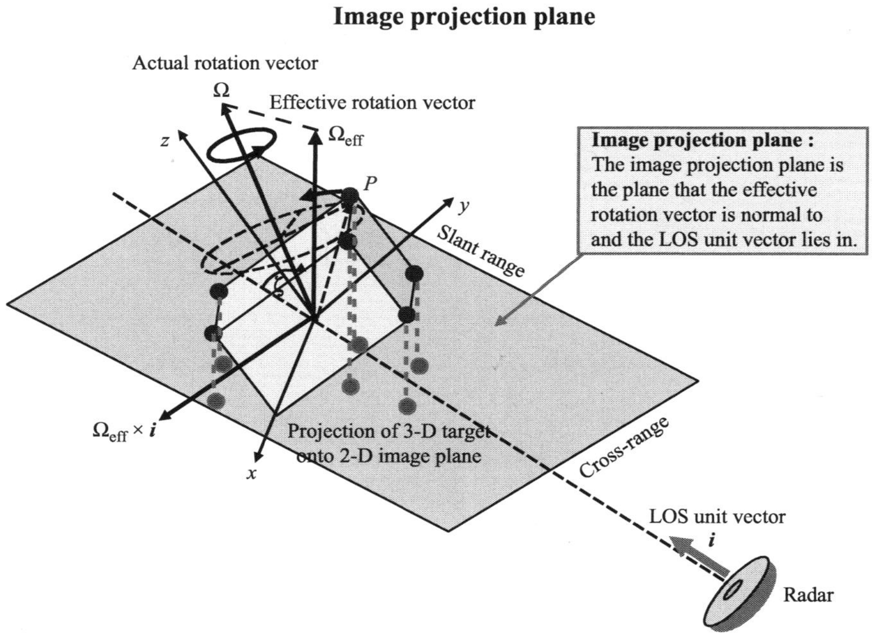

Figure 2.19 Image projection plane.

When a target has roll, pitch, and yaw motion, the combined actual rotation vector, $\boldsymbol{\Omega}$, determines the Doppler shift of a given scatterer in the target. The effective rotation vector, $\boldsymbol{\Omega}_{\text{eff}}$, is a vector that is normal to the LOS unit vector, $\boldsymbol{i}$, and on the plane on which $\boldsymbol{\Omega}$ and $\boldsymbol{i}$ lie, as indicated in Figure 2.19. Therefore, the image projection plane is defined as the plane that $\boldsymbol{\Omega}_{\text{eff}}$ is normal to and in which $\boldsymbol{i}$ lies, as shown in Figure 2.19 [19–21].

When the target's roll, pitch, and yaw motion changes with time, the effective rotation vector may vary with time, $\boldsymbol{\Omega}_{\text{eff}}(t)$. Therefore, an ISAR image of a target appears as a time-varying range-Doppler image on a constantly changing 2-D image projection plane.

2.6.3 Effect of Roll, Pitch, and Yaw Motion on ISAR Range-Doppler Imaging

If the target has roll, pitch, and yaw motion, the returned signal from the target can be expressed as

$$s_R(t) = \iiint \rho(\boldsymbol{r}) \exp\left\{ -j\frac{4\pi f_c}{c} \boldsymbol{r} \cdot \boldsymbol{i} \right\} d\boldsymbol{r}, \tag{2.68}$$

where the translational motion has been compensated, $\boldsymbol{r}$ is the location vector of a scatterer in the target, $\boldsymbol{i}$ is the unit vector along the radar LOS, and $\rho(\boldsymbol{r})$ is the target's reflectivity at $\boldsymbol{r}$. The reconstructed image can be expressed as

$$\rho(\boldsymbol{r}) = IFT\left[s_R(t) \exp\left\{ j\frac{4\pi f_c}{c} \boldsymbol{r} \cdot \boldsymbol{i} \right\} \right]. \tag{2.69}$$

When the target is rotating, $\boldsymbol{r}$ can be expressed in terms of the rotation matrix:

$$\boldsymbol{r} = \Re(\theta_r, \theta_p, \theta_y)\boldsymbol{r}_0, \tag{2.70}$$

where $\boldsymbol{r}_0$ is the location vector before rotating as indicated in Figure 2.18. Then,

$$\rho(\boldsymbol{r}) = IFT\left[s_R(t)\exp\left\{j\frac{4\pi f_c}{c}\left[\Re(\theta_r, \theta_p, \theta_y)\boldsymbol{r}_0\right]\cdot \boldsymbol{i}\right\}\right], \tag{2.71}$$

and the Doppler shift becomes

$$f_D = \frac{2f_c}{c}\frac{d}{dt}\left[\Re(\theta_r, \theta_p, \theta_y)\boldsymbol{r}_0\right]\cdot \boldsymbol{i}, \tag{2.72}$$

where $\frac{d}{dt}\left[\Re(\theta_r, \theta_p, \theta_y)\boldsymbol{r}_0\right]$ determines the effect of the roll, pitch, and yaw on Doppler shift of the scatterer at the location $\boldsymbol{r}_0 = (x_0, y_0)$ [20,21].

For a simple case of yaw motion only, that is, $\theta_r = 0$, $\theta_p = 0$, and $\theta_y = \Omega_y t$, the Doppler shift due to pure yaw motion becomes

$$f_{D_{yaw}} = \frac{2f_c}{c}(-x_0\Omega_y^2 t - y_0\Omega_y). \tag{2.73}$$

2.7 ISAR Image Processing

ISAR can generate high-resolution range-Doppler images. The basic concept of how ISAR generates a range-Doppler image of a target is illustrated in Figure 2.20.

The range resolution is achieved using short pulse signals, which can be obtained by pulse compressing wideband signal waveforms, such as LFM chirp or stepped-frequency waveforms. With a range resolution, the returned signals from different scatterer centers of the target can be resolved into different range cells. Target scatterers located in a same range cell can have different cross-ranges. The cross-range direction is perpendicular to the LOS and also perpendicular to the effective rotation vector, Ω_{eff}. A target such as an aircraft can be imaged via its roll, pitch, and yaw rotations. Target scatterers at different cross-range are resolved through their different Doppler shifts.

2.7.1 Background

Theoretical formulation of ISAR imaging is described by Walker's model [22–24]. We will focus on analyzing factors that make an ISAR image defocused and on image processing techniques that compensate the effect of target motion and rotation for refocusing an ISAR image.

ISAR image processing includes three steps: preprocessing, range processing, and cross-range processing.

1. PREPROCESSING: The preprocessing step consists of processing received raw ISAR data to remove amplitude and phase errors introduced during the data collection, to filter out unwanted modulations and interferences, and to eliminate artifacts.

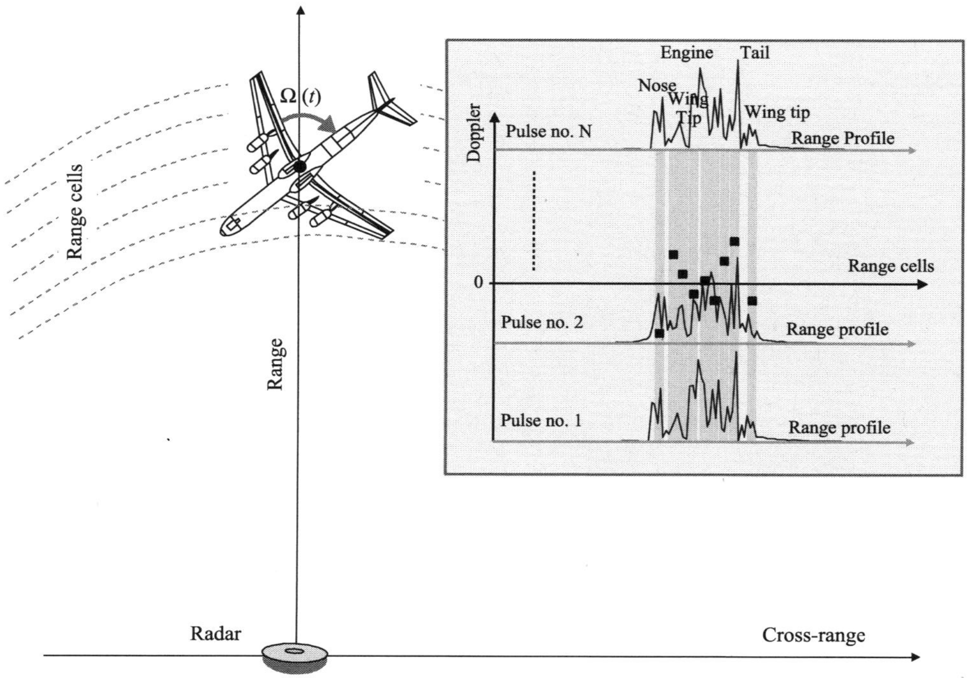

Figure 2.20 Basic concept of ISAR range-Doppler imaging.

2. RANGE PROCESSING: Range processing consists of compensating the translational motion of the target. Preprocessed and uncompensated range profiles are shown in Figure 2.21a, where the target's rapid shift in range can be seen. The procedure of the range processing includes a coarse range compensation and a fine range compensation.

   A simple method for range compensation is to apply the cross-correlation between two successive range profiles. However, after the cross-correlation the phase function may become nonlinear as shown in Figure 2.21b. Therefore, a phase correction must be applied. After phase correction processing, the result becomes a set of range-aligned range profiles as shown in Figure 2.21c. From the range-aligned range profiles, at each range cells we should apply the Fourier transform across pulses, and then an ISAR image of the target can be formed as shown in Figure 2.21d.

   However, if the target has rapid roll, pitch, and yaw rotation, the range processed ISAR image can still be seriously defocused. Thus, a further cross-range processing must be performed.
3. CROSS-RANGE PROCESSING: Range processing can compensate the translational motion only at the center of a target. If the target has rotation about its center of rotation, it may lead to time-varying Doppler drifts, $f_D(t)$, that induces additional defocusing. The time varying of the effective rotation rate, $\mathbf{\Omega}_{\text{eff}}(t)$ can also cause the cross-range defocusing. Therefore, the main goal of the cross-range processing is to estimate these two time-varying functions that cause the image defocusing.

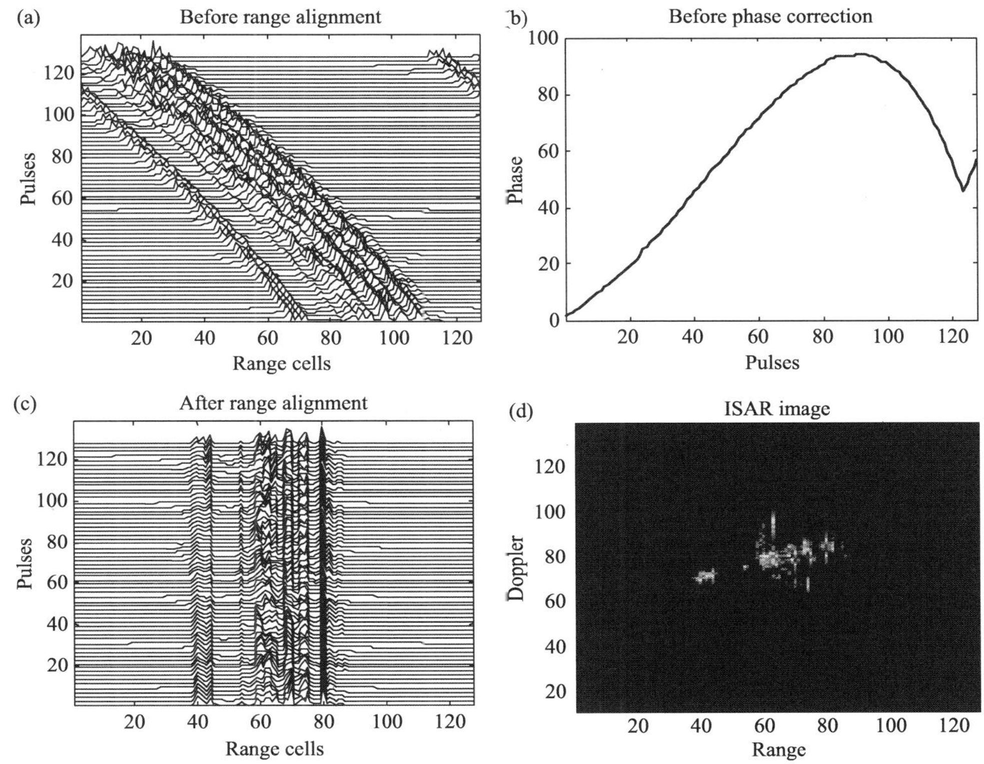

Figure 2.21 (a) Preprocessed and uncompensated range profiles; (b) The phase function after cross-correlation processing; (c) The result of range alignment; (d) ISAR image of the target.

Another issue in the cross-range processing is the cross-range scaling. In fact, to superpose a geometrical projection of the target onto the image projection plane, we need to convert the Doppler shift scale (Hz) to a range scale (meter). The cross-range scaling will be discussed in detail in Chapter 7.

2.7.2 Cross-range Focusing

After ISAR range processing, target's dominant scatterers will remain positioned at fixed range. Thus, ISAR image defocusing is concentrated on the cross-range processing. If a number of dominant scatterers on the target are located in a specific range cell as illustrated in Figure 2.22a, when the Doppler drift is zero, $f_D(t) = 0$, and the effective rotation rate is zero, $\Omega_{eff}(t) = 0$, the Doppler history of such scatterers does not change with time. Thus, the received signal from this specific range cell is in focused condition.

When there is a Doppler drift, $f_D(t) \neq 0$, all Doppler histories for scatterers on the target are subject to the same time variation as illustrated in Figure 2.22b. In this case, the ISAR image is suffering the same defocusing for all of the scatterers. If the target has a time-varying

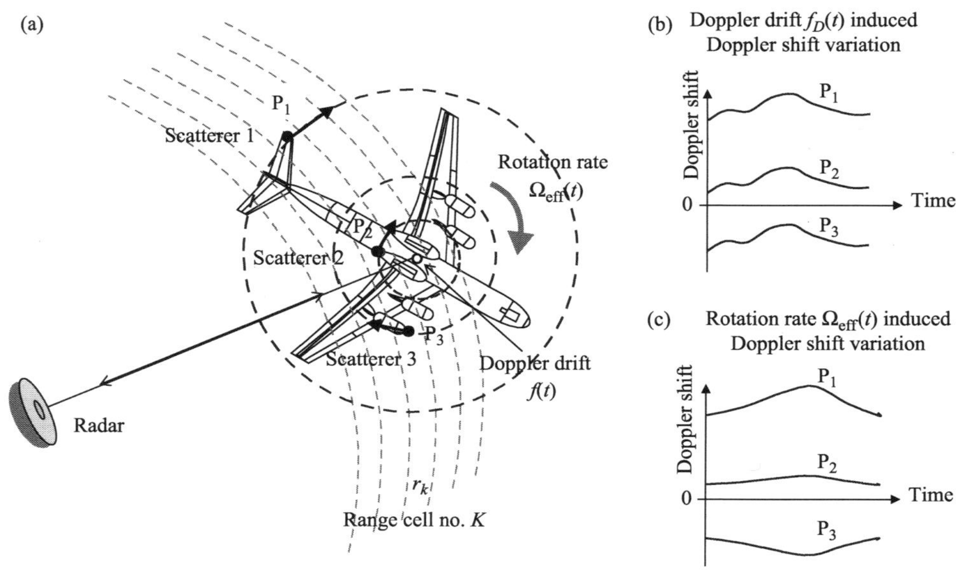

Figure 2.22 (a) Illustration of three dominant scatterers P_1, P_2, and P_3, located at the same range cell r_k. (b) If the target suffers from time-varying Doppler drift, $f_D(t)$, all Doppler histories for scatterers on the target are subject to the same variation. (c) If the target has a time-varying rotation rate, $\Omega_{eff}(t)$, all Doppler histories for scatterers on the target are subject to different time variation depending on the scatterer's distance from the center of rotation. Scatterers at longer distances from the rotation center suffer from larger Doppler variations.

rotation rate $\Omega_{eff}(t)$, all Doppler histories for scatterers on the target are subject to different time variation. The expansion of such variation depends on the scatterer distance from the center of rotation. Scatterers at longer distances suffer from larger Doppler variations as illustrated in Figure 2.22c.

In general, based on (1.9) and (2.67), the defocused baseband signal can be expressed by

$$s_B(t) = \sum_{n=1}^{N} A_n \exp\{-j2\pi f_D(t,n)t\} = \sum_{n=1}^{N} A_n \exp\left\{-j2\pi\left[\frac{2f}{c}\Omega_{eff}(t)r_{cr}(n)\right]t\right\}, \tag{2.74}$$

where N is the total number of scatterers on the target, $r_{cr}(n)$ is the distance from the target rotation center to the n-th scatter, and $f_D(t, n)$ is the Doppler drift of the n-th scatterer.

In cross-range processing, a method for compensation of Doppler drifts is to simply apply a phase correction function, $\Phi(t) = f_D(t)$, to the defocused baseband signal if we can estimate the Doppler drift function, $f_D(t, n) = f_D(t)$ since all Doppler histories of scatterers on the target are subject to the same variation:

$$s_{f_D}(t) = \exp\{j\Phi(t)t\} \sum_{n=1}^{N} A_n \exp\{-j2\pi f_D(t,n)t\}. \tag{2.75}$$

However, the compensation for time-varying rotation rate, $\Omega_{\text{eff}}(t)$, is more complicated. Generally speaking, ISAR image formation and autofocus algorithms are available for cross-range focusing, including the range-Doppler algorithm, the minimum variance algorithm, the phase difference algorithm, the prominent point processing algorithm, the phase gradient algorithm, the entropy minimization algorithm, the contrast optimization algorithm, and the time-frequency algorithm. Most of these algorithms will be detailed in the next chapters.

2.8 Bistatic ISAR

Some elements of bistatic ISAR are provided in this section. (More details will be found in Chapter 10, which is entirely dedicated to bistatic ISAR imaging). As will be detailed in Chapter 10, a configuration with separated transmitter and receiver (bistatic) offers more degrees of freedom for acquiring complementary target's information and for defeating low monostatic RCS, which are typical of monostatic radar. In bistatic radar systems [6], to determine a target's location the timing of the transmitted signal must be known exactly. Synchronization between the receiver and transmitter must therefore be ensured. The range and Doppler resolutions are all dependent on the bistatic angle. As in the monostatic case, the target's rotational motions may induce bistatic Doppler shifts. Such shifts, in the bistatic case, would depend on the bistatic geometry, that is, the position of the target, transmitter, and receiver.

Different from the pseudo-monostatic case or forward scattering case, in our case the term *bistatic* refers to the geometry, where the transmitter and the receiver are separated by a baseline distance that is comparable with the maximum target range. Such separation brings up an issue of synchronization between the two sites and requires phase synchronization between the local oscillator in the transmitter and the one in the receiver to accurately measure the target location and to perform range and Doppler processing.

To have a better understanding of bistatic ISAR imaging, we will now define the bistatic ISAR geometry.

The coordinate systems include global space-fixed coordinates and target local body-fixed coordinates as shown in Figure 2.23 (in a three-dimensional case). The bistatic

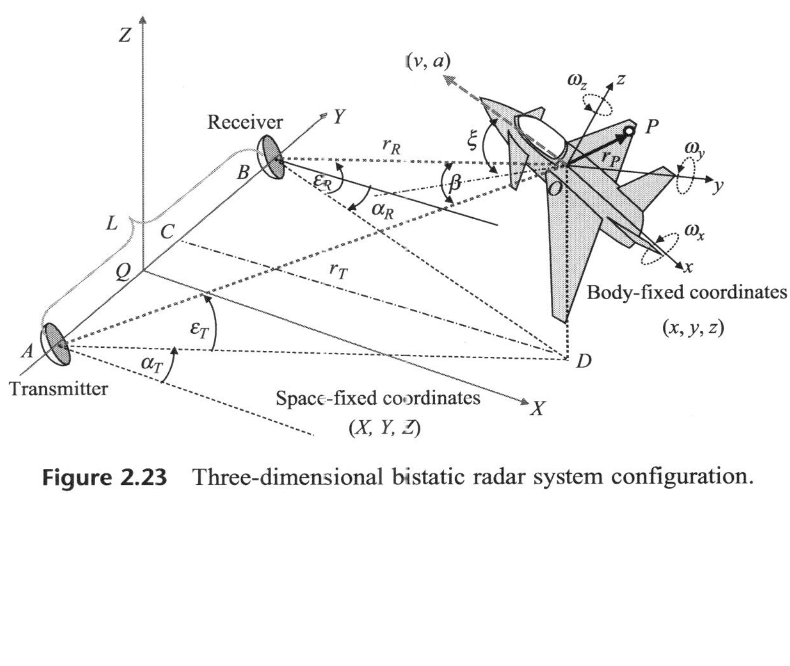

Figure 2.23 Three-dimensional bistatic radar system configuration.

plane is the one containing the transmitter, the receiver, and the target. The baseline, L, is the distance between the transmitter and receiver. The range from the transmitter to the target is a vector, $\boldsymbol{r}_T$, and the range from the receiver to the target is a vector, $\boldsymbol{r}_R$. The bistatic angle is the angle between the transmitter-to-target line and the receiver-to-target line. Assume the transmitter look angles (azimuth and elevation) are $(\alpha_T, \varepsilon_T)$, the receiver look angles $(\alpha_R, \varepsilon_R)$ can be obtained from the baseline distance, the target range, and looking angles relative to the transmitter. The positive angle is defined in a counterclockwise direction. Thus, the distance from the receiver to the target becomes

$$r_R = |\boldsymbol{r}_R| = (L^2 + r_T^2 \cos^2 \varepsilon_T - 2 r_T L \cos \varepsilon_T \sin \alpha_T)^{1/2}, \tag{2.76}$$

and the receiver look angles are

$$\alpha_R = \tan^{-1}\left(\frac{L - r_T \cos \varepsilon_T \sin \alpha_T}{r_T \cos \varepsilon_T \cos \alpha_T}\right), \tag{2.77}$$

$$\varepsilon_R = \tan^{-1}\left(\frac{r_T \sin \varepsilon_T}{(L^2 + r_T^2 \cos^2 \varepsilon_T - 2 r_T L \cos \varepsilon_T \sin \alpha_T)^{1/2}}\right). \tag{2.78}$$

The received signal from a point target, P, can be modeled as

$$s_r^P(t) = \rho(\boldsymbol{r}_P) \exp\left\{ j2\pi f \frac{|\boldsymbol{r}_T(t) + \boldsymbol{r}_P(t)| + |\boldsymbol{r}_R(t) + \boldsymbol{r}_P(t)|}{c} \right\}, \tag{2.79}$$

where c is the speed of wave propagation. Then, the received signal from a volume target is a volume integration of the point return over the whole target:

$$s_r(t) = \iiint \rho(\boldsymbol{r}_P) \exp\{j2\pi f[|\boldsymbol{r}_T(t) + \boldsymbol{r}_P(t)| + |\boldsymbol{r}_R(t) + \boldsymbol{r}_P(t)|]/c\} d\boldsymbol{r}_P \tag{2.80}$$

The phase term in the received signal is

$$\begin{aligned} \Phi_P(t) &= 2\pi f \frac{\{|\boldsymbol{r}_T(t) + \boldsymbol{r}_P(t)| + |\boldsymbol{r}_R(t) + \boldsymbol{r}_P(t)|\}}{c} \\ &= \frac{2\pi f \{r_T(t) + r_R(t) + [\boldsymbol{r}_T(t) + \boldsymbol{r}_R(t)] \cdot \boldsymbol{r}_P(t)\}}{c}, \\ &= \Phi_V(t) + \Phi_{\Omega,P}(t) \end{aligned} \tag{2.81}$$

where

$$\Phi_V(t) = 2\pi f \{r_T(t) + r_R(t)\}/c \tag{2.82}$$

is the phase term induced by translational motions and

$$\Phi_{\Omega,P}(t) = 2\pi f \{[\boldsymbol{r}_T(t) + \boldsymbol{r}_R(t)] \cdot \boldsymbol{r}_P(t)\}/c \tag{2.83}$$

is the phase term induced by rotational motions of the scatterer, P.

While the target has translational motion with a velocity, $\boldsymbol{V}$, and rotation with an angular velocity vector, $\Omega = (\omega_x, \omega_y, \omega_z)^T$, rotating about the target local-fixed axes x, y, and z, then the Doppler frequency shift induced by the translation and rotation can be obtained by means of a time derivative of the phase function.

The Doppler shift consists of two parts: one is induced by the translation and the other is induced by the rotation,

$$f_{D_{Bi}} = f_{D_{Tran}} + f_{D_{Rot}}, \tag{2.84}$$

where

$$f_{D_{Tran}} = \frac{f}{c}\frac{d}{dt}[r_T(t) + r_R(t)] \tag{2.85}$$

is the translational Doppler shift and

$$f_{D_{Rot}} = \frac{2f}{c}[\boldsymbol{\Omega} \times \boldsymbol{r_P}(t)] \tag{2.86}$$

is the rotation induced micro-Doppler, which is usually a periodic frequency function of time and distributed around the translational Doppler frequency.

If the target moves with $\boldsymbol{V}$ and an acceleration, $\boldsymbol{a}$, their components along the direction from the transmitter to the target are

$$V_T = \boldsymbol{V} \cdot \frac{\boldsymbol{r_T}}{|\boldsymbol{r_T}|}; \quad a_T = \boldsymbol{a} \cdot \frac{\boldsymbol{r_T}}{|\boldsymbol{r_T}|} \tag{2.87}$$

and components along the direction from the receiver to the target are

$$V_R = \boldsymbol{V} \cdot \frac{\boldsymbol{r_R}}{|\boldsymbol{r_R}|}; \quad a_R = \boldsymbol{a} \cdot \frac{\boldsymbol{r_R}}{|\boldsymbol{r_R}|}. \tag{2.88}$$

Then the range from the transmitter to the moving target becomes a continuous and differentiable function that can be approximated with a polynomial:

$$r_T = r_{T_0} + V_T t + \frac{1}{2} a_T t^2 + \cdots. \tag{2.89}$$

Equivalently, the range from the receiver to the target may be approximated as follows:

$$r_R = r_{R_0} + V_R t + \frac{1}{2} a_R t^2 + \cdots. \tag{2.90}$$

When the target has rotational motion, its rotation angles are determined by its initial angles, $\boldsymbol{\theta}_0$, and rotation rate, $\boldsymbol{\Omega}$:

$$\boldsymbol{\theta} = \boldsymbol{\theta}_0 + \boldsymbol{\Omega} t + \cdots, \tag{2.91}$$

where

$$\boldsymbol{\Omega} = (\omega_x, \omega_y, \omega_z)^T. \tag{2.92}$$

Therefore, for a second-order polynomial approximation the Doppler shift induced by the translational motion is

$$f_{D_{Tran}} = \frac{f}{c}\frac{d}{dt}[r_T(t) + r_R(t)] = \frac{f}{c}[V_T + V_R + (a_T + a_R)t] \tag{2.93}$$

and that induced by the rotation becomes

$$f_{D_{Rot}} = \frac{2\pi f}{c}\frac{d}{dt}\{[\boldsymbol{r_T}(t) + \boldsymbol{r_R}(t)] \cdot \boldsymbol{r_P}(t)\}, \tag{2.94}$$

Thus, the translational Doppler shift of a bistatic radar depends on three factors [6]. The first factor is the maximum Doppler shift that is produced when a target moves with a velocity, $\boldsymbol{V}$:

$$f_{D_{\text{Max}}} = \frac{2f}{c}|\boldsymbol{V}|. \tag{2.95}$$

The second factor is related to the bistatic triangulation factor

$$D = \cos\left(\frac{\alpha_R - \alpha_T}{2}\right) = \cos\left(\frac{\beta}{2}\right), \tag{2.96}$$

where $\beta = \alpha_R - \alpha_T$ is the bistatic angle.

The third factor is related to angle ξ, which is the angle between the moving direction of the target and the direction of the bisector as indicated in Figure 2.23:

$$C = \cos\xi. \tag{2.97}$$

Thus, the bistatic Doppler shift becomes

$$f_{D_{Bi}} = f_{D_{\text{Max}}} D \cdot C = \frac{2f}{c}|\boldsymbol{V}|\cos\left(\frac{\beta}{2}\right)\cos\xi. \tag{2.98}$$

Similar to the Doppler shift induced by the radial velocity in the monostatic radar case, for the bistatic radar case the Doppler shift is induced by the target's bisector velocity. The bistatic Doppler shift is always smaller than the maximum monostatic Doppler because the term $\cos(\beta/2)$ is always less than 1.

Compared with the range resolution, Δr_{Mono}, and Doppler resolution, $\Delta f_{D_{\text{Mono}}}$, of a monostatic ISAR, the range resolution, Δr_{Bi}, and Doppler resolution, Δf_{DBi}, of a bistatic ISAR also depends on the bistatic angle, β:

$$\Delta r_{Bi} = \Delta r_{\text{Mono}}/\cos\left(\frac{\beta}{2}\right), \tag{2.99}$$

and

$$\Delta f_{D_{Bi}} = \cos\left(\frac{\beta}{2}\right)\Delta f_{D_{\text{Mono}}}. \tag{2.100}$$

It is quite clear that there is a significant similarity between the characteristics of the received signal in the monostatic and bistatic cases, the same as that found in some system parameters, such as the range and Doppler resolution. In Chapter 10, we will introduce a virtual monostatic geometry that will replace the bistatic geometry and will allow for monostatic ISAR image processing to be used also in the bistatic case under some conditions. Such geometry will be based on a bistatically equivalent monostatic (BEM) ISAR configuration and will allow the use of monostatic ISAR signal modeling and consequently monostatic ISAR signal processing to form ISAR images in the bistatic case.

References

[1] Knott, E.F., J.F. Schaffer, and M.T. Tuley, *Radar Cross Section*, 2d ed., Artech House, 1993.

[2] Ruck, G.T., et al., *Radar Cross Section Handbook*, Plenum Press, 1970.

[3] Jenn, D., "Radar Cross-Section," in *Encyclopedia of RF and Microwave Engineering*, K. Chang, ed., John Wiley & Sons, 2005.

[4] Skolnik, M.I., *Introduction to Radar Systems*, 3d ed., McGraw-Hill, 2001.

[5] Crispin, J.W. Jr., and K.M. Siegel, *Methods of Radar Cross-Section Analysis*, Chap. 5, Academic Press, 1968.

[6] Willis, N.J., *Bistatic Radar*, 2d ed., SciTech Publishing, 2005.

[7] Ling, H., K. Chou, and S. Lee, "Shooting and Bouncing Rays: Calculating the RCS of an Arbitrarily Shaped Cavity," *IEEE Trans. Anten. Propag.*, vol. 37, no. 2, pp. 194–205, 1989.

[8] Kunz, K., and R. Luebbers, *The Finite-Difference Time Domain Method for Electromagnetics*, CRC Press, 1993.

[9] Chatzigeorgiadis, F., and D. Jenn, "A MATLAB Physical-Optics RCS Prediction Code," *IEEE Antenna and Propagation Magazine*, vol. 46, no. 4, pp. 137–139, 2004.

[10] Chatzigeorgiadis, F., *Development of Code for Physical Optics Radar Cross Section Prediction and Analysis Application*, Master's thesis, Naval Postgraduate School, Monterey, CA, September 2004.

[11] Shirman, Y.D. (ed.), *Computer Simulation of Aerial Target Radar Scattering, Recognition, Detection, and Tracking*, Artech House, 2002.

[12] Gabor, D., "Theory of Communication," *J.IEE (London)*, vol. 93, no. 3, pp. 429–457, November 1946.

[13] Woodward, P.W., *Probability and Information Theory, with Applications to Radar*, Pergamon Press, 1953.

[14] Gill, G.S., and J.C. Huang, "The Ambiguity Function of the Step Frequency Radar Signal Processor," *Proceedings of 1996 CIE International Conference on Radar*, pp. 375–380, 1996.

[15] North, D.O., "Analysis of Factors which Determine Signal-to-Noise Discrimination in Pulsed Carrier Systems," *Proc. IEEE*, vol. 51, pp. 1015–1028, 1963.

[16] Kelly, E.J., and R.P. Wishner, "Matched-Filter Theory for High-Velocity Targets," *IEEE Trans. Military Electr.*, vol. 9, pp. 56–69, 1965.

[17] Chen, C.C., and H. Andrews, "Multifrequency Imaging of Radar Turntable Data," *IEEE Trans. on AES*, vol. 16, no. 1, pp. 15–22, 1980.

[18] Chen, C.C., and H.C. Andrews, "Target Motion Induced Radar Imaging," *IEEE Trans. on AES*, vol. 16, no. 1, pp. 1–14, 1980.

[19] Wehner, D.R., *High-Resolution Radar*, 2d ed., Artech House, 1994.

[20] Chen V.C., and W.J. Miceli, "The Effect of Roll, Pitch and Yaw Motions on ISAR Imaging," *Proceedings of SPIE on Radar Processing, Technology, and Applications*, 1999.

[21] Chen V.C., and W.J. Miceli, "Analysis of Rotation Effects on ISAR Imaging Based on TIRA Radar and ARDS Data," *NATO RTO Symposium on High Resolution Radar Techniques*, Spain, 1999.

[22] Walker, J.L., "Range-Doppler Imaging of Rotating Objects," *IEEE Trans. on Aerospace Electronic Systems*, vol. 16, no. 1, pp. 23–52, 1980.

[23] Brown, W.M., "Walker Model for Radar Sensing of Rigid Targets Fields," *IEEE Trans. on AES*, vol. 16, no. 1, pp. 104–107, 1980.

[24] Ausherman, D.A., A. Kozma, J.L. Walker, H.M. Jones, and E.C. Poggio, "Developments in Radar Imaging," *IEEE Trans. on Aerospace and Electronic Systems*, vol. 20, no. 4, pp. 363–400, 1984.

CHAPTER 3

ISAR Image Formation

As we discussed in Chapter 1, for increasing radar image resolution in cross-range a larger antenna aperture is required. Antenna aperture can be synthesized through the relative rotation between a target to be imaged and the radar. From the point of view of the synthesizing aperture, for a stationary radar and a rotating target (Figure 3.1a) inverse synthetic aperture radar (ISAR) image of the target is equivalent to a spotlight synthetic aperture radar (SAR) image of a stationary target (Figure 3.1b). If the target has sufficient rotation, the coherent processing interval (CPI) in ISAR can be significantly shorter than that in SAR while keeping the same cross-range accuracy. This will be explained more quantitatively later on.

In this chapter, we will introduce the most commonly used ISAR range-Doppler image formation in Section 3.1. However, if the Doppler spectrum generated by a rotating target has severe time variation, ISAR range-Doppler image will become smeared in the Doppler domain. In these cases, in addition to conventional rotational motion compensation (RMC) methods such as the polar formatting algorithm (PFA), the time-frequency–based image formation can be used. We will introduce the time-frequency image formation algorithm in Section 3.2. For better display of two-dimensional (2-D) ISAR imagery, in Section 3.3 we suggest a useful windowing and zero-padding method for suppressing sidelobes.

3.1 ISAR Range-Doppler Image Formation

To generate an ISAR range-Doppler image of a moving target, if the target's effective rotation angle (defined in Section 2.6.2) is greater than the angle $[(\lambda/l_{\text{LOS}})^{1/2}/2]$ (where λ is the wavelength, and l_{LOS} is the projected target size along the radar LOS), the generated range-Doppler image will be defocused in the range domain [1]. However, in the Doppler domain, based on (1.30), the Doppler shift, f_D, of any scatterer on the target is determined by the rotation rate, Ω, the scatterer's cross-range displacement from the center of rotation, r_c, and the wavelength, λ:

$$f_D = \frac{2\Omega}{\lambda} r_c. \tag{3.1}$$

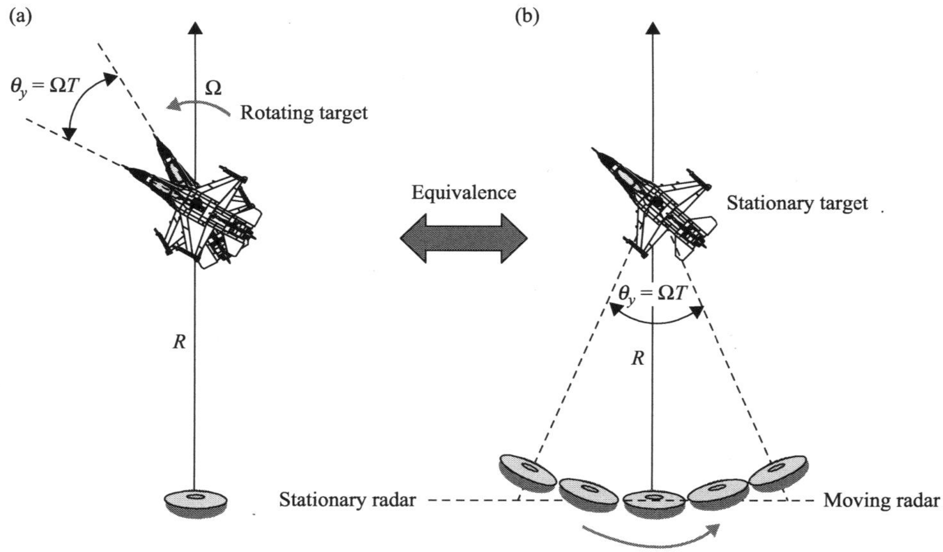

Figure 3.1 (a) A stationary radar and a rotating target is equivalent to (b) a spotlight SAR imaging of a stationary target.

As discussed in the Chapter 1, according to (1.21) the radar received signal can be modeled by

$$s_R(t) = \exp\left[-j4\pi f \frac{R(t)}{c}\right] \iint_{-\infty}^{\infty} \rho(x, y) \exp\{-j2[xk_x(t) - yk_y(t)]\} dxdy, \tag{3.2}$$

where

$$k_x(t) = k \cos\theta(t) \tag{3.3}$$

and

$$k_y(t) = k \sin\theta(t). \tag{3.4}$$

The instantaneous range and rotation angle can be written in terms of target motion history:

$$\begin{cases} R(t) = R_0 + v_0 t + \frac{1}{2} a_0 t^2 + \cdots \\ \theta(t) = \theta_0 + \Omega_0 t + \frac{1}{2} \gamma_0 t^2 + \cdots \end{cases}, \tag{3.5}$$

where the translational motion parameters are initial range R_0, velocity v_0, and acceleration a_0, and the angular rotation parameters are initial angle θ_0, angular velocity Ω_0, and angular acceleration γ_0. If these translational motion parameters can be accurately estimated, the extraneous phase term $\exp\left[-j4\pi f \frac{R(t)}{c}\right]$ can be completely removed. Thus, the target's reflectivity density function $\rho(x, y)$ can be reconstructed exactly by taking a 2-D inverse Fourier transform.

Thus, for ISAR range-Doppler image formation, the first step is to carry out the translational motion compensation (TMC). It estimates the target's translational motion parameters and removes the extra phase term, such that the target's range is no longer varying with time. Then, by taking the Fourier transform along the pulses (slow-time) domain as indicated in Figure 1.7, the range-Doppler image of the target can be reconstructed. In many cases, however, the target may also rotate about an axis. The rotational motion can make Doppler shifts to be time varying; thus, by using the Fourier transform the reconstructed ISAR image will be smeared in the Doppler domain. In this case, the RMC must be carried out to correct for the rotation. After taking the TMC and RMC, ISAR range-Doppler images can be correctly reconstructed by a 2-D Fourier transform. Therefore, the previously described image formation is typically addressed as the ISAR range-Doppler image formation.

As we know, in SAR image formation algorithms, there is a range-Doppler algorithm (RDA), although it is different from the ISAR range-Doppler image formation. The SAR RDA was introduced in 1976 to solve the problem of range migration in a satellite SAR [2] and since then has become the standard. Here, we first recall the SAR RDA to provide groundwork to understand the ISAR RDA.

3.1.1 SAR and ISAR Range-Doppler Algorithms

SAR is used to generate radar image of stationary targets, such as buildings and bridges. To achieve high cross-range resolution, SAR synthesizes a large-sized antenna aperture through a radar platform's motion in the along-track direction. The process of SAR imaging of a point target is depicted in Figure 3.2. SAR collects data from the target space, then converts data in the data space, and finally forms a SAR image in the image space.

On the other hand, ISAR synthesizes a larger antenna aperture through the relative rotation between radar and target. In a ground-based ISAR, a radar image of a moving target is generated through the target's relative rotation with respect to the radar. In an airborne ISAR, a radar image of a moving target is generated through the relative rotation contributed by both the target and the radar motions. Figure 3.3 illustrates the difference between SAR imaging of stationary targets and ISAR imaging of moving targets.

The collected SAR raw data are similar to that of ISAR. Typically, the pulse number in ISAR data is replaced by the azimuth-cell number in SAR. Thus, the collected SAR data are 2-D in the delay-time and azimuth-cell domains (Figure 3.4), similar to the ISAR raw data arrangement in Figure 1.6. The difference in terms of platform movement in SAR and target movement in ISAR is that the time-varying azimuth change due to radar platform movement corresponds to the time-varying view angle change due to target movement.

A block diagram of the SAR RDA is shown in Figure 3.5 [3]. The SAR data are in the range-cell (fast-time) and azimuth-cell (slow-time) domains. A matched filter is used in the range domain for range compression. Because the range-Doppler algorithm works in the range and azimuth-frequency (Doppler) domains, the Fourier transform in the azimuth domain must be taken to convert to the Doppler domain. Then, the range cell migration correction (RCMC) is performed in the range and Doppler domains to compensate for range cell migration. After the RCMC, an azimuth-matched filter is applied. Finally, by taking the inverse fast Fourier transform (IFFT) in the azimuth domain, the SAR image is formed.

Similarly, a block diagram of the ISAR RDA is shown in Figure 3.6, where the raw ISAR data are arranged in the range cells (fast-time) and number of pulses (slow-time) domains.

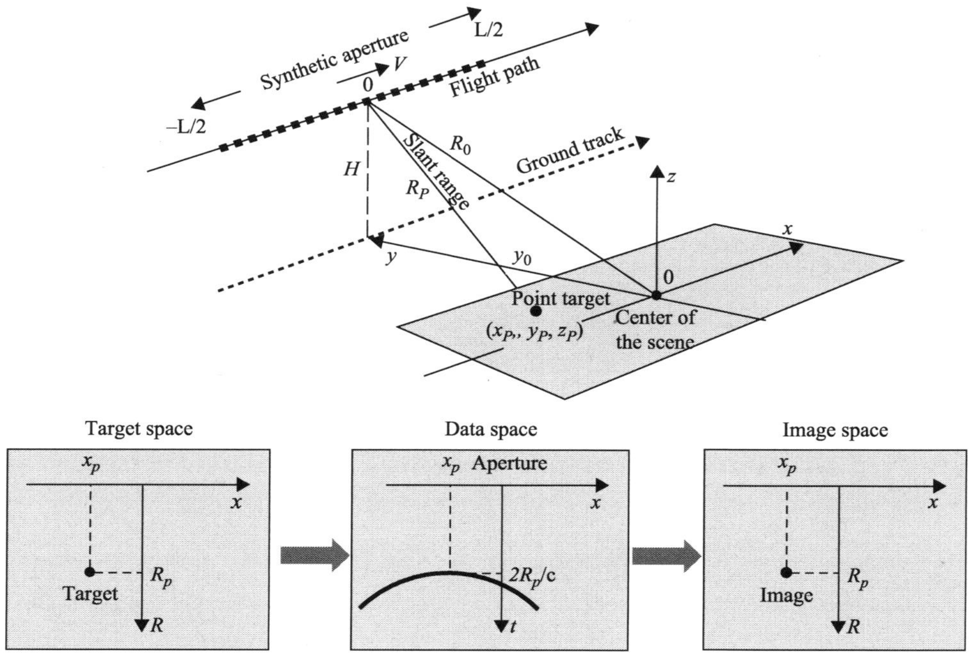

Figure 3.2 The process of SAR imaging of a stationary point target.

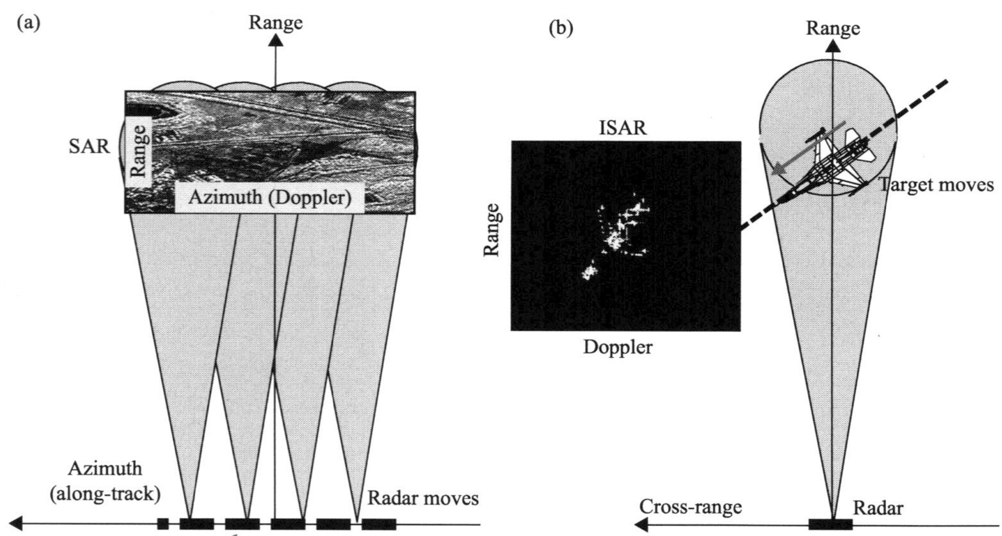

Figure 3.3 (a) SAR imaging of stationary objects in range and azimuth domain and (b) ISAR imaging of moving targets in range and cross-range domain.

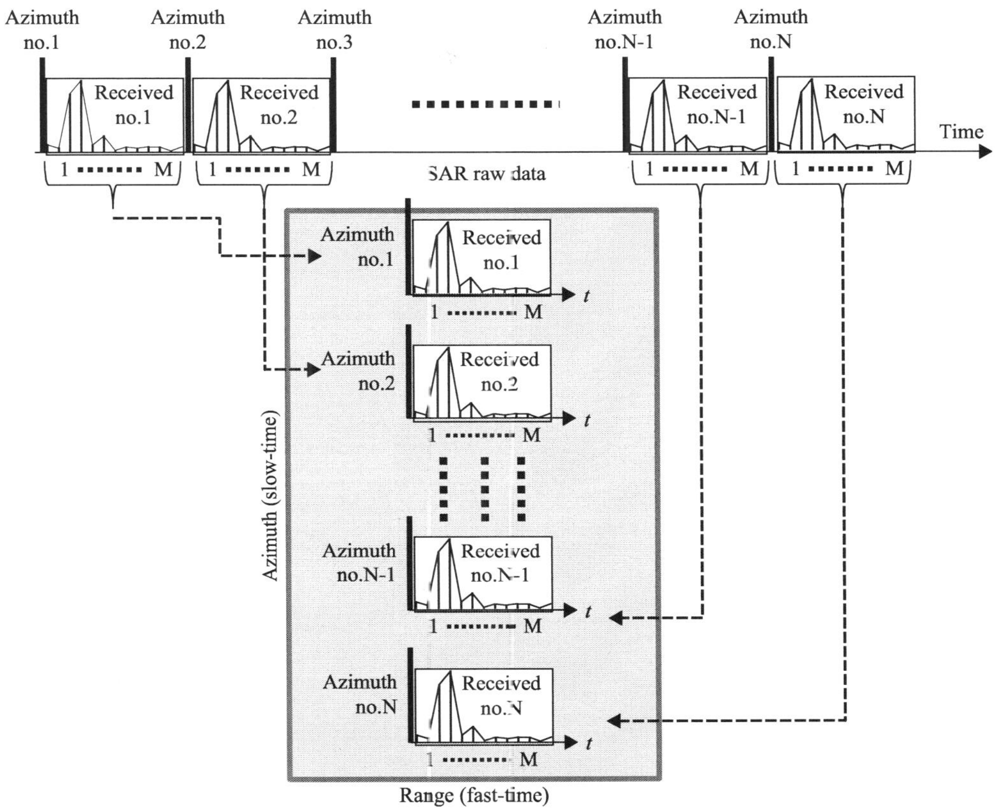

Figure 3.4 SAR raw data arrangement in 2-D delay-time and azimuth-cell domains.

To reconstruct an ISAR range-Doppler image, we first take range compression to obtain ISAR range profiles, after which we apply TMC to remove target's translational motion. The common process of TMC includes two stages: range alignment and phase adjustment.

If the target has more significant rotational motion during the CPI time, the formed ISAR range-Doppler image can still be unfocused and smeared due to the rotation-induced time-varying Doppler spectrum. In these cases, additional image-focusing algorithms for correcting rotation errors, such as the PFA, must be applied. After removing translational and rotational motion, ISAR range-Doppler image is finally generated by taking the Fourier transform in the pulses domain. To display ISAR image in the range and cross-range domain, cross-range scaling is also needed to convert Doppler shift to cross-range domain.

In ISAR range-Doppler image formation, the most important step is the motion compensation. In Chapter 4, ISAR motion compensation methods and related algorithms will be discussed in detail.

3.1.2 ISAR Range Compression

ISAR data are arranged in the delay-time and pulses domains. Because the delay-time is related to the range by $R(t) = c \cdot t/2$, the ISAR data can also be displayed in the range cells and

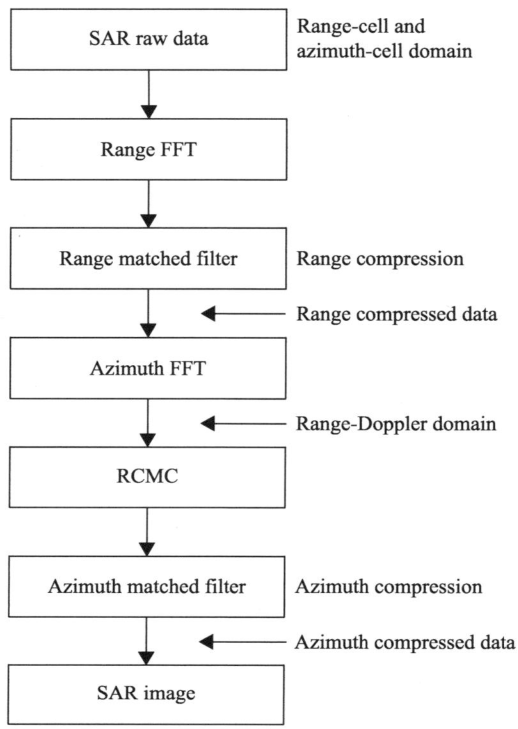

Figure 3.5 Block diagram of SAR range-Doppler algorithm.

pulses domain. Generally, for a given signal waveform, matched filter must be applied to generate pulse compressed range profiles. Figure 3.7a shows an example of ISAR range profiles, where magnitude peaks of the signal indicate range locations of dominant scatterers. Figure 3.7b is the phase function at the range cell no. 50 after the range compression.

Dynamic range is an important specification in radar receivers. It is defined by the ratio between the maximum and minimum values of the capable received signal intensity and is formulated by

$$\text{Dynamic range} = 20 \cdot \log_{10} A_{\max} - 20 \cdot \log_{10} A_{\min} = 20 \cdot \log_{10}\left(\frac{A_{\max}}{A_{\min}}\right), \tag{3.6}$$

where $A_{\max}$ and $A_{\min}$ are the maximum and minimum intensity values in linear scales, respectively. For example, to get 60 dB dynamic range, the ratio between the maximum and minimum values of capable received signal intensity ($A_{\max}/A_{\min}$) must be 1000.

3.1.3 Range Alignment—Coarse Translational Motion Compensation

In ISAR data, the effect of target translational motion can usually be compensated for by aligning range profiles, such that the radar returned signals from the same scatterer are always kept in the same range cell. Range alignment is called *coarse motion compensation*.

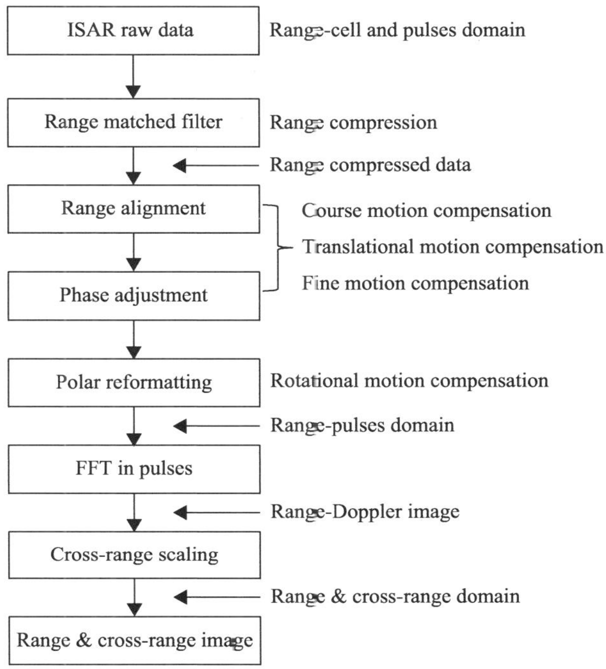

Figure 3.6 Block diagram of ISAR range-Doppler algorithm.

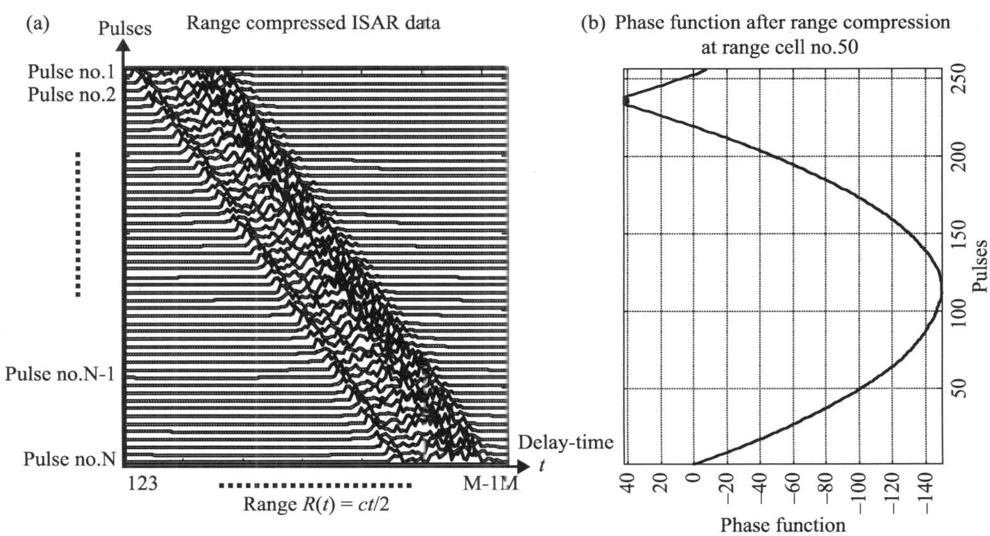

Figure 3.7 ISAR data arranged in 2-D range cells and pulses domain.

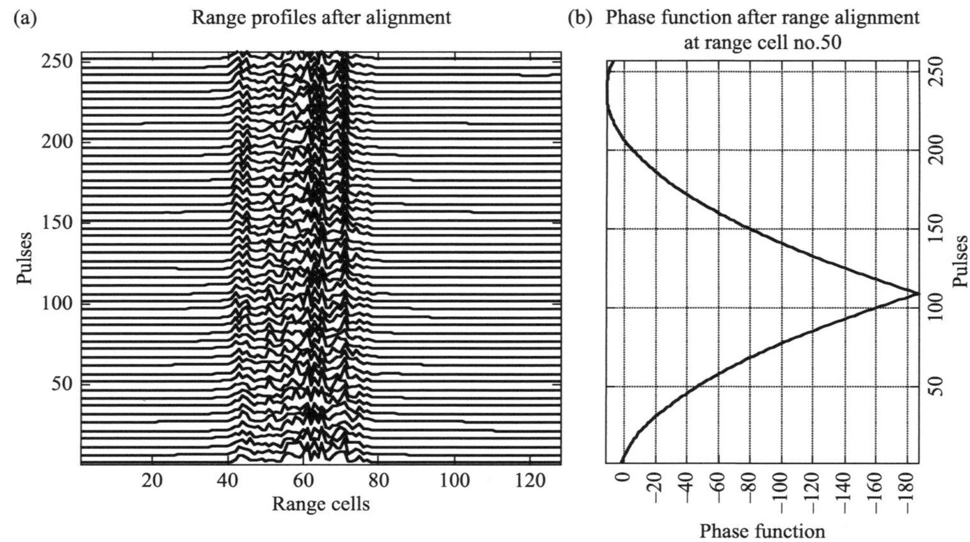

Figure 3.8 Aligned range profiles after applying envelope cross-correlation.

The range-cell alignment process is usually implemented by aligning a strongest magnitude peak in each range profiles. An envelope cross-correlation method between two range profiles is commonly used to estimate the range cell shift between two profiles. Figure 3.8a shows that, after the range alignment, the ISAR range profiles in Figure 3.7 become aligned. Figure 3.8b is the phase function at range cell no. 50 after the range alignment, which is still nonlinear. Thus, a phase adjustment procedure is required.

3.1.4 Phase Adjustment—Fine Translational Motion Compensation

Aligning the range can also cause phase drifts. Figure 3.9a shows a resulting nonlinear phase function across pulses at a selected range cell caused by the range alignment process. To remove the phase drifts and makes linear phase functions at range cells where the target occupies, we must apply a phase adjustment processing, called *fine motion compensation.* An example of phase adjustment method is the minimum variance method [4], which will be introduced in Chapter 4.

3.1.5 Rotational Motion Compensation

As we described in Chapter 1, in ISAR range-Doppler images Doppler shifts are induced by the target's rotation. If a target rotates too fast or the CPI is too long, after range alignment and phase adjustment Doppler frequency shifts can still be time varying. In these cases, the final reconstructed ISAR range-Doppler image can still be smeared. Therefore, we must correct the result due to fast rotation of the target. PFA is a well-known technique for compensating for rotational motion [5].

PFA is based on the tomography developed in medical imaging, which has been used to reconstruct a spatial domain object. To do this, we must take a series of observations through

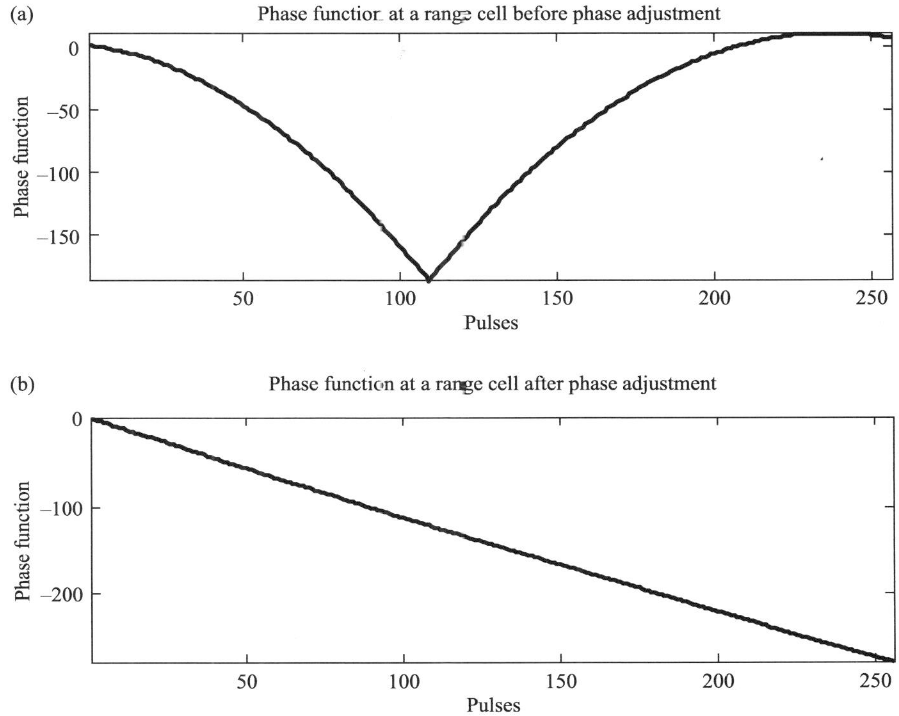

Figure 3.9 Phase function at a range cell (a) before phase adjustment and (b) after phase adjustment, when it becomes linear.

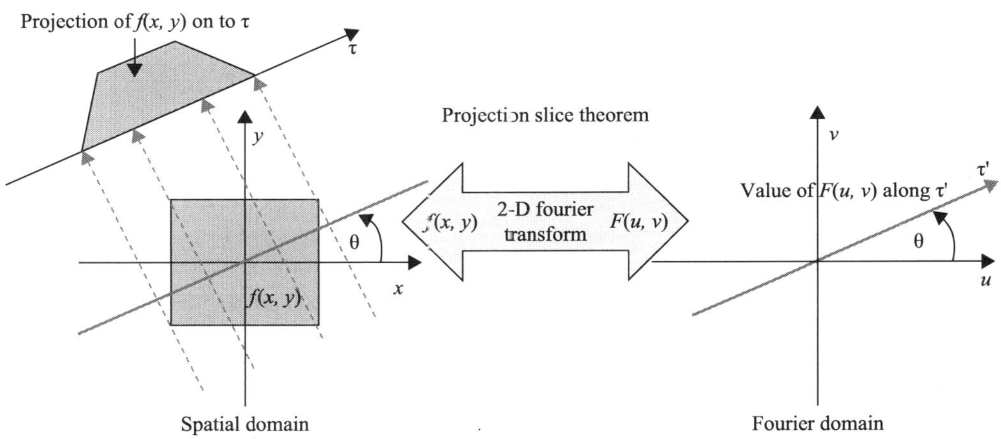

Figure 3.10 Projection slice theorem.

the object. According to the projection slice theorem [5], observation is a projection of the object onto a line, as illustrated on the left side of Figure 3.10. Thus, by applying the Fourier transform to a set of observations over a series of aspect angles (shown on the left side of Figure 3.11), the series of observations populates a region of Fourier space on the right side of

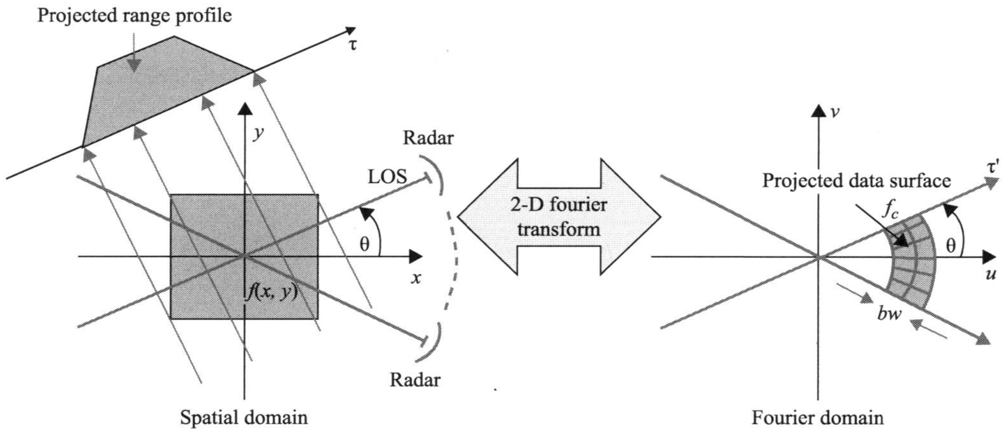

Figure 3.11 Relationship between the radar range profiles in the spatial domain and the projected radar data surface in the Fourier domain.

Figure 3.11. This projected data surface is used to reconstruct an image of the object through the inverse Fourier transform. A 2-D Fourier transform of a spatial function $f(x, y)$ is defined as

$$F(u, v) = \iint_{-\infty}^{\infty} f(x, y) \exp\{-j2\pi(ux + vy)\}dxdy. \tag{3.7}$$

For a projection of $f(x, y)$ at an angle θ, its Fourier transform becomes a slice line through $F(u, v)$ at an angle θ as illustrated on the right side of Figure 3.10.

Because a radar-received pulse signal can be seen as the projection of electromagnetic scattering from an object onto the radar line of sight (LOS), the PFA is suitable for radar image formation. In Figure 3.11, if the angle of a radar LOS is θ, then the projection of a target $f(x, y)$ onto τ becomes a projected range profile. On the other hand, in the Fourier domain, Fourier transform of the radar signal, $F(u, v)$, will produce a line segment that offsets from the origin ($u = 0$, $v = 0$) by the amount of carrier frequency and with the same angle θ as the radar LOS. The length of the line segment is determined by the bandwidth of the radar signal. When radar LOS angle sweeps, the swept line segments become the projected radar data surface in the Fourier space. The relationship between the radar range profiles in the spatial domain and the projected radar data surface in the Fourier domain is shown in Figure 3.11.

In principle, the ISAR PFA is similar to the spotlight SAR PFA [2]. However, in ISAR the target's aspect angle is changed by the target's motion and thus is unknown and uncontrollable. In spotlight SAR, the radar motion determines the aspect angle. Because the aspect change defines the data surface, the target rotation must be estimated before applying the PFA in ISAR. To implement the PFA in ISAR, we must measure target motion parameters from the received radar data for modeling the data surface, projecting the data surface onto a planer surface, interpolating the data into equally spaced samples, and performing the inverse Fourier transform.

Details on ISAR PFA will be discussed in Chapter 4. Here, we only summarize the algorithm we used as follows.

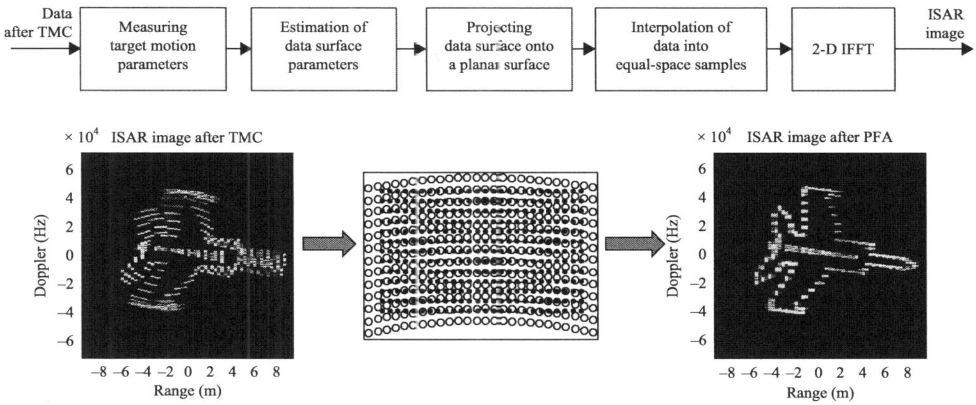

Figure 3.12 ISAR PFA block diagram and imagery improvement.

We take three locations on a performed image for measuring the motion parameters, where one location is used as a reference point and the other two are used to estimate the surface parameters. Each location provides us with a range, velocity (Doppler), and translational acceleration and a set of simultaneous equations relating these values to the quadratic terms in the model that are used for estimating the model parameters. The projection and interpolation steps are done simultaneously for each point on a rectangular grid by back-projecting the grid onto the data surface and performing the interpolation onto the data surface. Finally, a 2-D Fourier transform of the interpolated data produces the final image.

The block diagram of the previously summarized ISAR PFA is depicted in Figure 3.12. A comparison of ISAR images produced by the conventional TMC and after the PFA shows a significant improvement in image quality.

3.2 Time-Frequency–Based Image Formation

Time-frequency analysis is a useful tool in signal analysis. By taking the time-frequency transform of a signal, we can gain insight into time-dependent frequency features in the signal. Features presented in the joint time and frequency domain are more informative and superior to features in either the time or frequency domain alone.

The mathematical basis of time-frequency analysis and its applications to radar imaging and signal analysis can be found in [6–9]. The advantages of using time-frequency analysis in radar range profile, target feature extraction, ISAR motion compensation, and ISAR image formation are remarkable [10–12].

As we defined earlier, radar range profile is a mapping of the target reflectivity onto range cells. Typically, range profile consists of a number of peaks in distinct range cells that correspond to the scattering centers of the target. These features can be used to identify targets. However, in the real world electromagnetic (EM) dispersion and diffraction often exist. Because time-varying scattering features cannot be observed in the time or frequency domain alone, the joint time-frequency analysis becomes a powerful technique to assess the dispersion and diffraction scattering phenomenology [13–16].

In addition to scattering from a stationary target, radar signals can also be reflected from a target that incorporates vibrating or rotating structures, such as propellers of a fixed-wing aircraft or rotor blades of a helicopter. Therefore, the reflected signals contain time-varying Doppler (called micro-Doppler) characteristics associated with these vibrating or rotating structures [17]. Micro-Doppler feature is a unique signature of a target through its motion structures and thus provides useful information for target recognition. To exploit the time-varying Doppler feature of a target, Fourier analysis is no longer suitable. Instead, the joint time-frequency analysis should be applied [9,10].

Time-frequency analysis for ISAR motion compensation can be found in [11], where an adaptive spectrogram proposed in [7] is used to select and extract the phase of multiple prominent point scatterers on the target. The extracted phase is then coupled with the multiple prominent point processing (PPP) model to eliminate undesirable motion errors in the radar raw data. In this manner, the phase of the focused image is preserved and the Doppler resolution offered by the full CPI can be achieved.

In ISAR image formation, Fourier transform is usually used to retrieve the Doppler or cross-range information. The underlying assumption for applying the Fourier transform is that, during the radar processing time interval, all scattterers of the target must remain in their range cells and their Doppler frequency shifts must remain constant. If there are time-varying Doppler shifts due to target's motion, the ISAR image formed by the use of the Fourier transform becomes smeared in the Doppler domain. Therefore, in these cases we should not simply apply the conventional Fourier transform. Instead, the joint time-frequency transform should be applied. As we know, the joint time-frequency transform brings an additional transform dimension (i.e., the time domain transformation).

By applying the joint time-frequency transform, a 2-D range-Doppler image becomes a 3-D time-range-Doppler cubic image [8,9], which we can use to effectively examine the ISAR 2-D range-Doppler image at each time instant to eliminate range drift and Doppler smearing.

3.2.1 Joint Time-Frequency Transforms

In practical applications in digital signal processing, the discrete Fourier transform (DFT) is used to perform the Fourier transform, which converts a time-domain digital signal into the frequency domain.

Since any signal to be analyzed must be in a limited time interval, T, if the digital sampling period of the signal is ΔT, the number of samples in the signal is $N = T/\Delta T$. In the corresponding frequency domain, after the DFT, the extent of the signal spectrum becomes $F = 1/\Delta T$ with N frequency samples at a frequency sampling period $\Delta F = 1/T$, where ΔF is the frequency resolution of the digital signal determined by the total time interval of the signal to be analyzed. Therefore, to have higher frequency resolution the time interval of the signal must be larger.

A spectrogram is a spectro-temporal representation method that provides the actual change of frequency contents of a signal over time, and it is calculated using the most well-known and simplest joint time-frequency transform: short-time Fourier transform (STFT). A spectrogram is represented by the squared magnitude of the STFT without keeping the signal's phase information: $\text{Spectrogram}(t, f) = |\text{STFT}(t, f)|^2$.

The STFT is a moving window Fourier transform. Examining the frequency content of the signal as the time window is moved generates a 2-D time-frequency transform, which contains information on the signal's frequency contents at different times. A drawback of the STFT is that its resolution limit is imposed by the length of the window function. A shorter time window results in better time resolution but leads to worse frequency resolution and vice versa. To overcome the resolution limit of the STFT, various bilinear time-frequency representations have been proposed.

The STFT is defined by

$$\mathrm{STFT}(t, f) = \int_{-\infty}^{\infty} w(\tau - t)s(\tau) \exp\{-j2\pi f\tau\}d\tau, \tag{3.8}$$

where $s(t)$ is a signal, and $w(t)$ is a short-time window function. With a time-limited window function, the resolution of the STFT is determined by the window size.

The Wigner-Ville distribution (WVD) is a typical bilinear time-frequency representation, which has the highest time-frequency resolution, but with cross-term interferences. This severely limits the usefulness of the WVD in its original form so numerous modifications were proposed. The smoothed pseudo WVD (SPWVD) is a good example [6,7].

According to the Wiener-Khinchin theorem, the power spectrum $P(f)$ of a signal $s(t)$ is the Fourier transform of the autocorrelation function $R(\tau)$ of the signal

$$P(f) = \int_{-\infty}^{\infty} R(\tau) \exp\{-j2\pi f\tau\}d\tau, \tag{3.9}$$

where the autocorrelation function is independent of time and is defined as

$$R(\tau) = \int_{-\infty}^{\infty} s(t)s^*(t - \tau)dt. \tag{3.10}$$

However, the power spectrum indicates only the frequency components contained over the entire time duration of the signal. It does not show how the frequency components are distributed in time. An approach to achieve high-energy concentration in both the time and frequency domains is to use the autocorrelation function and make it time dependent [7].

The Fourier transform of a time-dependent autocorrelation function $R(t, \tau)$ is a time-dependent power spectrum of the signal, which is a function of time and frequency:

$$P(t, f) = \int_{-\infty}^{\infty} R(t, \tau) \exp\{-j2\pi f\tau\}d\tau. \tag{3.11}$$

The choice of $R(t, \tau)$ is not arbitrary because $P(t, f)$ must satisfy the frequency and time marginal conditions, and the time-dependent spectrum should be real valued, that is, $P(t, f) = P^*(t, f)$, and nonnegative [7].

When the time-dependent autocorrelation function is defined as

$$R(t, \tau) = s\left(t + \frac{\tau}{2}\right)s^*\left(t - \frac{\tau}{2}\right), \tag{3.12}$$

the time-dependent power spectrum becomes

$$\mathrm{WVD}(t, f) = \int_{-\infty}^{\infty} s\left(t + \frac{\tau}{2}\right) s\left(t - \frac{\tau}{2}\right) \exp\{-j2\pi f\tau\}d\tau, \tag{3.13}$$

which is named the WVD.

Unlike the STFT, there is no short-time window function involved in the WVD. In fact, the time-reversed signal, as shown in (3.13), can be considered as a matched filtering. The frequency resolution of the WVD is very close to that of the Fourier transform [6,7].

The bilinear WVD has better joint time-frequency resolution than any linear transform, such as the STFT. However, it suffers from the problem of cross-term interference; that is, the WVD of the sum of two signals is not the sum of their individual WVDs. If a signal contains more than one component in the joint time-frequency domain, its WVD will contain cross-terms that occur halfway between each pair of auto-terms. The magnitude of these oscillatory cross-terms can be twice as large as the auto-terms. To reduce the cross-term interference, filtered WVDs have been used to preserve the useful properties of the time-frequency transform with a slightly reduced time-frequency resolution and largely reduced cross-term interference. The WVD with a linear low-pass filter belongs to Cohen's class [6].

The general form of Cohen's class is defined by

$$\text{Cohen}(t, f) = \iint s\left(u + \frac{\tau}{2}\right) s^*\left(u - \frac{\tau}{2}\right) \phi(t - u, \tau) \exp\{-j2\pi f\tau\} du d\tau, \tag{3.14}$$

where $\phi(t, \tau)$ is a low-pass filter, and its Fourier transform $\Phi(\theta, \tau)$ is called the kernel function. If $\Phi(\theta, \tau) = 1$, then $\phi(t, \tau) = \delta(t)$, and Cohen class reduces to the WVD. The Cohen's class with different kernel functions, such as the SPWVD, can be used to largely reduce the cross-term interference in the WVD.

The SPWVD is defined as

$$\text{SPWVD}(t, f) = \int_{-\infty}^{\infty} h(\tau) \int_{-\infty}^{\infty} w(t - u) s\left(u + \frac{\tau}{2}\right) s^*(u - \frac{\tau}{2})\, du \cdot \exp\{-j2\pi f\tau\} d\tau, \tag{3.15}$$

where $s(t)$ is the signal to be analyzed, $w(t)$ is the smoothing time window, and $h(t)$ is the smoothing frequency window.

A publicly available time-frequency toolbox from CNRS of France provides a variety of time-frequency representation, including the STFT and the SPWVD [18].

3.2.2 Image Formation Based on Joint Time-Frequency Transform

As we discussed earlier, after compensating for translational motion in the range domain, an ISAR image of a target can be formed by taking one-dimensional (1-D) Fourier transform along pulses (slow time). However, if Doppler shifts are time varying, the ISAR image using the Fourier transform is smeared in the Doppler domain. Thus, we have to apply the RMC, such the PFA.

As an alternative, for noncooperative targets with severe rotatinal motion the joint time-frequency transform can be easily applied for ISAR image formation [9]. Figure 3.13 illustrates the time-frequency–based image formation. The standard translational motion compensation procedure is still needed prior to the time-frequency transform.

If the translational motion compensated data consist of N range profiles $G(r_m, t_n)$ and each range profile has M range cells, then by taking the joint time-frequency transform at each range cell and along pulses, an $N \times N$ time-varying Doppler spectrum is obtained.

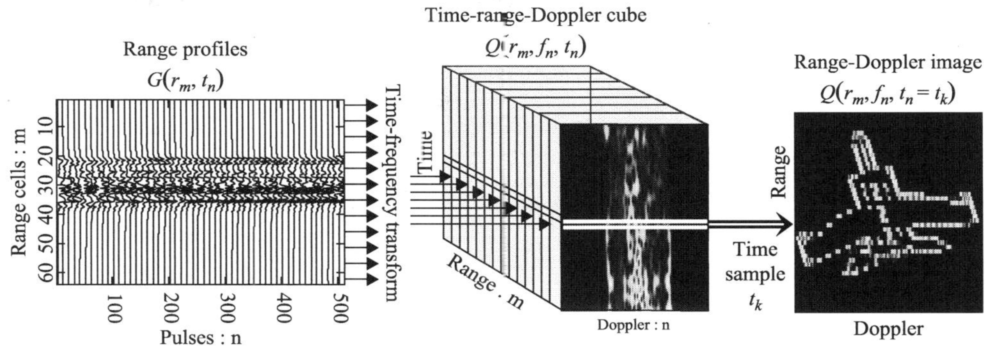

Figure 3.13 Illustration of the time-frequency–based image formation.

Then, by combining the $N \times N$ time-varying Doppler spectrum at M range cells, an $N \times M \times N$ time-range-Doppler cube $Q(r_m, f_n, t_n)$ can be formed

$$Q(r_m, f_n, t_n) = TF\{G(r_m, t_n)\}_n, \tag{3.16}$$

where TF denotes the time-frequency operation with respect to n. At a particular time instant, t_k, only one range-Doppler image frame $Q(r_m, f_n, t_n = t_k)$ can be extracted from the cubic image. Thus, a total of N image frames are available, and every one represents a complete range-Doppler image extracted at a particular time instant. Therefore, by replacing the Fourier transform with the time-frequency transform, a 2-D range-Doppler Fourier image frame becomes a 3-D time-range-Doppler cubic image. Actually, by integrating the N time-range-Doppler frames, the resulting image is exactly the 2-D range-Doppler Fourier image. After time sampling the cubic image, a sequence of 2-D range-Doppler images can be viewed. Each individual time-sampled range-Doppler image frame is a well-focused image with superior resolution. The N image sequence also provides the temporal changing information from time to time (Figure 3.14).

Figure 3.15a shows simulation result of a conventional ISAR Fourier-based image for a fast rotating Mig-25 aircraft. The radar is a stepped-frequency continuous wave (SFCW) radar operating at 9.0 GHz with 64 frequency steps and 7.5 MHz step size. For a total number of bursts of 512 and the burst repetition frequency of 281 Hz, the total CPI time is 1.82 sec. Because of fast rotating, the Fourier-based image of the aircraft is smeared in the Doppler domain, as shown in Figure 3.15a, even after applying translational motion compensation. For comparison, we also applied the time-frequency–based image formation using the STFT and the SPWVD. Figure 3.15b is one frame of the time-frequency–based images using the STFT, and Figure 3.15c is one frame using the SPWVD.

From the simulation results in Figure 3.15, we can see that without applying sophisticated rotational motion compensation algorithms ISAR images of a fast rotating target can still be reconstructed using the time-frequency–based image formation. Compared with the SPWVD-based method, the ISAR image reconstructed by the STFT-based method shows lower Doppler resolution because the frequency resolution of the STFT method depends on the window size. A suitable window size must be selected before applying the STFT-based image formation.

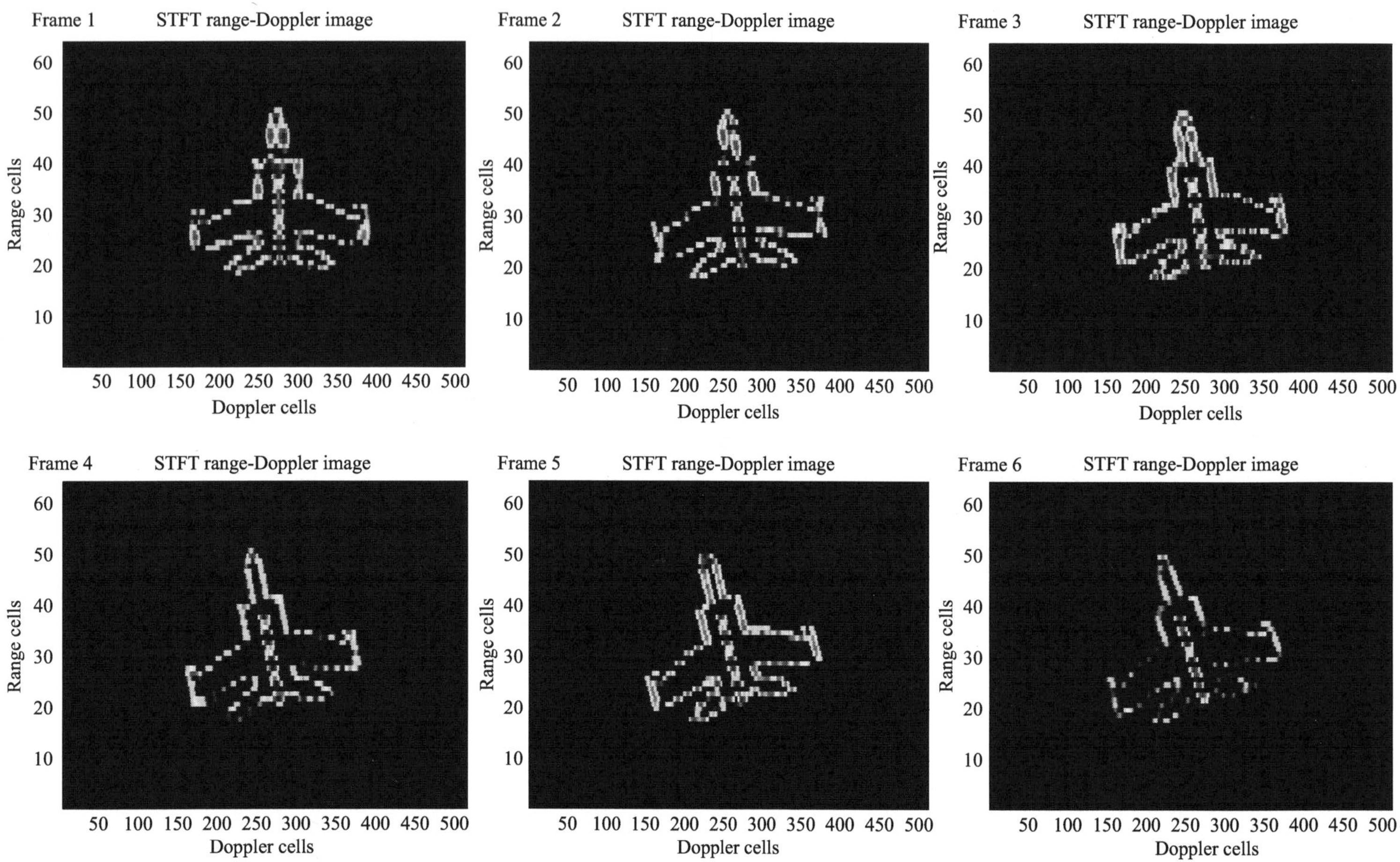

Figure 3.14 Selected a number of individual time-sampled frame from the time-range-Doppler cubic image, where each image provides not only superior resolution but also temporal changing information from one time to another.

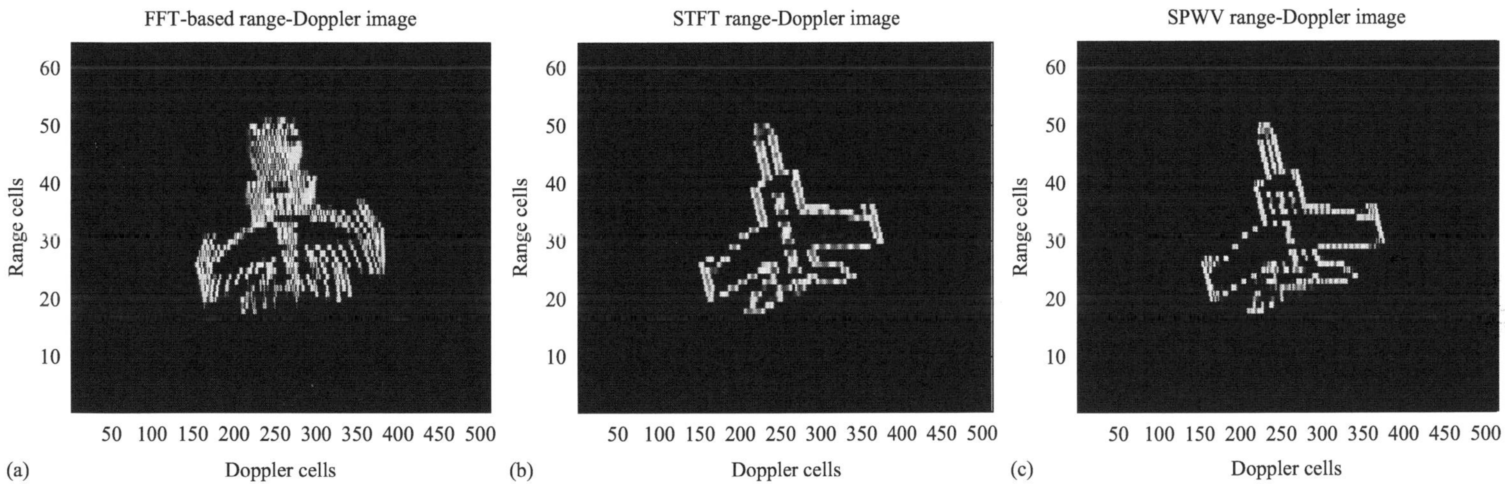

Figure 3.15 (a) Smeared FFT-based range-Doppler image of a fast rotating Mig-25; (b) after applying the time-frequency–based image algorithm using the STFT; (c) the time-frequency–based image using the SPWVD.

3.3 Display 2-D ISAR Imagery—Windowing and Zero Padding for Sidelobe Suppression

After ISAR image formation, a suitable method for displaying 2-D ISAR imagery is instructional for the end of this chapter. For display 2-D ISAR imagery, windowing and zero-padding processing are very useful for suppressing sidelobes in display 2-D ISAR imagery according to [19]. Without windowing the motion compensated data, in the reconstructed ISAR image high sidelobes next to the main peaks will appear as shown in Figure 3.16a. Thus, the windowing function should be applied to reduce the sidelobes at the cost of a slightly reduced resolution. Commonly used Hamming windows can be used for this purpose and is defined by

$$w_k = 0.54 - 0.46\cos\left(\frac{2\pi}{N-1}\right)k; \quad (k = 0, 1, \ldots, N-1), \tag{3.17}$$

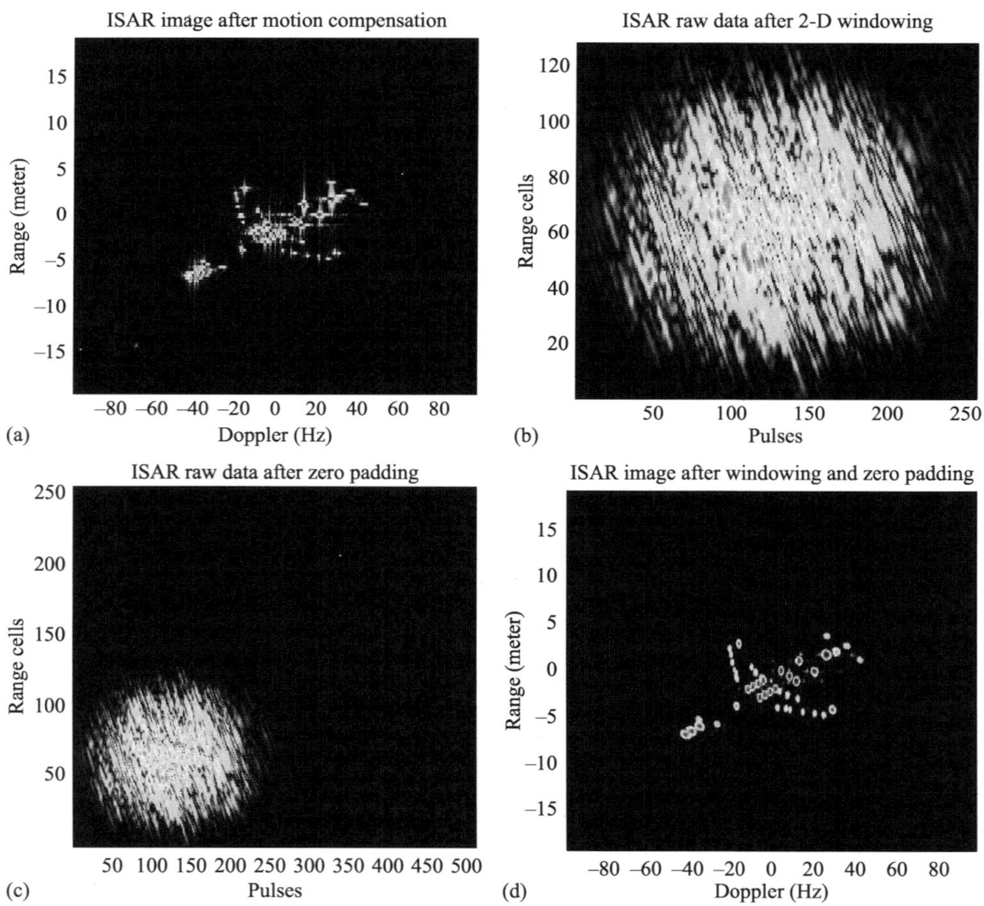

Figure 3.16 (a) ISAR image after motion compensation; (b) Raw data after applying 2-D Hamming windowing; (c) Raw data after zero padding; (d) Final displayed ISAR imagery after windowing and zero padding.

where w_k is the weighting coefficients of the Hamming window function, and N is the length of the window. For ISAR imagery, the window function should be applied in both the range cells and the pulses direction. After applying 2-D Hamming windowing, the raw data are shown in Figure 3.16b.

ISAR imagery is a discrete representation of the target reflectivity distribution. Because scattering centers may not be located exactly at the discrete grid points, interpolation must be taken before final display of the ISAR imagery. To do this, we need to do 2-D zero padding, after which the original raw data should occupy only a quarter of area in the extended zero-padded raw data (Figure 3.16c).

After zero padding, the 2-D Fourier transform is applied to generate ISAR imagery. The vertical direction of the imagery should be the range domain, and the horizontal direction should be the Doppler (cross-range) domain, with a dynamic range of 40 dB maximum for a normalized ISAR imagery. Thus, the final displayed ISAR imagery is shown in Figure 3.16d.

References

[1] Wehner, D.R., *High-Resolution Radar*, 2d ed., Artech House, 1995.

[2] Carrara, W.G., R.S. Goodman, and R.M. Majewski, *Spotlight Synthetic Aperture Radar: Signal Processing Algorithms*, Artech House, 1995.

[3] Wu, C., "A Digital System to Produce Imagery from SAR," Proc. of AIAA Conference on Systems Design Driven by Sensors, paper 76-968, 1976.

[4] Haywood, B., and R. Evens, "Motion Compensation for ISAR imaging," Proc. of ASSPA, Adelaide, Australia, pp. 113–117, 1989.

[5] Kak, A.C., and M. Slaney, *Principles of Computerized Tomographic Imaging*, Society of Industrial and Applied Mathematics, 2001.

[6] Cohen, L., *Time-Frequency Analysis*, Prentice-Hall, 1995.

[7] Qian, S., and D. Chen, *Joint Time-Frequency Analysis-Methods and Applications*, Prentice-Hall, 1996.

[8] Chen, V.C., "Reconstruction of Inverse Synthetic Aperture Images Using Adaptive Time-Frequency Wavelet Transforms," *SPIE Proc. on Wavelet Applications*, vol. 2491, pp. 373–386, Apr. 1995.

[9] Chen, V.C., and S. Qian, "Joint Time-Frequency Transform for Radar Range-Doppler Imaging," *IEEE Trans. Aerospace Electronic Systems*, vol. AES-34, no. 2, pp. 486–499, 1998.

[10] Chen, V.C., and H. Ling, *Time-Frequency Transforms for Radar Imaging and Signal Analysis*, Artech House, 2002.

[11] Wang, Y., H. Ling, and V.C. Chen, "ISAR Motion Compensation via Adaptive Joint Time-Frequency Technique," *IEEE Trans. Aerospace Electronic Systems*, vol. AES-34, pp. 670–677, Apr. 1998.

[12] Ling, H., Y. Wang, and V. Chen, "ISAR Image Formation and Feature Extraction Using Adaptive Joint Time-Frequency Processing," *SPIE Proc. on Wavelet Applications*, vol. 3708, pp. 424–432, Apr. 1997.

[13] Moghaddar, A., and E.K. Walton, "Time-Frequency-Distribution Analysis of Scattering from Waveguide Cavities," *IEEE Trans. Antennas Propagat.*, vol. AP-41, pp. 677–680, May 1993.

[14] Ling, H., J. Moore, D. Bouche, and V. Saavedra, "Time-Frequency Analysis of Backscattered Data from a Coated Strip with a Gap," *IEEE Trans. Antennas Propagat.*, vol. AP-41, pp. 1147–1150, August 1993.

[15] Moore, J., and H. Ling, "Time-Frequency Analysis of the Scattering Phenomenology in Finite Dielectric Gratings," *Microwave Optical Tech. Lett.*, vol. 6, pp. 597–600, August 1993.

[16] Trintinalia, L.C., and H. Ling, "Interpretation of Scattering Phenomenology in Slotted Waveguide Structures via Time-Frequency Processing," *IEEE Trans. Antennas Propagat.*, vol. AP-43, pp. 1253–1261, November 1995.

[17] Chen, V.C., *The Micro-Doppler Effect in Radar*, Artech House, 2011.

[18] Time-Frequency Toolbox: http://tftb.nongnu.org

[19] *Consolidated SET 040 TG-22 Standard for Pre-processing and Display of 2D-ISAR Imagery*, NATO unclassified, NC3A-NL/ACD/2000/741:3/NAT/20 (SET), Nov. 22, 2000.

CHAPTER 4

ISAR Motion Compensation

As introduced in Chapter 3, inverse synthetic aperture radar (ISAR) data are arranged in a 2-D matrix, where the number of range cells is in its row and the number of pulses is in its column, or vice versa. To reconstruct an ISAR range-Doppler image, we first take range compression to obtain ISAR range profiles, which is a sequence of consecutive range profiles over the coherent processing interval (CPI). Range compression is usually performed through a matched filter.

From the range profiles, we can see motion of targets; that is, targets appear at different positions in different range profiles. Then, motion compensation will be carried out, including translational motion compensation (TMC) and rotational motion compensation (RMC). Finally, after removing translational motion and rotational motion, by taking the Fourier transform along the number of pulses (slow-time domain), an ISAR range-Doppler image can be formed. In this chapter, we will introduce ISAR motion compensation methods, including the cross-correlation, range centroid, and minimum-entropy methods for range alignment and the minimum variance, Doppler centroid, phase gradient, and entropy methods for phase adjustment.

Many motion compensation algorithms were developed for solving the image-smearing problem [1–12]. TMC includes range alignment and phase adjustment or phase correction. Range alignment is accomplished by tracking the movement of a prominent scatterer with strong peak in range profiles. This is called the coarse range alignment, which allows the prominent scatterer to be sorted into the same range cell across the range profiles. The accuracy of the alignment is limited by the range resolution cell. However, only the coarse range alignment is not sufficient for removing phase drift errors in the range profiles. Consequently, a suitable phase adjustment procedure must be carried out to remove the residual phase errors and drifts.

4.1 Translational Motion Compensation

Translational motion can be compensated for by aligning range profiles through shifting a necessary number of range cells, such that the received signals from the same scatterer are always kept in the same range cell. This is called the range-cell alignment or range tracking, which is the coarse motion compensation. However, due to the shifting in range cells, it also causes phase drifts. Thus, a phase adjustment procedure must be applied to remove the phase drifts and makes phase functions as linear as possible within the target occupied range cells. The phase adjustment is called the fine motion compensation.

A simple method for range-cell alignment is to align the strong peak of a prominent scatterer in range profiles as proposed in [1]. Envelope cross-correlation between the two range profiles is calculated to estimate the shift of range cells. Other methods for range alignment include range centroid, minimum entropy, and maximum contrast [2–4]. Methods for phase adjustment include minimum variance method, Doppler centroid method, phase gradient method, and entropy, contrast, or other methods that use a cost function [1,5–7]. We will give examples regarding how to perform motion compensation and will provide MATLAB source codes for reproducing the results.

4.1.1 Range Alignment and Phase Adjustment

Cross-correlation is a straightforward method for range alignment. In [1], descriptions on performing cross-correlation between two range profiles in the spatial domain and in the frequency domain can be found. The cross-correlation method is actually performed between a range profile and a preselected reference range profile. The peak location of the cross-correlation is used to estimate the shifted number of range cells from the reference range profile. By shifting back the estimated number of range cells, the misaligned range profile becomes aligned with the reference range profile. The previous procedure is repeated for each successive range profile until the dominant scatterer in all range profiles is aligned.

Other range alignment methods include minimizing entropy of averaged range profiles and global range alignment [8,9].

Figure 4.1 shows the geometry of radar and a moving target in a flat earth model. Assume that the radar is located at the origin ($X = 0$, $Y = 0$, $Z = 0$) in the global radar coordinate system and a target at (X_0, Y_0, Z_0) is moving toward a heading direction. The target-heading

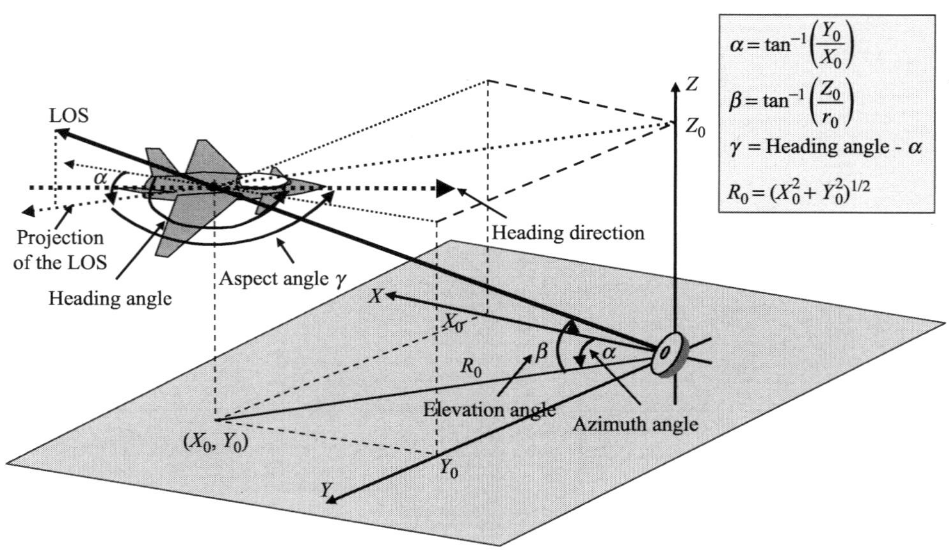

Figure 4.1 ISAR and target geometry.

angle is defined by the angle from the direction of the positive X-axis to the heading direction as indicated in Figure 4.1. The target azimuth angle viewed from the radar is

$$\alpha = \tan^{-1}(Y_0/X_0), \tag{4.1}$$

and the elevation angle is

$$\beta = \tan^{-1}(Z_0/R_0), \tag{4.2}$$

where $R_0 = (X_0^2 + Y_0^2)^{1/2}$ is the ground range.

As shown in Figure 4.1, the target aspect angle, γ, is the angle between the projection of the radar line of sight (LOS) and the target-heading direction, which can be calculated by the target-heading angle minus the radar azimuth angle, α. To understand how to count the aspect angle for a given heading direction, it may be much clearer with two-dimensional (2-D) ISAR and target geometry as depicted in Figure 4.2, which shows three target aspect angles (45°, –45°, and –135°) and their respective heading directions.

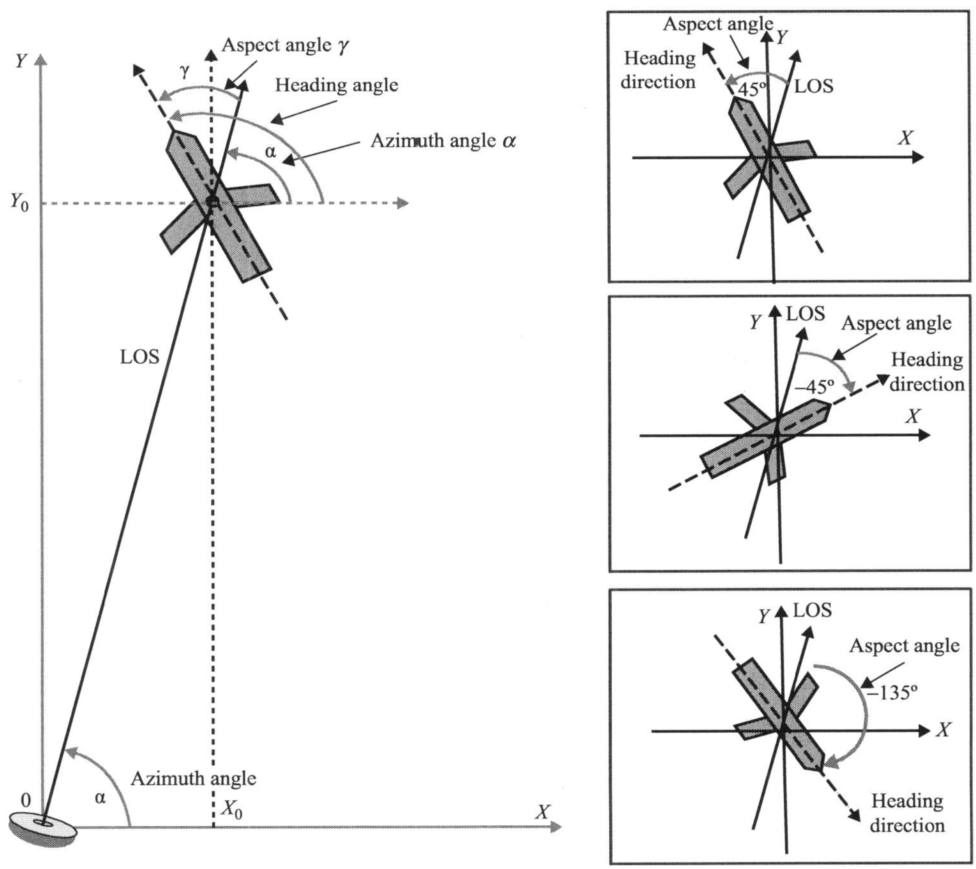

Figure 4.2 Examples of three target aspect angles (45°, –45°, and –135°) with their respective heading directions.

After discussing the aspect angle, we now demonstrate the detailed procedure for ISAR range alignment. Given a radar located at ($X = 0$, $Y = 0$, $Z = 0$) and a target at an initial position ($X_0 = 500$ m, $Y_0 = 2{,}000$ m, $Z_0 = 200$ m), for simplicity we use 2-D geometry as shown in Figure 4.3. Thus, the radar azimuth angle, α, is 76°. If the target is heading to a direction at 235° anticlockwise angle from the positive X-axis, according to definition the target aspect angle should be 159°. Table 4.1 lists the aforementioned geometric parameters of the radar and target.

Figure 4.4 is a block diagram of the range alignment based on the cross-correlation method. To align range profiles, we first select a strong peak from same prominent scatterer in each range profile and keep tracking it from one profile to the next. Then we align the peak by shifting it to the same range cell for each range profile.

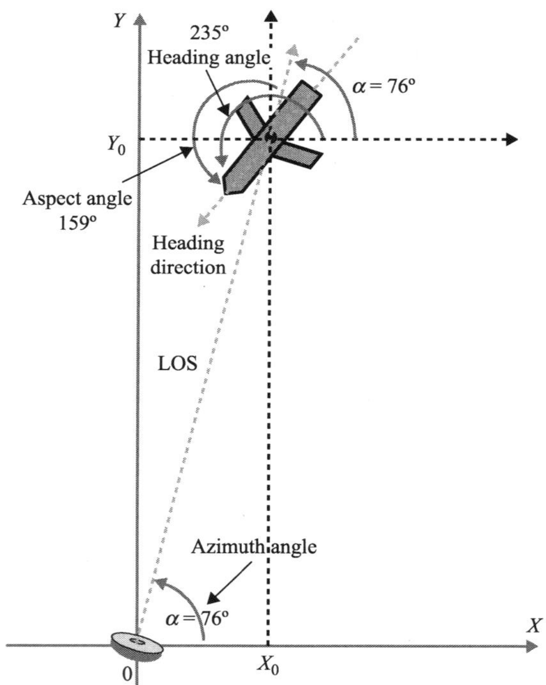

Figure 4.3 2-D ISAR geometry with a moving target.

Table 4.1 Radar and Target Geometric Parameters

Radar location	$X = 0$ m; $Y = 0$ m; $Z = 0$ m
Target initial location	$X_0 = 500$ m; $Y_0 = 2{,}000$ m; $Z_0 = 200$ m
Target initial azimuth angle	76°
Target-heading direction angle	235° (anticlockwise from positive X-axis)
	125° (clockwise from positive X-axis)
Target aspect angle	159°

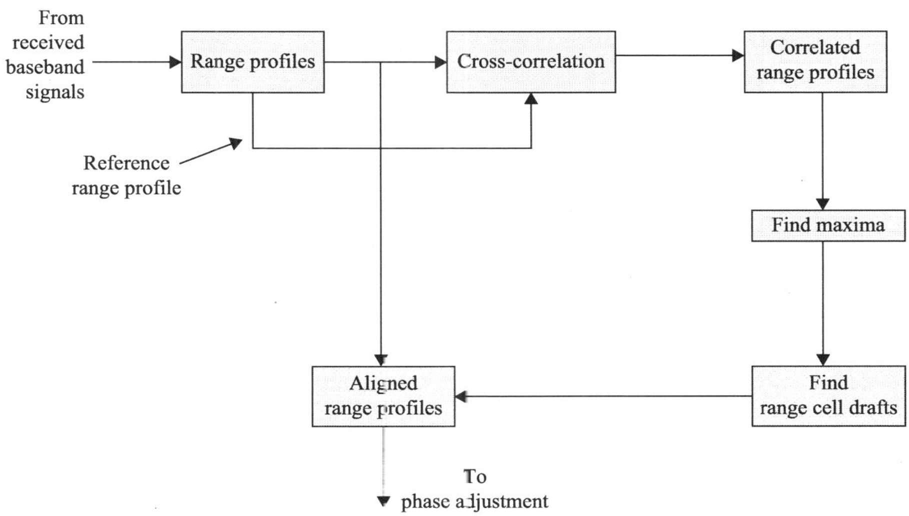

Figure 4.4 Block diagram of a range alignment procedure.

As we know, the cross-correlation of two same-shaped signals has a highest peak value at zero time delay between them. Thus, to find a strong peak of a target in range profiles, we assign a reference range profile to perform cross-correlation with other range profiles. From the location of the strong peak, we can count its shifted cells from the reference range profile. Thus, we can pull it back.

To select a suitable reference range profile, we should choose the one that shows higher target peaks near the center of its range cells. Therefore, the target image can be guaranteed to be displayed near the center of the range domain. By calculating the cross-correlation between the reference range profile and each of other range profiles, a 2-D cross-correlation function can be generated. Then by finding the maxima we can estimate the drifted number of range cells for each individual range profile. Finally, after shifting back the estimated number of range cells for these range profiles, the misaligned range profiles become aligned.

To demonstrate this procedure, we use the geometric parameters of radar and target as listed in Table 4.1. The target has only translational motion at 100 m/sec and heading at 125° clockwise or 235° anticlockwise from the *X*-axis. Thus, we can see that in the range profiles a target is moving from range cells to range cells as shown in Figure 4.5a.

Because of the motion, the target position is drifting from one range profile to the next. Then, a simple cross-correlation method is used to find the drifted number of range cells for each individual range profile. As we know, the cross-correlation can be performed either in the time domain or in the frequency domain. Cross-correlation in the frequency domain is simply multiplication, which is faster than the time-domain correlation. Thus, we apply the cross-correlation in the frequency domain. Therefore, after the cross-correlation processing, the range profiles must be transformed back to the spatial frequency domain using 1-D Fourier transform along the range cells as shown in Figure 4.5b.

To select the reference range profile, we pick up the one at the center of the number of pulses. The cross-correlation operation in the spatial frequency domain is the frequency-domain range profiles multiplied by the conjugate of the frequency-domain reference range

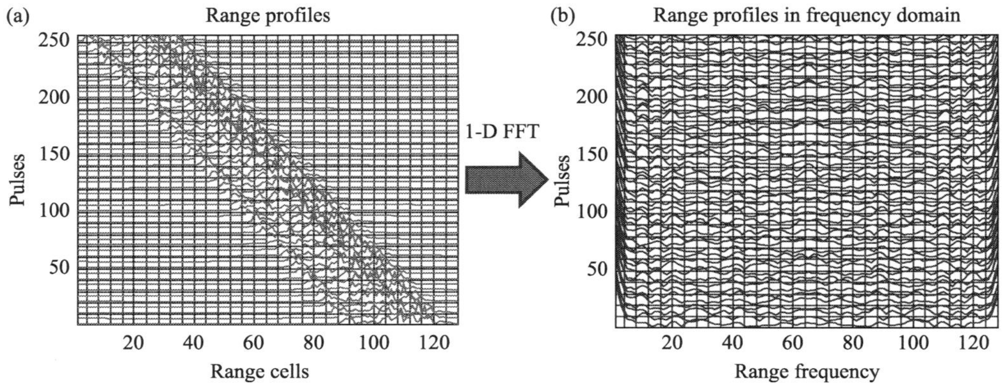

Figure 4.5 Range profiles (a) in a range cells and pulses domain and (b) in a range frequency and pulses domain.

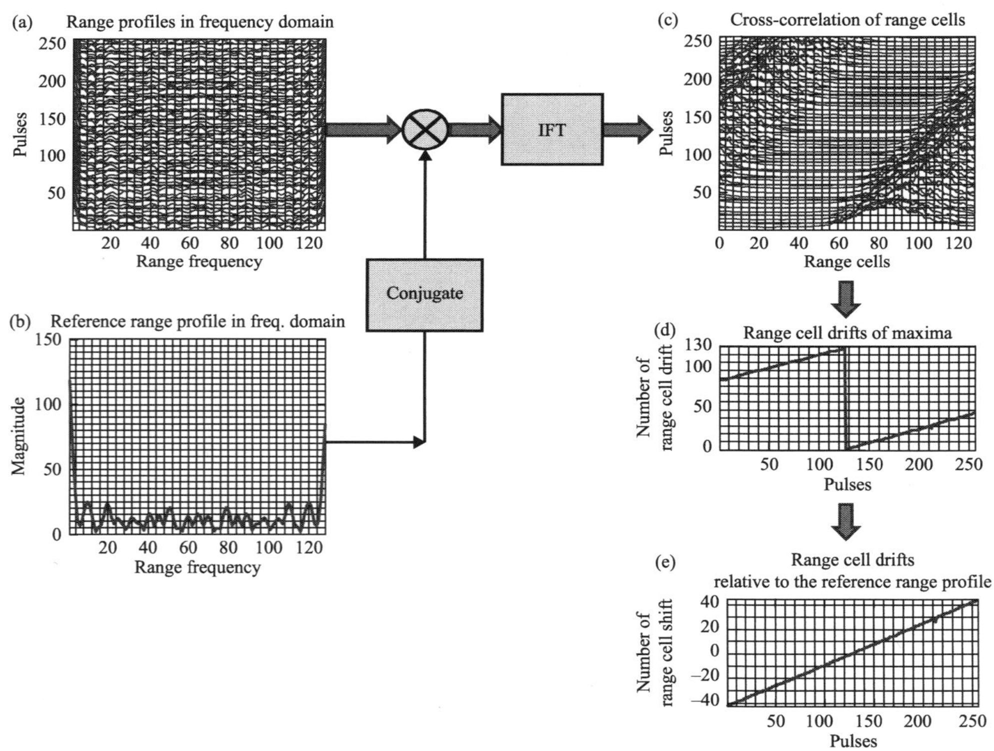

Figure 4.6 (a) The range profiles in the spatial frequency domain; (b) the reference range profile in the spatial frequency domain; (c) the cross-correlation function of range cells; (d) the range cell drifts found from the maxima in the cross-correlation function; and (e) the found range drifts vs. pulses.

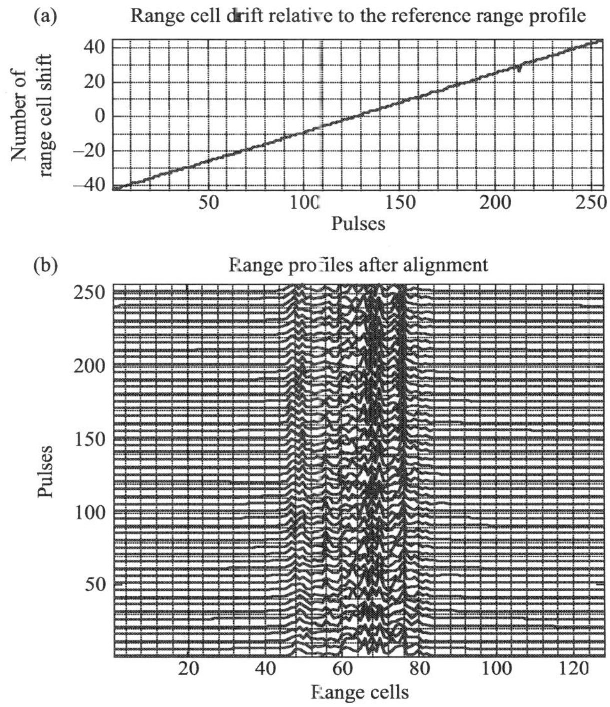

Figure 4.7 (a) The drifts of a range cell relative to the reference range profile; (b) the range profiles after alignment by phase shifting.

profile (Figure 4.6a and Figure 4.6b). Finally, by taking the inverse Fourier transform the cross-correlation is obtained (Figure 4.6c). As indicated in Figure 4.6, after having the cross-correlation function drifts of range cells can be found from the maxima in the cross-correlation function (Figure 4.6d). The number of range cells drifted from the reference range profile are illustrated in Figure 4.6e.

Based on the found drifted range cell with respect to the reference range profile shown in Figure 4.7a, we can shift back to the number of drifted range cells. The range profiles will be aligned with the reference range profile as demonstrated in Figure 4.7b.

However, after the range cell alignment the phase function at each range cell along the number of pulses becomes nonlinear (Figure 4.8a) because of phase errors caused by shifting range cells. Phase adjustment compensates for the phase errors such that phase functions become linear and ISAR image is focused. Figure 4.8b shows the aligned range profiles before phase adjustment. After applying phase adjustment, the phase function becomes linear (Figure 4.8c). Therefore, by taking 1-D Foureir transform, the final ISAR range-Doppler image of the target becomes focused (Figure 4.8d).

The phase adjustment can be implemented using many approaches such as minimum variance, Doppler centroid, phase gradient, and entropy, contrast, or other cost-function methods [4–10]. Here, we introduce the very common minimum variance method for phase adjustment [10].

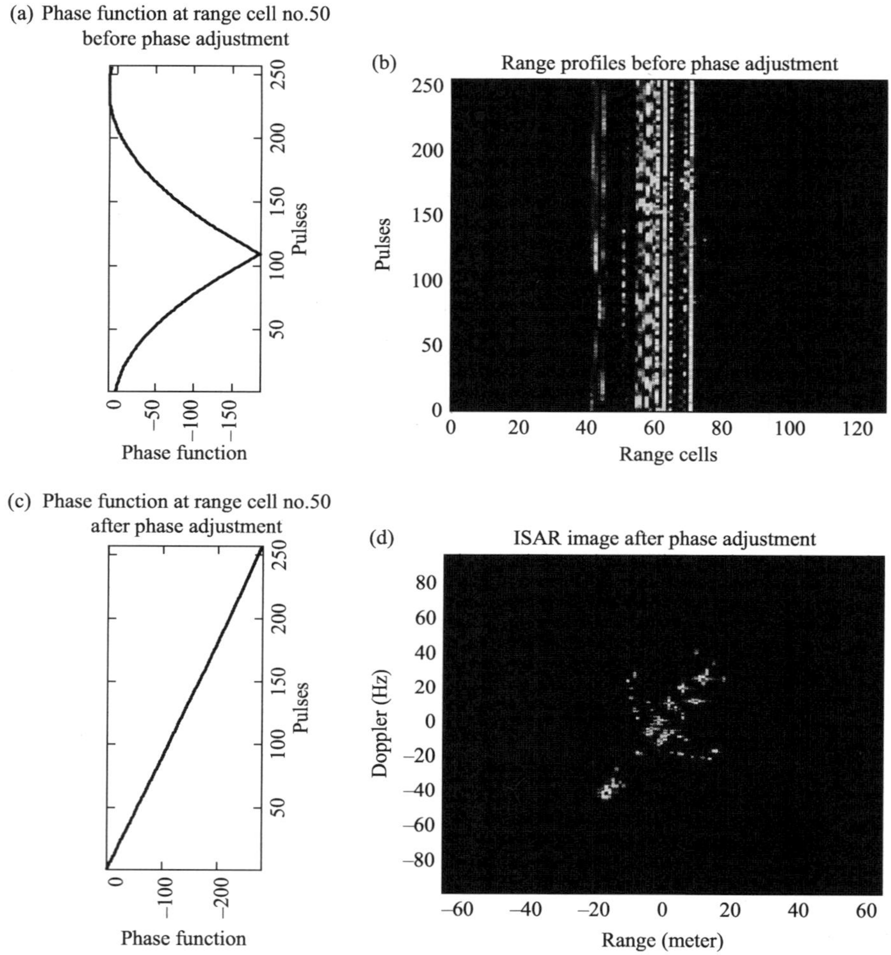

Figure 4.8 (a) The phase function at the range cell no. 45 before phase adjustment; (b) ISAR image before phase adjustment; (c) the linearized phase function at the same range cell after phase adjustment; (d) final ISAR image of the target after range alignment and phase adjustment.

The minimum variance algorithm is used to find a small number of scatterers (e.g., 3) that have the most minimum variances in their envelope range profiles. Figure 4.9 shows the range profiles after the range alignment. For this special case, three minimum variance scatterers at range cell 50, 68, and 76 can be found.

After finding theses scatterers, we can calculate a phase correction function across pulses and averaged from the range cells of these scatterers with the most minimum variances (Figure 4.10). Then this phase correction function is used to compensate for phase functions at each range cell.

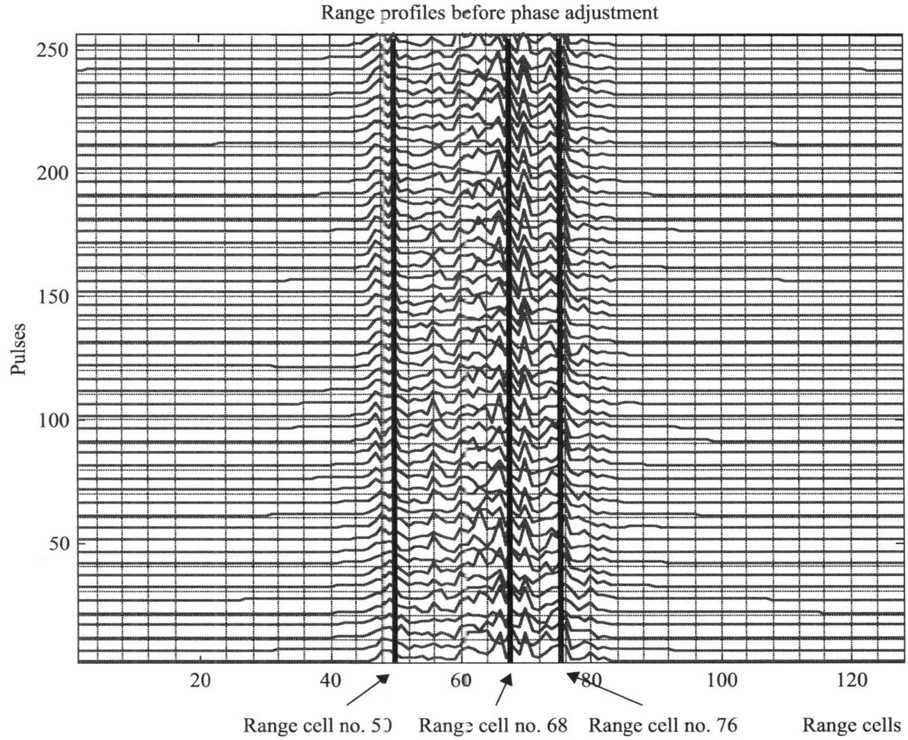

Figure 4.9 Range profiles after range alignment but before phase adjustment. Three scatterers with minimum variances at range cell no. 50, 68, and 76 can be found.

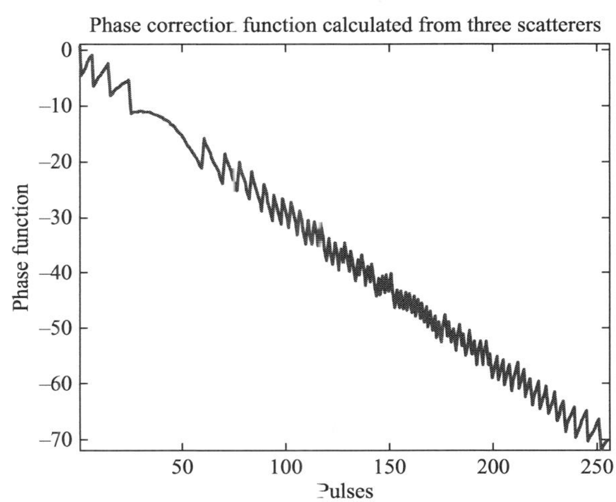

Figure 4.10 Phase correction function calculated from three scatterers with most minimum variances.

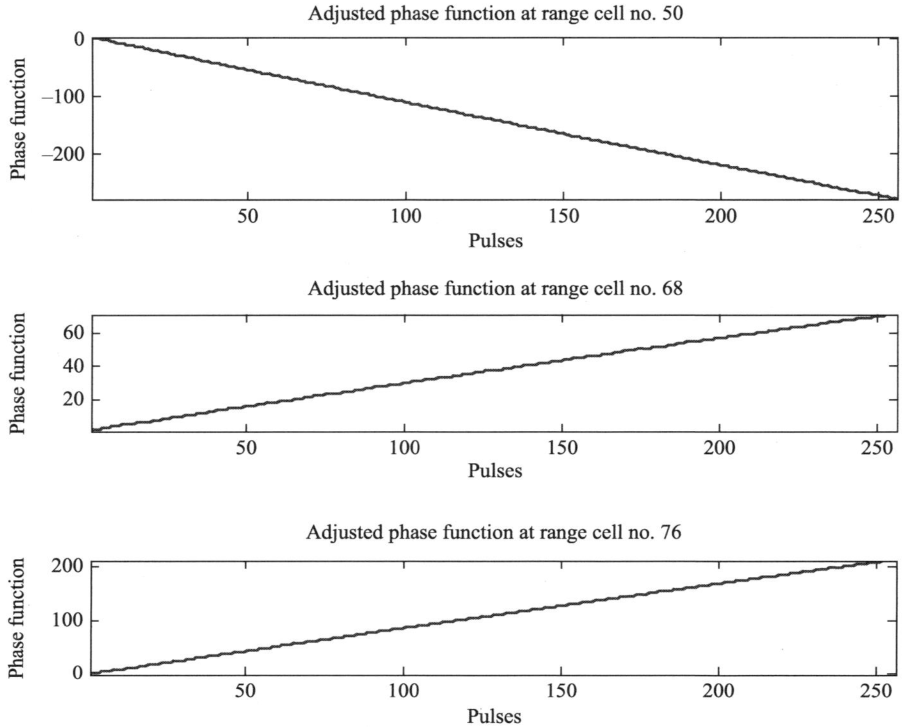

Figure 4.11 Phase functions perfectly adjusted at the three scatterers.

The estimated phase is a function of time, $\psi(t)$, which is applied to compensate the phase function across pulses for each range cell, such that the nonlinearity of the phase function can be corrected. After the phase compensation, the linearized phase functions at range cell no. 50, 68, and 76 in the range profiles are shown in Figure 4.11. Thus, ISAR image of the target can be obtained using inverse Fourier transform of the range aligned and phase adjusted range profiles:

$$\mathrm{IFT}\left\{ s_R(t) \cdot \exp\left[j4\pi f \frac{R(t)}{c} \right] \cdot \exp[j\psi(t)] \right\}, \tag{4.3}$$

where s_R(t) is the radar received baseband signal, $\exp\left[j4\pi f \frac{R(t)}{c}\right]$ is the term for removing range cell shifts, and $\exp[j\varphi(t)]$ is the phase adjustment term.

4.1.2 Range and Doppler Centroid Method

Range centroid and Doppler centroid methods can be used for range alignment and phase adjustment in the TMC and have been applied to ISAR motion compensation for a long time [11,12]. In this section, we introduce these methods as alternatives for TMC.

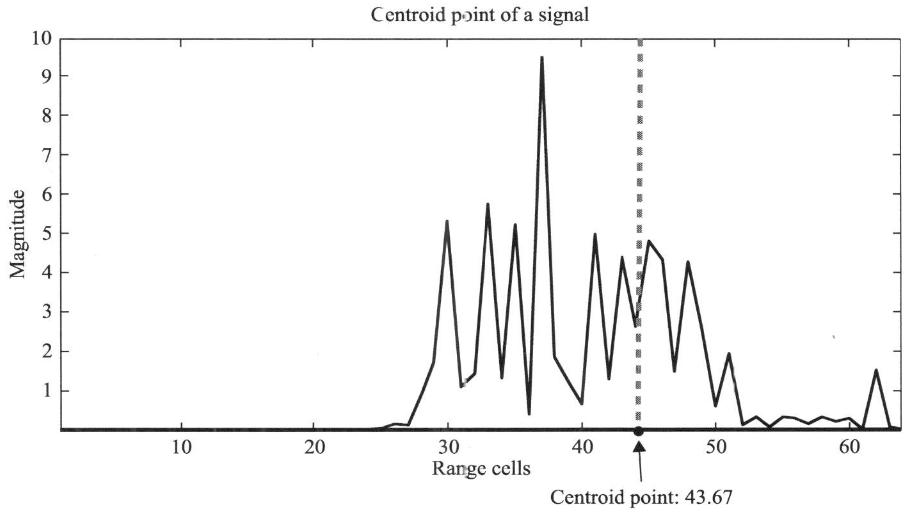

Figure 4.12 The calculated centroid point of a most centralized signal.

The centroid is a measure of the center of mass. For example, for a discrete signal $S = [s_1, s_2, \ldots, s_K]$, the centroid is defined by

$$\text{Centroid} = \sum_{k=1}^{K} k \cdot s_k \Big/ \sum_{k=1}^{K} s_k, \tag{4.4}$$

where the signal is expressed by a series of number K samples.

Figure 4.12 shows that the centroid point of a normal magnitude signal is at range cell no. 44, calculated from (4.4). However, a centroid point may not always fall in a higher energy region. It may be located in a zero-energy region, in which case the signal's energy is not centralized but is split (Figure 4.13a) and is falling in a zero-energy region at range cell no. 35.

In these cases, the centroid location is not suitable for range alignment. Therefore, when we use the centroid method for the TMC, we first need to determine whether the signal is centralized. If it is not, it should be circularly shifted to make it being centralized (Figure 4.13b). Thus, we calculate the centroid point for the circular shifted signal.

In the example given in Figure 4.13, the centroid point of the circular shifted signal is located at range cell no. 29. Then, the next step is to find its corresponding centroid point location at the original signal in Figure 4.13a. This is determined by comparing the variance of the centroid between the original signal and the circular shifted signal:

$$\text{variance} = \sum_{k=1}^{K} (k - \text{centroid})^2 \cdot s_k. \tag{4.5}$$

Then, the correct centroid point location at the original signal can be found to be at range cell no. 61 (Figure 4.13c).

The following example shows how to use the range centroid and Doppler centroid method for ISAR motion compensation. We use a point scatterer model of a Mig-25 aircraft and plot

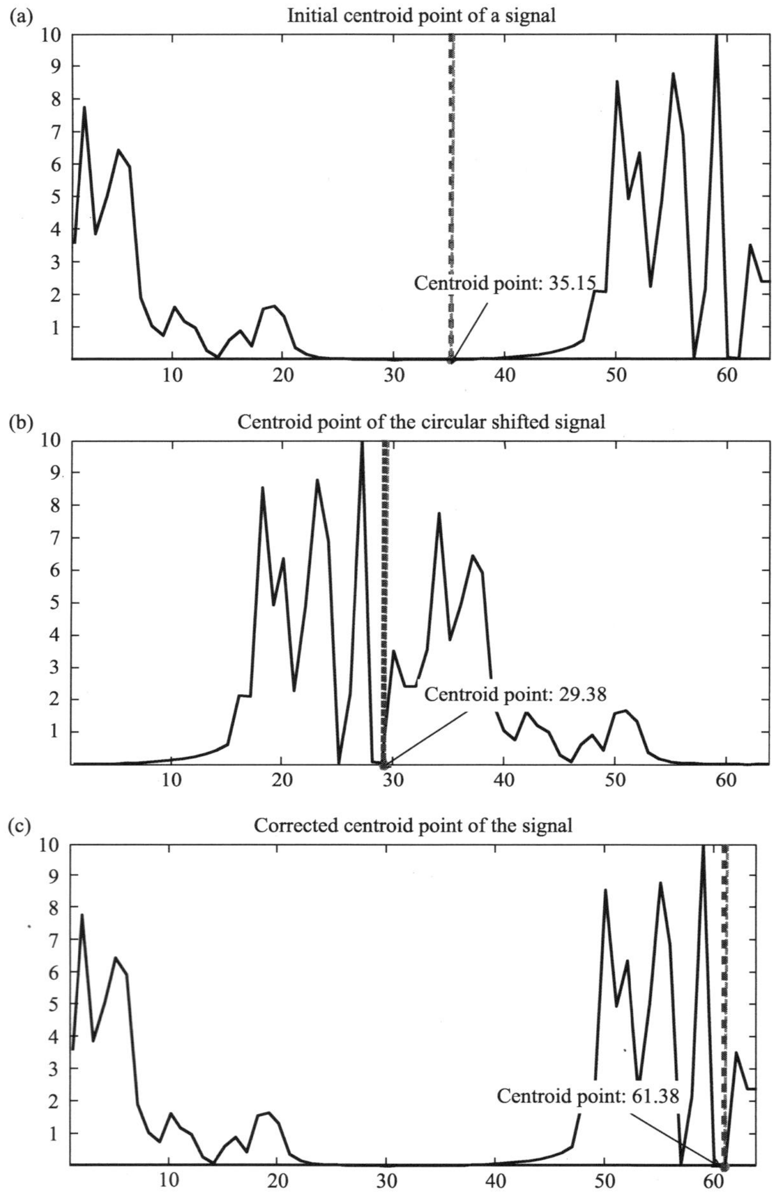

Figure 4.13 (a) The calculated centroid point at 35.15 for a noncentralized signal; (b) the centroid point at 29.38 for the same signal after circular shift; (c) the corresponding centroid point located at the original signal in (a) is 61.38.

the 2-D geometry of the radar and the aircraft in Figure 4.14. The radar is located at $(X = 0, Y = 0)$, and the aircraft is moving with velocity V_0 starting from (X_0, Y_0). Thus, the initial radial velocity of the aircraft is $(V_0)_R = V_0 \cdot X_0/R_0$ and the initial velocity perpendicular to the radial velocity is $(V_0)_\perp = V_0 \cdot Y_0/R_0$. Then, the initial angular velocity is $\omega_0 = (V_0)_\perp/R_0 = V_0 Y_0/(R_0)^2$. If the aircraft has rotation Ω at the same time, the

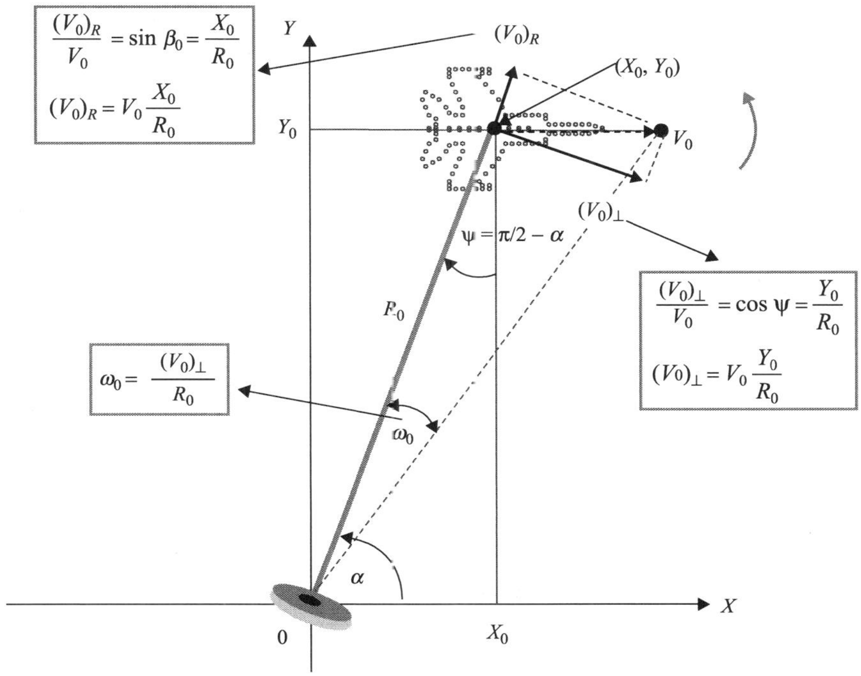

Figure 4.14 A point scatterer model of Mig-25 aircraft and 2-D geometry of the radar and aircraft.

additional angular velocity of the maneuvering must be added to the angular velocity: $\omega_0 = V_0Y_0/(R_0)^2 + \Omega$.

Assume the initial aircraft location is ($X_0 = 350$ m, $Y_0 = 3{,}500$ m) with an initial velocity $V_0 = 100$ m/s. The radar operated at X-band of $f_0 = 9.0$ GHz is a stepped-frequency radar with frequency step of $\Delta f = 7.5$ MHz and $M = 64$ frequency steps in each burst. If the PRF is 18 KHz, the CPI with $N = 512$ bursts and $M = 64$ frequency steps is $T = 1.82$ sec. The aircraft translational motion with a speed V_0 can induce angular velocity $\omega_T = 1.62°$/sec because of the LOS angle's change. On the other hand, the aircraft rotating can also produce angular velocity, which is assumed $\omega_R = 8.02°$/sec in the simulation study. Thus, the combined angular velocity $\omega = \omega_T + \omega_R$ is 8.64°/sec, which is relatively large. Therefore, the ISAR image generated during the CPI becomes smeared in the Doppler (or cross-range) direction.

Figure 4.15 demonstrates the range centroid method for range alignment. Figure 4.15a shows the range profiles before applying the range centroid processing, and 4.15b illustrates the range curve estimated from the centroid range points in the range profiles. The estimated target's range is a function of time, $R(t)$. This range function is used to compensate the range shift from pulse to pulse in the range profiles, $s_R(t)$, such that the extraneous phase term $\exp[-j4\pi f\frac{R(t)}{c}]$ due to target's motion can be exactly removed. Therefore, ISAR image of the target can be obtained simply by taking the inverse Fourier transform of the range compensated signal:

$$\mathrm{IFT}\left\{s_R(t)\cdot\exp\left[j4\pi f\frac{R(t)}{c}\right]\right\}.$$

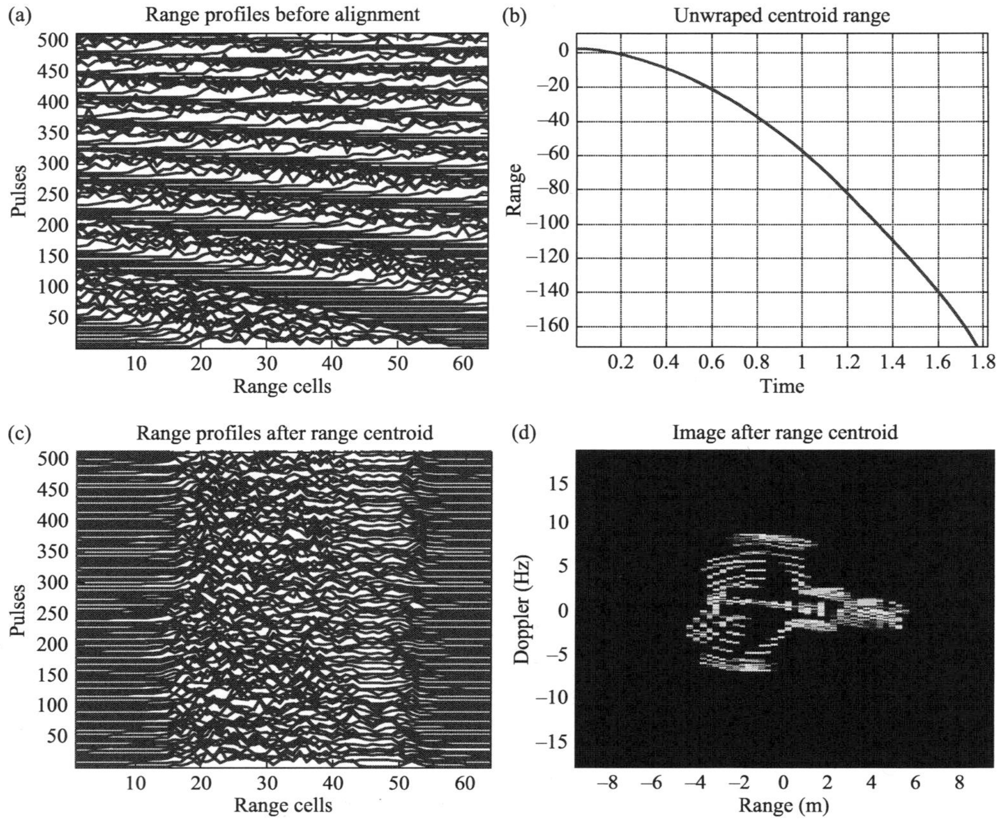

Figure 4.15 Demonstration of range centroid method for range alignment.

Figure 4.15c shows the range profiles after range alignment by using the range centroid method, and 4.15d is the ISAR image of the Mig-25 model after range centroid processing, which is still smearing. Thus, a Doppler centroid processing is needed.

After range centroid processing, the range profiles become aligned. However, the Doppler profiles are still not aligned and are time varying (see the Doppler center line in Figure 4.16a). Thus, additional Doppler centroid processing in the Doppler domain should be applied. After Doppler centroid, the Doppler profiles become aligned (Figure 4.16b).

However, after the Doppler centroid processing, the range profiles may become unaligned. Thus, a refining processing may be necessary to realign the range profiles and keep the Doppler profiles aligned. The final refined ISAR image compared with the image after range centroid processing is shown in Figure 4.17.

After the range centroid and Doppler centroid, we still need to apply the conventional RMC for correcting the rotational motion in Figure 4.17b.

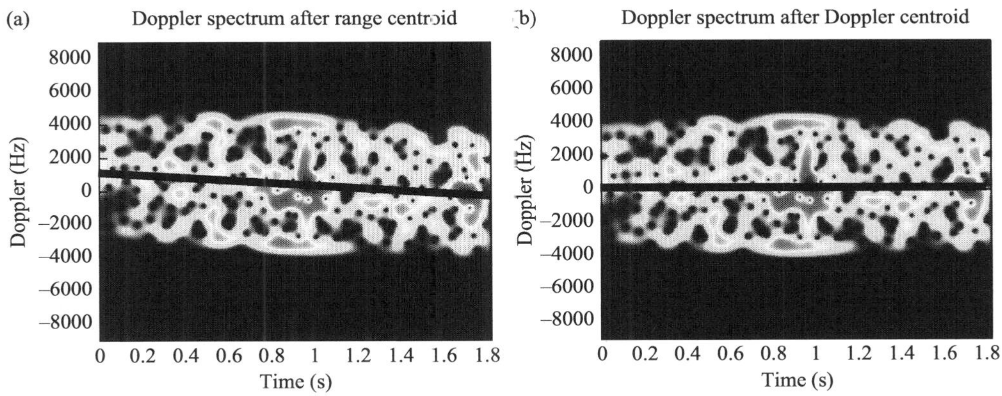

Figure 4.16 (a) Doppler spectrum calculated from pulse to pulse at one range cell after the range centroid processing; (b) Doppler spectrum after the Doppler centroid processing.

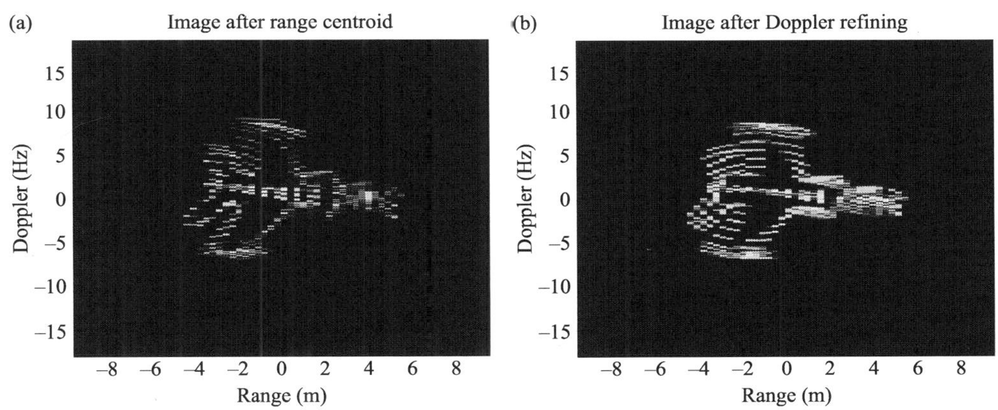

Figure 4.17 (a) ISAR image after the range centroid processing; (b) ISAR image after the Doppler centroid and refining processing.

4.1.3 Entropy Minimization Method for Range Alignment and Phase Adjustment

The entropy minimization method has been used more often than the centroid methods for the TMC [8,9,14,15]. Entropy function can be used as a cost function to align range profiles and adjust phase functions in ISAR. In statistical thermodynamics, entropy is a measure of unpredictability. Let S be a discrete random variable $S = [s_1, \ldots, s_N]$ with its probability distribution function $p(s_n) \geq 0$ and mean value $E(S) = \sum_{n=1}^{N} s_n p(s_n)$. The Shannon entropy, $H(S)$, of S is defined by [13]

$$H(S) = -\sum_{n=1}^{N} p(s_n) \log p(s_n), \tag{4.6}$$

which quantifies the unevenness of the probability distribution function, $p(s_n)$. When $p(s_n)$ is certain or impossible, that is, $p(s_n) = 1$ or 0, the Shannon entropy is minimum. When all random variables are equally likely, that is, $p(s_n) = 0.5$, it is maximum.

The entropy can be used to evaluate the result of range cell alignment. If range profiles are accurately aligned, the summation function of all envelope range profiles should have sharp peaks at dominant scatterer centers. In other words, the entropy function should reach the minimum value. Let S_n be the n-th range profile and S_{n+1} be the successive $(n+1)$-th range profile. Then the Shannon entropy function can be rewritten as

$$H(S_n, S_{n+1}) = -\sum_{m=1}^{M} p(k,m) \cdot \log\{p(k,m)\}, \tag{4.7}$$

where M is the total number of range cells in a range profile, and the probability distribution function, $p(k, m)$, is defined as

$$p(k,m) = \frac{|S_n(m)| + |S_{n+1}(m-k)|}{\sum\limits_{m=0}^{M-1} [|S_n(m)| + |S_{n+1}(m-k)|]}, \tag{4.8}$$

where m is the index of range cell, and k is the relative range cell shift between the two range profiles.

Thus, the relative range cell shift k between the two range profiles can be estimated by

$$\hat{k} = \arg\min\{H(S_n, S_{n+1})\}. \tag{4.9}$$

The entropy algorithm allows us find such a number $\hat{k}$ that minimizes the entropy function; therefore, by shifting $\hat{k}$ range cells the n-th range profile will align with the $(n+1)$-th range profile. To reduce the accumulate error while doing range cell alignment, a method using averaging range profiles instead of two range profiles was proposed [15].

The following example demonstrates how to use the entropy minimization method for range alignment. A point scatterer model of a B-727 aircraft and its radar and aircraft 2-D geometry are depicted in Figure 4.18. The radar and target geometric parameters are the same as those listed in Table 4.1. The radar operates at 9.0 GHz with bandwidth of 125 MHz and uses a stepped-frequency waveform. The stepped-frequency radar has 128 frequencies with 1.17 MHz frequency step and 256 bursts at 25 KHz burst repetition frequency.

For range alignment, based on radar received signal $s_R(t)$, we use 1-D entropy function to estimate the target's range function. Let $S_n = [s_{R_n}(t_1), s_{R_n}(t_2), \ldots s_{R_n}(t_M)]$ be the n-th range profile and $S_{n+1} = [s_{R_{n+1}}(t_1), s_{R_{n+1}}(t_2), \ldots s_{R_{n+1}}(t_M)]$ be the successive $(n+1)$-th range profile; then equations (4.7) through (4.9) are used to estimate the relative range cell shift, $\hat{k}$. Based on the estimated shift, $\hat{k}$, the target's range function of time, $R(t)$, can be estimated as shown in Figure 4.19. Thus, the range function is used to compensate shifts of range cells, such that the target motion induced extraneous phase term $\exp[-j4\pi fR(t)/c]$ can be totally removed. Finally, ISAR image of the target can be obtained simply by taking the inverse Fourier transform of the range compensated signal:

$$\text{IFT}\{s_R(t) \cdot \exp[j4\pi fR(t)/c]\}. \tag{4.10}$$

After range alignment, phase adjustment is also needed. Again, the entropy minimization method is used to do phase adjustment. In [16], a parametric method was developed to estimate target motion parameters.

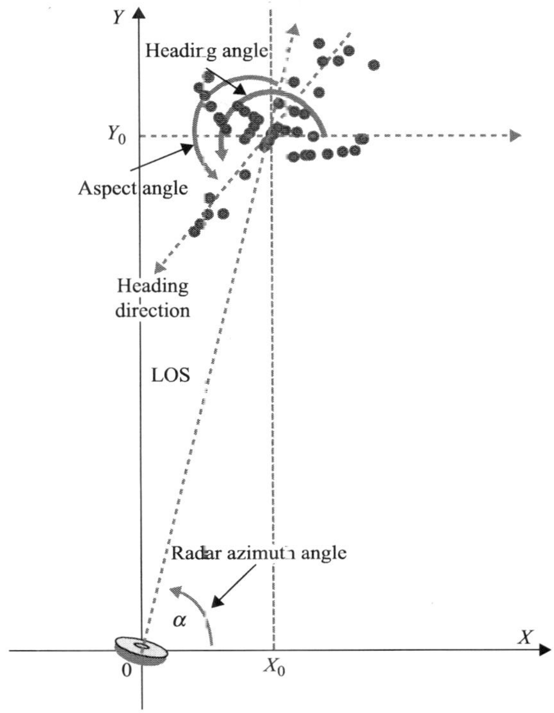

Figure 4.18 Geometry of stepped-frequency radar and B-727 aircraft point scatterer model.

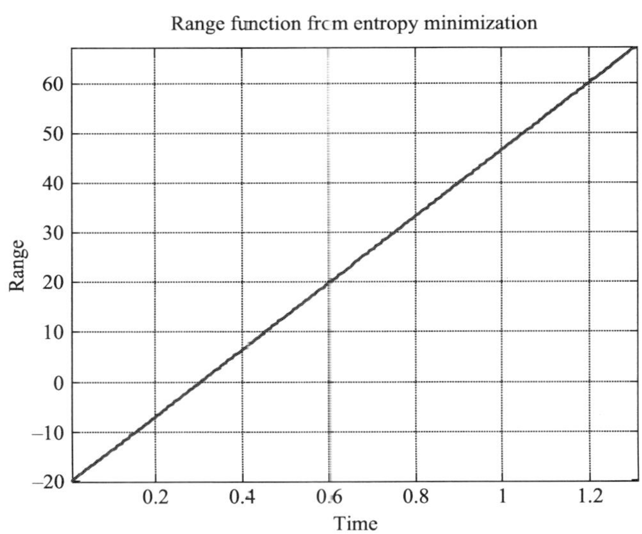

Figure 4.19 Estimated target range function using entropy minimization method.

Before using the entropy minimization, we first use a parametric method to estimate target's velocity and acceleration parameters. If the estimated parameter is not right and the range-Doppler image is not well focused, a 2-D entropy minimization method can help for the estimation of the mostly correct parameters. According to the estimated target motion parameters (velocity, v, and acceleration a) the phase function due to target's motion can be calculated by

$$\varphi(t) = vt + \frac{1}{2}at^2. \tag{4.11}$$

Then, the phase adjustment term $\exp\{j4\pi f\varphi(t)/c\}$ is applied to the aligned range profiles. After taking a 2-D Fourier transform, the phase-adjusted range-Doppler image can be generated. If the estimated motion parameter is not correct, the phase-adjusted range-Doppler image will not be well focused. Thus, we can use the 2-D entropy minimization method to reach the mostly correct motion parameters.

The 2-D entropy function is defined as follows. Let S_{mn} $(m = 1, 2, \ldots, M; n = 1, 2, \ldots, N)$ be the $M \times N$ matrix; then the entropy function is

$$H(S_{mn}) = -\sum_{m=1}^{M}\sum_{n=1}^{N} p(k,m) \cdot \log\{p(k,m)\}, \tag{4.12}$$

where the probability distribution function is

$$p(m,n) = S_{mn} \bigg/ \sum_{m=1}^{M}\sum_{n=1}^{N} S_{mn}. \tag{4.13}$$

The 2-D entropy method is to estimate m and n by

$$(\hat{m}, \hat{n}) = \arg\min\{H(S_{mn})\}. \tag{4.14}$$

The estimated velocity and acceleration parameters using the 2-D entropy method are shown in Figure 4.20, where the minima is at (vel = –2.3 m/s, acc = 2.64 m/s^2). Thus, the phase function can be calculated as shown in Figure 4.21a, and the adjusted phase functions at every range cell are shown in Figure 4.21b and are mostly linearized. Then, the ISAR image of the target can be obtained simply by taking the inverse Fourier transform of the range aligned and phase-adjusted signal (Figure 4.21c):

$$\text{IFT}\left\{s_R(t) \cdot \exp\left[j4\pi f\frac{R(t)}{c}\right] \cdot \exp[j\varphi(t)]\right\}. \tag{4.15}$$

4.2 Rotational Motion Compensation

After carrying out the TMC and removing the translational motion, the reconstructed ISAR range-Doppler image may be still unfocused. This is because of the time-varying quadratic term in the Doppler frequency shift, which makes the target scatterers drift from one range cell to the next. Therefore, the RMC is needed to compensate the angular rotation rate, Ω.

A useful scheme to deal with target's rotational motion is the polar formatting algorithm (PFA) as introduced in Chapter 3, Section 3.1.5, which reformats data from the polar

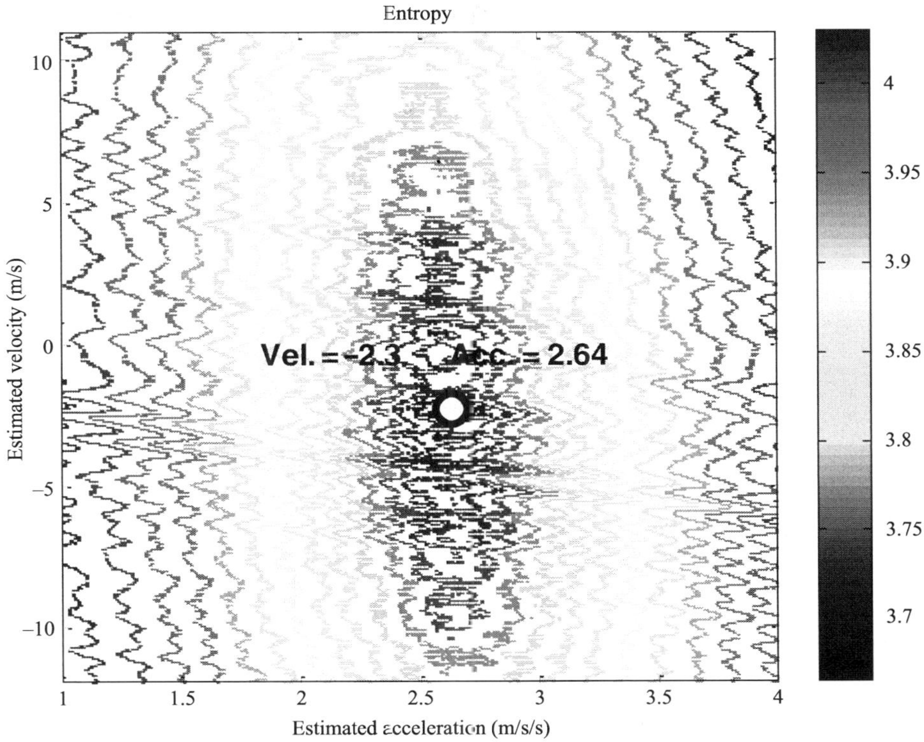

Figure 4.20 Estimated velocity and acceleration parameters using 2-D entropy minimization.

coordinates to rectangular coordinates [2]. As introduced in Chapter 3, Section 3.2, the joint time-frequency–based image formation proposed in [17] can handle the problem caused by the target's fast rotation. Here, we discuss how to handle the compensation of rotational motion using the ISAR PFA.

4.2.1 Introduction

As discussed in Chapter 1, the radar-received signal is expressed by

$$s_R(t) = \exp\left[-j4\pi f \frac{R(t)}{c}\right] \iint_{-\infty}^{\infty} \rho(x,y)\exp\{-j2\pi[xf_x(t) - yf_y(t)]\}dxdy, \tag{4.16}$$

where $f_x(t) = 2f\cos\theta(t)/c$, and $f_y(t) = 2f\sin\theta(t)/c$. The range function, $R(t)$, and rotation angle function, $\theta(t)$, can be modeled in terms of motion parameters:

$$\begin{cases} R(t) = R_0 + v_0 t + \frac{1}{2}a_0 t^2 + \cdots \\ \theta(t) = \vartheta_0 + \Omega_C t + \frac{1}{2}\gamma_0 t^2 + \cdots \end{cases}, \tag{4.17}$$

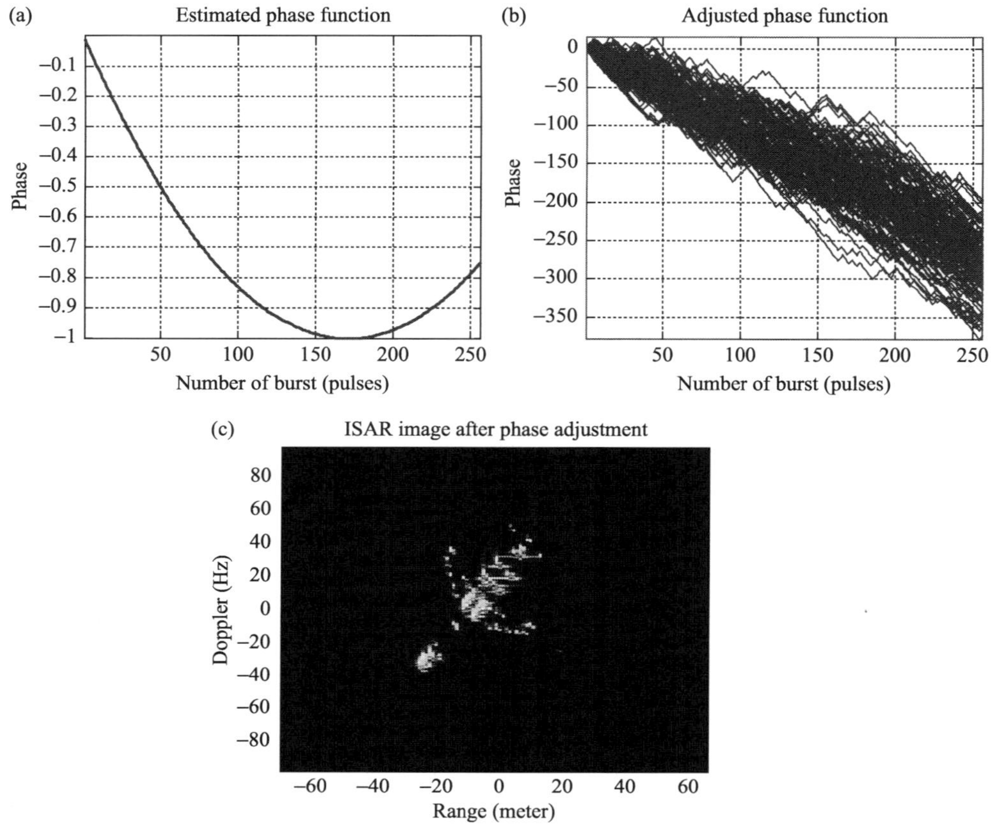

Figure 4.21 (a) Phase function obtained from the estimated motion parameters; (b) adjusted phase functions at every range cell; (c) focused ISAR image after phase adjustment.

where the translational motion parameters are initial range, R_0, velocity, v_0, and acceleration a_0; and the angular rotation parameters re initial angle, θ_0, angular velocity, Ω_0, and angular acceleration γ_0.

After removing the translational motion, the target has only rotation about the center at a constant range, *R,* as depicted in Figure 4.22. Thus, the range of a scatterer, *P,* located at (x, y) in the target local coordinates is given by

$$R_P = R + x\cos\theta(t) - y\sin\theta(t). \tag{4.18}$$

Without showing the time variable, *t*, (4.16) can be rewritten as

$$s_R(f_x, f_y) = \exp[-j4\pi fR/c] \iint_{-\infty}^{\infty} \rho(x,y)\exp\{-j2\pi[xf_x - yf_y]\}dxdy, \tag{4.19}$$

where the time variable, *t*, is embedded in f_x and f_y. The components of the spatial frequency are

$$f_x = 2f\cos\theta/c \tag{4.20}$$

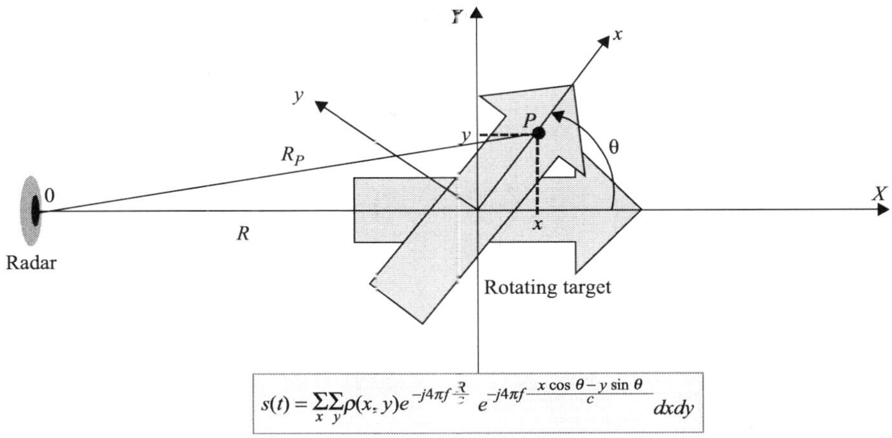

Figure 4.22 Radar and a rotating target.

and

$$f_y = 2f \sin\theta / c. \tag{4.21}$$

The two Cartesian coordinates f_x and f_y and two polar coordinates f and θ are related by the transform

$$\begin{cases} f = \sqrt{f_x^2 + f_y^2} \\ \theta = \tan^{-1}\left(\dfrac{f_y}{f_x}\right) \end{cases}. \tag{4.22}$$

If the range function, R, is known exactly, the phase term in (4.19), $\exp[-j4\pi fR/c]$, can be removed. Thus, the baseband signal becomes

$$s_B(f_x, f_y) = \iint_{-\infty}^{\infty} \rho(x, y) \exp\{-j2\pi [xf_x - yf_y]\} dxdy. \tag{4.23}$$

Then, the target reflectivity density function, $\rho(x, y)$, can be reconstructed by taking the inverse Fourier transform

$$\rho(x, y) = \iint_{-\infty}^{\infty} s_B(f_x, f_y) \exp\{j2\pi [xf_x - yf_y]\} df_x df_y. \tag{4.24}$$

Because the returned radar signal from a rotating target can be represented in the spatial frequency (f_x, f_y) domain as in (4.23), if the target rotation angle and radar frequency bandwidth are large, the radar data $s_B(f_x, f_y)$ on the (f_x, f_y) grid distributed in the Cartesian coordinates (f_x, f_y) can be mapped onto the (f, θ)-grid distributed in the polar coordinates (f, θ) as illustrated in Figure 4.23.

However, based on the relationship between the polar coordinates and the Cartesian coordinates, by using the PFA any data in the polar (f, θ) domain can be remapped into the Cartesian (f_x, f_y) domain as illustrated in Figure 4.24. If all scatterers of a target are

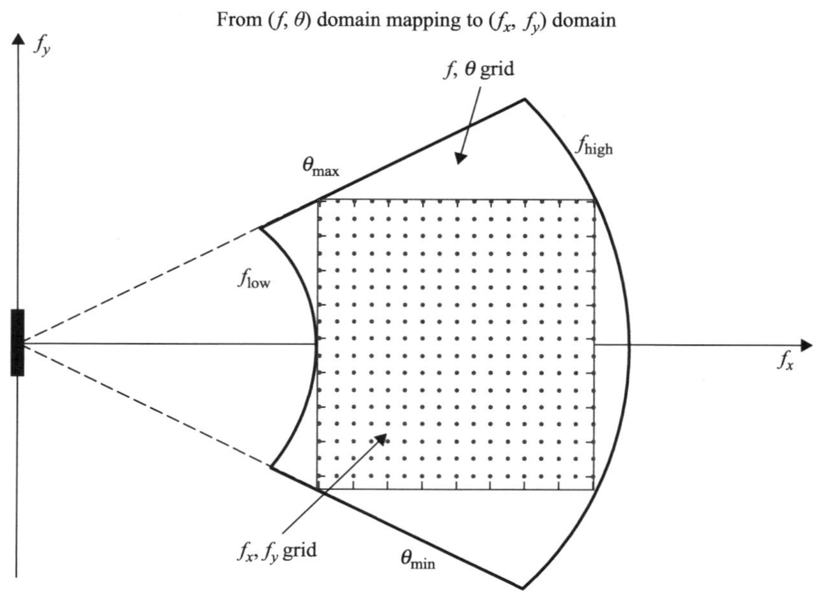

Figure 4.23 Polar reformation mapping from (f, θ) domain to (f_x, f_y) domain.

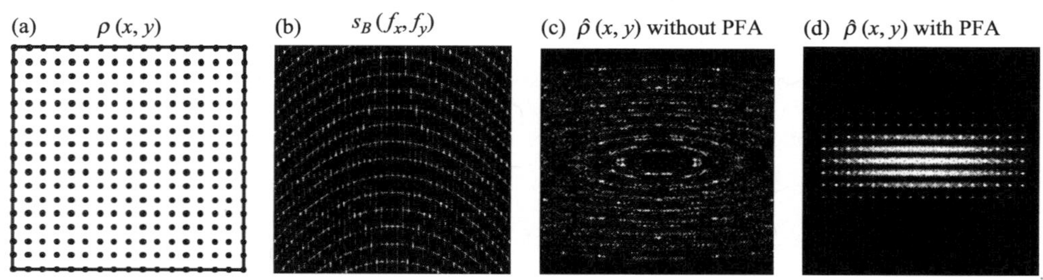

Figure 4.24 (a) Original target shape; (b) received data from the target; (c) reconstructed image of the target without using polar formatting algorithm; and (d) reconstructed image after using the polar formatting algorithm.

uniformly distributed inside a rectangle (Figure 4.24a), because of fast rotation of the target or long image integration time the received radar data from the target are affected by the rotation (Figure 4.24b). Using the data $s_B(f_x, f_y)$ and without using polar reformatting method, the reconstructed ISAR image $\hat{\rho}(x, y)$ of the rectangular target is smeared (Figure 4.24c). After applying the polar formatting algorithm, the reconstructed ISAR image $\hat{\rho}(x, y)$ of the rectangular target becomes focused and rectangular shaped (Figure 4.24d).

4.2.2 ISAR Polar Formatting

Polar formatting is an image formation method based on tomography. It was originally developed for spotlight SAR imaging [2,18] and was then adapted for use in ISAR image formation [19].

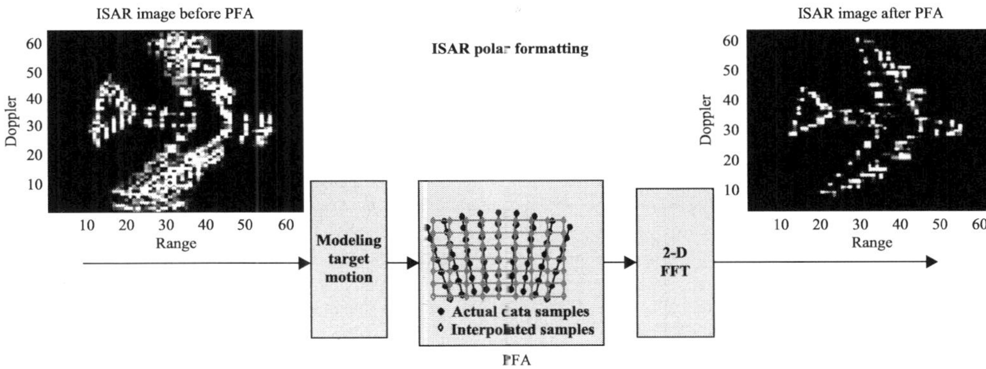

Figure 4.25 ISAR image formed by polar formatting algorithm.

In principle, the ISAR PFA is similar to that for spotlight SAR. However, in ISAR the motion of the target provides the change in aspect angle necessary for Doppler processing, whereas in spotlight SAR the change in aspect angle is from the motion of the radar. In both of them, the aspect angle change between the radar and the target is unknown. In ISAR especially the aspect angle change is also uncontrollable. Therefore, target's rotation parameters must be measured before the ISAR PFA is applied.

The relative rotation between the radar and the target makes the collected data distributed in polar format. To form a focused image using the Fourier transform, a polar-to-rectangular remapping processing must be applied. The implementation of the ISAR PFA involves (1) measuring target motion parameters, (2) estimating parameters of data surface model, (3) projecting the data surface onto a planar surface, (4) interpolating the data into equally spaced samples, and (5) performing the inverse Fourier transform. The image formed by the PFA is illustrated in Figure 4.25. The comparison between images formed by the conventional method shown on the left side of the figure and by the polar reformatting method shown on the right side of the figure can clearly prove the improvement in image focusing.

Because ISAR is for imaging of noncooperative moving targets, the target motion parameter must be measured directly from the received radar data. Let us recall parameter models of the range and the rotation angle in (4.17), which is based on a rigid-body target with point scatterers modeling and assuming the target contains both the translation and rotation. Thus, the radar-received signal in the baseband can be expressed by

$$\begin{aligned}
s_B(t) &= \exp\left[-j4\pi f\frac{R(t)}{c}\right]\iint_{-\infty}^{\infty}\rho(x,y)\exp\{-j2\pi[xf_x(t)-yf_y(t)]\}\,dxdy \\
&= \iint_{-\infty}^{\infty}\rho(x,y)\exp\left\{-j\frac{4\pi f}{c}[R(t)+x\cos\theta(t)-y\sin\theta(t)]\right\}dxdy \\
&\cong \iint_{-\infty}^{\infty}\rho(x,y)\exp\left\{-j\frac{4\pi f}{c}\left[R_0+v_0t+\frac{1}{2}a_0t^2+x-\Omega_0yt+\frac{1}{2}\Omega_0^2xt^2-\frac{1}{2}\gamma_0yt^2\right]\right\}dxdy \\
&= \iint_{-\infty}^{\infty}\rho(x,y)\exp\left\{-j\frac{4\pi f}{c}\left[(R_0+x)+(v_0-\Omega_0y)t+\frac{1}{2}(a_0+\Omega_0^2x-\gamma_0y)t^2\right]\right\}dxdy,
\end{aligned} \tag{4.25}$$

where the initial rotation angle $\theta_0 = 0$. The constant phase term $\exp[-j4\pi f R_0/c]$ is not relevant to the image formation and can be ignored. However, other terms related to the velocity, acceleration, angular velocity, and angular acceleration can lead to image smearing. Thus, after removing the translational motion–related phase terms, we must estimate the rotational motion parameters before applying the PFA.

One approach to estimate the rotation parameters is the time-frequency–based analysis as described in [20,21]. We notice that in (4.25) there are two quadratic phase terms: $\exp\left\{-j\frac{4\pi f}{c}\left[\frac{1}{2}a_0t^2\right]\right\}$, which represents the translation motion error and is independent of the cross-range, y; and the cross-range, y, dependent term, $\exp\left\{-j\frac{4\pi f}{c}\left[\frac{1}{2}\gamma_0 y t^2\right]\right\}$, which represents the rotational motion generated phase error. By using the time-frequency–based approach, we first remove the translational motion error term $\exp\left\{-j\frac{4\pi f}{c}\left[\frac{1}{2}a_0t^2\right]\right\}$ by extracting the phase function of a selected prominent point scatterer located at (x_1, y_1) using the time-frequency analysis. Then, the uncompensated phase error terms are reduced to

$$s_B(t) \cong \iint_{-\infty}^{\infty} \rho(x, y)\exp\left\{-j\frac{4\pi f}{c}\left[x - y\left(\Omega_0 t - \frac{1}{2}\gamma_0 t^2\right)\right]\right\}dxdy, \tag{4.26}$$

where only the rotational motion error remains.

Next, we can extract the phase function related to target's rotation using another prominent point scatterer at (x_2, y_2) and the time-frequency analysis. From the time-varying Doppler spectrum of the rotating prominent point scatterer, we can find the Doppler shift, f_D, and Doppler shift rate, df_D/dt, which are related to rotation velocity and rotation acceleration, respectively. Therefore, the rotation rate, Ω_0, and rotation acceleration, γ_0, can be estimated. Once these rotation parameters are estimated, we are ready to perform the ISAR PFA.

Instead of modeling the target motion, a method proposed in [19] is directly modeling the data surface. This method uses measurable scatterer motion quantities, such as range, velocity, and acceleration, to estimate the data surface model parameters. The data surface can be modeled by a quadratic surface, which requires fewer parameters but still allows us to compensate for the majority of nonlinear rotational motion in the target. Within the surface, the spacing between the data line segments is also modeled by a quadratic function. In the modeling, there are two parameters of interest: the quadratic term representing the curvature of the data surface, called the out-of-plane acceleration; and the quadratic term representing the line segment spacing, called the in-plane acceleration.

Now we are ready to apply the ISAR PFA to imaging the noncooperative moving targets.

4.2.3 Application of ISAR Polar Formatting Algorithm

Here we use an example to demonstrate how to apply the ISAR PFA to generate a focused ISAR image of a rotating target. MATLAB source codes are also provided for testing the ISAR PFA.

For simplicity, we use 2-D Cartesian coordinates. Assume a stationary radar is located at $(X = 0,\ Y = 0)$ and a target initially located at $(X = X_0,\ Y = Y_0)$ has only translational motion with initial velocity, v_0. Because the target aspect angle changes when it is moving, the translational motion can also produce target rotation relative to the radar. The relative

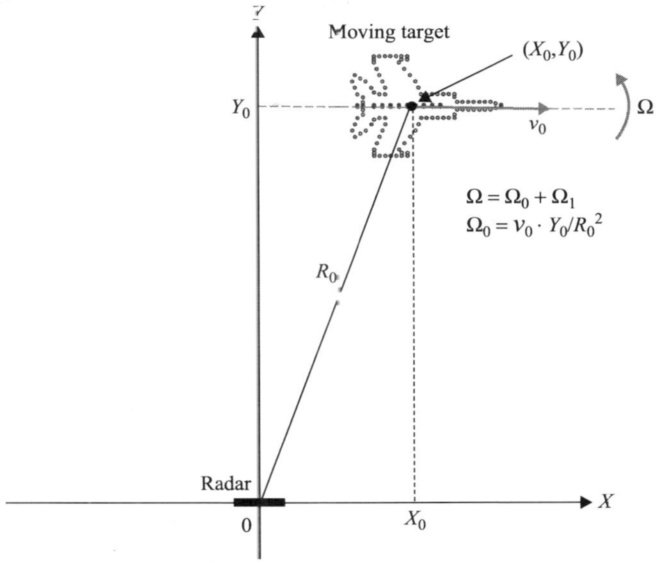

Figure 4.26 Geometry of radar and a moving target.

rotation rate is determined by $\Omega_0 = v_0 Y_0/R_0^2$ as indicated in Figure 4.26. In addition to the translational motion produced rotation rate, the target itself can also has an additional rotation rate, Ω_1. Thus, the total rotation is $\Omega = \Omega_0 + \Omega_1$.

The following example shows how to apply PFA to focus an ISAR image of rotating targets. The point scatterer model of a Mig-25 is used, and 2-D geometry of the radar and the aircraft is depicted in Figure 4.26, where the radar is located at ($X = 0$, $Y = 0$) and the aircraft is moving with velocity $v_0 = 100$ m/s starting from ($X_0 = 350$ m, $Y_0 = 7{,}000$ m) at a heading angle of 0° with respect to the X-axis. Thus, the translation-generated relative rotation rate is $\Omega_0 = v_0 Y_0/R_0^2 = 0.014(\text{rad/sec}) = 0.82°/\text{sec}$. We assume a stepped frequency radar operating at $f_c = 9.0$ GHz with 512 MHz bandwidth, 64 frequency steps, 512 bursts, and 8.0 MHz step frequency. We also assume the burst repetition frequency (BPF) to be 18 KHz, and thus the total CPI is 1.82 sec. The range resolution then becomes $\Delta r_r = c/(2BW) = 0.29$m, and cross-range resolution is $\Delta r_{cr} = c \cdot BRF/(2f_c \Omega MN) = 0.056$ m, where we assume additional rotation rate is $\Omega_1 = 0.15$ (rad/sec) $= 8.6°$/sec and thus the total rotation rate is $\Omega = 0.164$ (rad/sec) $= 9.4°$/sec. The total rotation rate is very high for the purpose of simulation of fast rotating targets but not for real-world target's rotation. Thus, from such simulated range profiles (Figure 4.27a), the range function can be estimated (Figure 4.27b).

After translational motion compensation, however, due to fast rotation the ISAR image is still smearing (Figure 4.28a). Therefore, we must apply ISAR PFA to compensate for for rotational motion. The ISAR image after the PFA is shown in Figure 4.28b. To apply the ISAR PFA, the rotational motion parameters must be estimated as demonstrated in the provided MATLAB source codes.

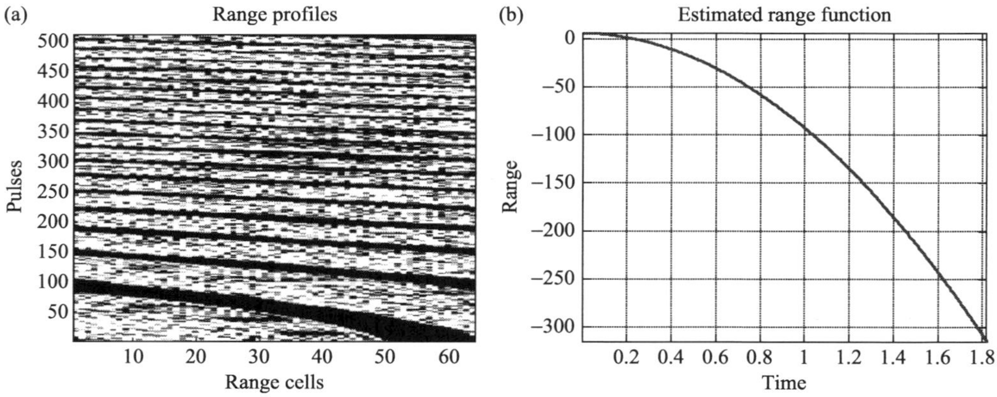

Figure 4.27 The simulated fast rotating target's (a) range profiles and (b) estimated range function.

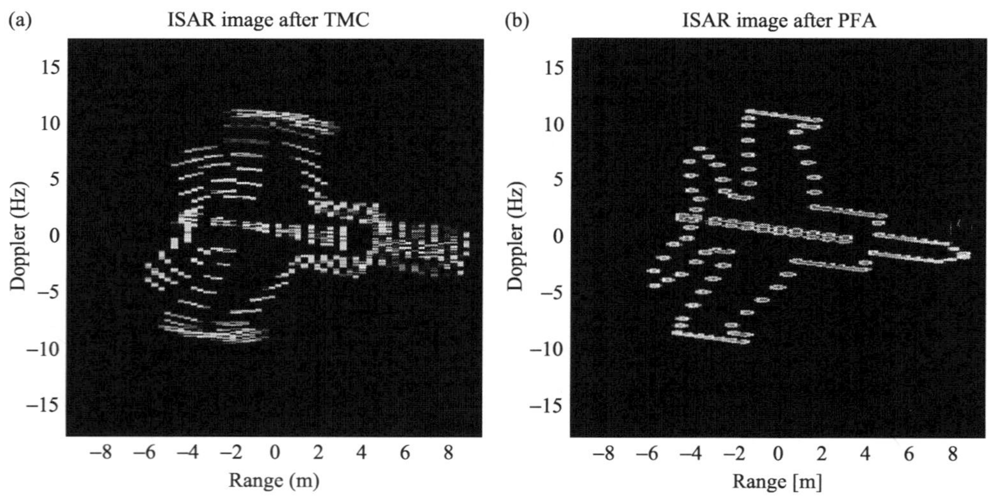

Figure 4.28 (a) ISAR image of a simulated fast rotating Mig-25 after translational motion compensation; (b) ISAR image of the Mig-25 after ISAR PFA.

References

[1] Chen, C.C., and H.C. Andrew, "Target-Motion Induced Radar Imaging," *IEEE Trans.*, AES-16, no.1, pp. 2–14, 1980.

[2] Walker, J.L., "Range-Doppler Imaging of Rotating Objects," *IEEE Trans.*, AES-16, no.1, pp. 23–52, 1980.

[3] Brown, W.M., and R.J. Fredricks, "Range-Doppler Imaging with Motion through Resolution Cells," *IEEE Trans.*, AES-5, no.1, pp. 98–102, 1969.

[4] Berizzi, F., and G. Corsini, "Focusing of Two-Dimensional ISAR Images of Objects by Contrast Maximization," *Proc. of 22nd European Microwave Conference*, vol. 2, Finland, pp. 951–956, 1992.

[5] Wahl, D.E., P.H. Eichel, D.C. Ghiglia, and C.V. Jakowatz, "Phase Gradient Autofocus—A Robust Tool for High Resolution SAR Phase Correction," *IEEE Trans.*, AES-30, no. 1 pp. 827–835, 1994.

[6] Delisle, G.Y., and H. Wu, "Moving Target Imaging and Trajectory Computation Using ISAR," *IEEE Trans.*, AES-30, no. 3, pp. 887–899, 1994.

[7] Berizzi, F., and G. Cosini, "Autofocusing of Inverse Synthetic Aperture Radar Images Using Contrast Optimization," *IEEE Trans.*, AES-32, no. 3, pp. 1185–1191, 1996.

[8] Li, X., G. Liu, and J. Ni, "Autofocusing of ISAR Images Based on Entropy Minimization," *IEEE Trans.*, AES-35, no. 4, pp. 1240–1251, 1999.

[9] Wang, J., and D. Kasilingam, "Global Range Alignment for ISAR," *IEEE Trans.*, AES-39, no. 1, pp. 351–357, 2003.

[10] Haywood, B., and R. Evens, "Motion Compensation for ISAR Imaging," *Proc. of ASSPA*, Adelaide, Australia, pp. 113–117, 1989.

[11] Prickett, M.J., and C.C. Chen, "Principle of Inverse Synthetic Aperture Radar (ISAR) Imaging," *IEEE EASCON Record*, Arlington, VA, pp. 340–345, 1980.

[12] Itoh, T., H. Sueda, and Y. Watanabe, "Motion Compensation for ISAR via Centroid Tracking," *IEEE Trans.*, AES-32, no. 3, pp. 1191–1197, 1996.

[13] C.E. Shannon, "A Mathematical Theory of Communication," *Bell System Technical Journal*, vol. 27, no. 3, pp. 379–423, 1948.

[14] Wang, J., and X. Liu, "Improved Global Range Alignment for ISAR," *IEEE Trans.*, AES-43, no. 3, pp. 1070–1074, 2007.

[15] Zhu, D., L. Wang, Y. Yu, Q. Tao, and Z. Zhu, "Robust ISAR Range Alignment via Minimizing the Entropy of the Average Range Profiles," *IEEE Trans. GRS Letters*, vol. 6, no. 2, pp. 204–208, 2009.

[16] Bocker, R.P., T.B. Henderson, S.A. Jones, and B.R. Frieden, "A New Inverse Synthetic Aperture Radar Algorithm for Translational Motion Compensation," *Proc. SPIE*, vol. 1569, pp. 298–310, 1991.

[17] Chen, V.C., and S. Qian, "Joint Time-Frequency Transform for Radar Range-Doppler Imaging," *IEEE Trans.*, AES-34, no. 2, pp. 486–499, 1998.

[18] Walker, J.L., W. Carrara, and I. Cindrich, "Optical Processing of Rotating-Object Radar Data Using a Polar Recording Format," Technical Report RADC-TR-73-136, AD 526 738, Rome Air Development Center, Rome, NY, May 1973.

[19] Lipps, R., and D. Kerr, "Polar Reformatting for ISAR Imaging," *Proc. of 1998 IEEE Radar Conference*, pp. 275–280, 1998.

[20] Chen, V.C., and H. Ling, *Time-Frequency Transforms for Radar Imaging and Signal Analysis* (Chapter 6), Artech House, 2002.

[21] Wang, Y., H. Ling, and V.C. Chen, "ISAR Motion Compensation via Adaptive Joint Time-Frequency Technique," *IEEE Trans.*, AES-34, no. 2, pp. 670–677, 1998.

CHAPTER 5

ISAR Autofocus Algorithms

Autofocus allows us to generate images with better quality by automatically adjusting the image focusing parameters. For radar imaging, autofocus means to automatically correct phase errors based on collected radar returns from targets.

As discussed in Chapter 4, inverse synthetic aperture radar (ISAR) motion compensation includes range alignment and phase adjustment. The phase adjustment process is for removing the residual translation error on phase terms. The phase errors are the causes of image defocusing. If a phase adjustment algorithm is based solely on the radar data itself, this is called the autofocus algorithm [1,2].

ISAR autofocus methods can be parametric and nonparametric. Parametric method uses a parametric model of the radar-received signal. Image contrast-based autofocus (ICBA) and entropy-based autofocus algorithm belong to the parametric method [3,4]. The prominent point processing (PPP) algorithm and the phase gradient autofocus (PGA) algorithm belong to the nonparametric ISAR autofocus method [1,2]. In this chapter, we will introduce the PPP, PGA, ICBA, and entropy-based autofocus methods. The issue on the keystone autofocus algorithm in ISAR will also be introduced.

5.1 Prominent Point Processing Autofocus

The PPP algorithm, initially used in resolving moving target imaging in synthetic aperture radar (SAR) [5], utilizes information from prominent points to correct phase errors and convert nonuniform rotation into a uniform rotation. Multiple PPP algorithm tracks multiple selected prominent scatterers in a target to extract motion parameters [1].

In spotlight SAR, a motion compensation method must be able to remove space variant and invariant errors. The multiple PPP algorithm can remove both the space-invariant and variant errors. In the multiple PPP, the first prominent point is usually selected for removing translational motion and adjusting the phase of the received signals so as to form a new image center. Then, a second prominent point, which is for correcting the phase error induced by nonuniform rotations, must be selected. If necessary, we can also select a third prominent point for estimating the rotation rate and the azimuth scale factor of the resulting image to achieve complete focusing [1].

However, in ISAR, after course motion compensation, the image may still be smeared due to phase errors induced by the target rotation and residual translation errors. To focus the image, we need to identify a prominent point at the rotation center of the target and track its phase variation. Then, an appropriate approach for searching an optimal phase function must be used. Finally, by applying the conjugate of the estimated optimal phase function, the image can be focused on the rotation center.

A simple approach to measure the optimal phase function is the exhaustive searching [6]. Because a phase function can be represented by a polynomial, we use the Taylor polynomial to approximate the phase function at the range cell that the rotation center falls in. The Taylor polynomial centered at a point, x_0, is defined by

$$f(x) = f(x_0) + f'(x_0)(x - x_0) + \frac{1}{2!}f''(x_0)(x - x_0)^2 + \frac{1}{3!}f'''(x_0)(x - x_0)^3 + \cdots. \tag{5.1}$$

Since the phase function is related to the range function by $\Phi(t) = 4\pi f R_P(t)/c$, we can simply represent the range function by the Taylor polynomial centered at zero:

$$R_P(t) = R_P(0) + R_P'(0)t + \frac{1}{2!}R_P''(0)t^2 + \frac{1}{3!}R_0'''(0)t^3, \tag{5.2}$$

where the higher order terms are omitted.

If we can have the polynomial coefficients $\left[R_P(0), R_P'(0), \frac{1}{2!}R_P''(0), \frac{1}{3!}R_0'''(0)\right]$, the range or phase function can be constructed. Thus, an exhausted search process can be used to find a set of coefficients for constructing the optimal phase function. However, to simplify the search process, we search only two coefficients instead of four. Because the coefficient $R_P(0)$ is a constant range that has been removed by translational motion compensation, we can set this coefficient to zero. The third-order coefficient $\frac{1}{3!}R_0'''(0)$ determines the jerk motion, which is relatively small compared with the velocity and the acceleration. We can set it to zero, too. Thus, we search only the velocity and acceleration coefficients $R_P'(0)$ and $\frac{1}{2!}R_P''(0)$ within their possible limits. By applying the conjugate of the searched optimal phase function, the rotation center of the image can be focused.

The next step is to remove the rotational phase error by estimating the target rotation parameters. To do this, we need to select a second prominent point scatterer, which should be far away from the center of rotation. After we have target rotation parameters, ISAR polar formatting algorithm can be applied and the entire ISAR image can be focused.

To estimate the target rotation parameters, a parametric approach may be used. As discussed in Chapter 1, for a target that has translational motion $R_0(t)$ and rotational motion $\theta(t)$ and by assuming the azimuth angle α is zero as shown in Figure 5.1, the radar received signal from the target can be represented by

$$s_R(t) = \iint_{-\infty}^{\infty} \rho(x,y) \exp\{-j2\pi f[2R_P(t)/c]\}dxdy, \tag{5.3}$$

where $R_P(t)$ is the range from the radar to the scatterer, P. From Chapter 1, Section 1.3.2, it can be expressed by

$$R_P(t) \cong R_0(t) + x_P \cos\theta(t) - y_P \sin\theta(t), \tag{5.4}$$

where assuming $R_0(t) \gg r_P$, and thus $r_P^2/[2R_0(t)] \approx 0$.

After course motion compensation, the geometric center of the target has no translation, that is, $R_0(t) = R_0$. The target can rotate only about the geometric center of the target. Thus, the

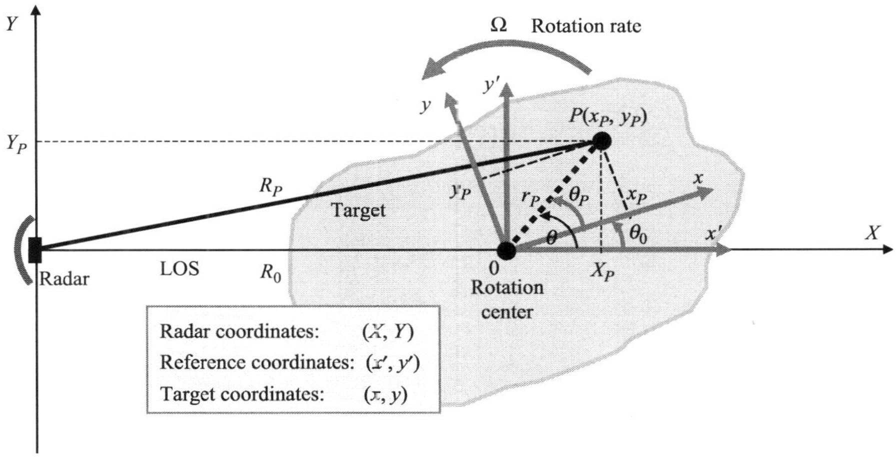

Figure 5.1 Geometry of a rotating target and radar at azimuthal angle = 0.

range from the radar to the point scatterer, P, becomes $R_P(t) \cong R_0 + x_P \cos\theta(t) - y_P \sin\theta(t)$, where R_0 is the range from the radar to the geometric center of the target.

If representing the point scatterer P in polar coordinates and assuming the range R_0 is much greater than the dimension of the target, the range equation can be rewritten as

$$R_P(t) = R_0 + r_P \cos\left(\theta_0 + \theta_P + \Omega t + \frac{1}{2}\gamma t^2\right), \tag{5.5}$$

where (r_P, θ_P) is polar coordinate of the point scatterer P, θ_0 is the initial aspect angle of the target, Ω is the angular velocity, and γ is the angular acceleration of the rotating target. Here, the order higher than 2nd rotations has been ignored.

Similarly, the velocity and acceleration equations can be derived as

$$\dot{R}_P(t) = \frac{d}{dt}R_P(t) = \Omega\, r_P \sin\left(\theta_0 + \theta_P + \Omega t + \frac{1}{2}\gamma t^2\right), \tag{5.6}$$

and

$$\ddot{R}_P(t) = \frac{d^2}{dt^2}R_P(t) = -\Omega^2\, r_P \cos\left(\theta_0 + \theta_P + \Omega t + \frac{1}{2}\gamma t^2\right). \tag{5.7}$$

A method based on the target motion equations was proposed in [7], which can be used to estimate motion parameters. The following is to introduce how to apply this method for ISAR image autofocusing.

If we count the range from the center of the rotation, the motion equations can be rewritten as

$$\begin{cases} R_{0,P}(t) = r_P \cos\left(\theta_0 + \theta_P + \Omega t + \frac{1}{2}\gamma t^2\right) \\ \dot{R}_{0,P}(t) = \Omega\, r_P \sin\left(\theta_0 + \theta_P + \Omega t + \frac{1}{2}\gamma t^2\right), \\ \ddot{R}_{0,P}(t) = -\Omega^2\, r_P \cos\left(\theta_0 + \theta_P + \Omega t + \frac{1}{2}\gamma t^2\right) \end{cases} \tag{5.8}$$

Figure 5.2 depicts the procedure of how to estimate angular velocity and acceleration from received data for the ISAR PFA.

To estimate the angular velocity, we select a prominent point whose Doppler shift (or velocity) is the same as that of the rotation center but is at a range cell different from the rotation center as illustrated in Figure 5.3.

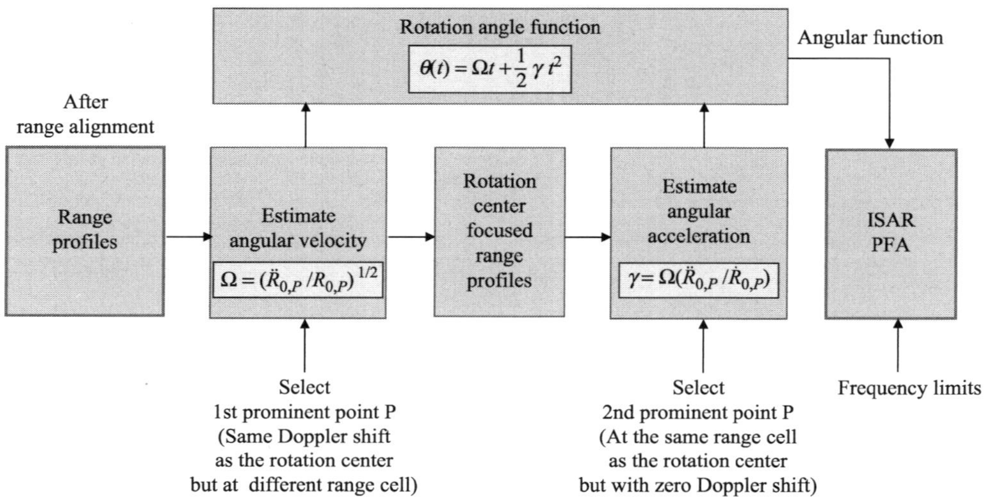

Figure 5.2 Estimation of angular velocity and acceleration from received data.

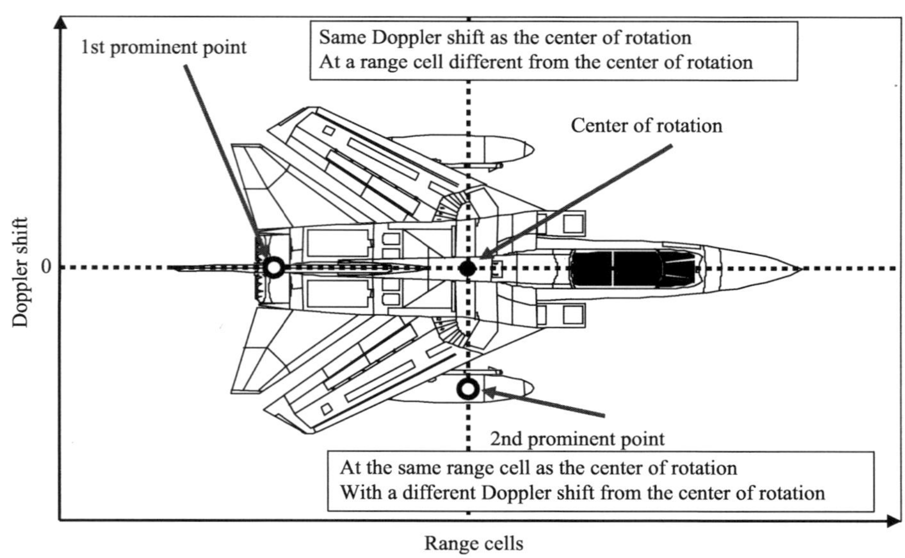

Figure 5.3 Illustration of the selection of prominent points.

Because the Doppler shift of the rotation center is zero, we have

$$\dot{R}_{0,P} = \Omega\, r_P \sin\hat{\theta}_P = 0, \tag{5.9}$$

where $\hat{\theta}_P$ represents the average rotation angle during the coherent processing interval (CPI). Thus, we have $\sin\hat{\theta}_P = 0$ or $\cos\hat{\theta}_P = 1$. In other words, average rotation angle should be $\hat{\theta}_P = 0$.

Then, the corresponding range and acceleration equations become

$$R_{0,P} = r_P \cos\hat{\theta}_P = r_P, \tag{5.10}$$

and

$$\ddot{R}_{0,P} = -\Omega^2\, r_P \cos\hat{\theta}_P + \gamma\, r_P \sin\hat{\theta}_P = -\Omega^2\, r_P. \tag{5.11}$$

From the updated motion equations (5.10) and (5.11), we can estimate the angular velocity by

$$\Omega = \left(\ddot{R}_{0,P}/R_{0,P}\right)^{1/2}. \tag{5.12}$$

Similarly, the acceleration parameter can be also estimated by selecting another prominent point whose range cell is the same as that of the rotation center, but its Doppler shift is not zero as indicated in Figure 5.3. Based on the selected prominent point, we have

$$R_{0,P} = r_P \cos\hat{\theta}_P = 0. \tag{5.13}$$

This leads to $\cos\hat{\theta}_P = 0$ or $\sin\hat{\theta}_P = 1$. In other words, average rotation angle should be $\hat{\theta}_P = \pm\frac{\pi}{2}$.

Thus, the corresponding velocity and acceleration equations become

$$\dot{R}_{0,P} = \Omega\, r_P \sin\hat{\theta}_P = \Omega\, r_P, \tag{5.14}$$

and

$$\ddot{R}_{0,P} = -\Omega^2\, r_P \cos\hat{\theta}_P + \gamma\, r_P \sin\hat{\theta}_P = \gamma\, r_P. \tag{5.15}$$

From motion equations (5.14) and (5.15), we can estimate the angular acceleration by

$$\gamma = \Omega\,(\ddot{R}_{0,P}/\dot{R}_{0,P}). \tag{5.16}$$

After estimating the angular velocity and acceleration parameters Ω and γ, we can calculate the rotation angle function $\theta(t) = \Omega t + \frac{1}{2}\gamma t^2$. Finally, with the estimated angular function and known frequency region of the selected signal waveform, we can apply the ISAR PFA to focus the entire image of the target.

The following is an example of selecting prominent points for removal of residual translational and rotational phase errors to apply ISAR PFA. MATLAB source codes are also provided in the book for exercises.

5.1.1 Removal of Residual Uncompensated Translation

As discussed in Section 4.2.3 and shown in Figure 4.26, a point scatterer model of a Mig-25 aircraft is moving with velocity $v_0 = 100$ m/s from its initial location at ($X_0 = 350$ m, $Y_0 = 7{,}000$ m) with a heading angle of 0°. A stepped-frequency radar operating at $f_c = 9.0$ GHz is located at ($X = 0$, $Y = 0$). Using 8.0 MHz frequency step and the total number of 64 frequency steps, the signal bandwidth can achieve 512 MHz. Thus, the range resolution is $\Delta r_r = c/(2BW) = 0.29$ m. With a burst repetition frequency of 18 KHz and a total number of bursts 512, the total CPI is 1.82 sec.

As indicated in Section 4.2.3, the target's rotation rate due to its translation is $\Omega_0 = v_0 Y_0/R_0^2 = 0.014$ (rad/sec) $= 0.82°$/sec. We assume additional rotation rate is $\Omega_1 = 0.15$ (rad/sec) $= 8.6°$/sec and, thus, the total rotation rate is $\Omega = 0.164$ (rad/sec) $=$ 9.4°/sec. Then the cross-range resolution is $\Delta r_{cr} = c \cdot BRF/(2f_c \Omega MN) = 0.056$ m.

Under these conditions, after applying TMC the ISAR image is shown in Figure 4.28a and redrawn in Figure 5.4, where we can see the image smearing due to residual uncompensated translation and fast rotation.

First of all, we want to remove the residual uncompensated translation displacement by selecting an appropriate prominent point. In this case, we should select a point whose location is solely determined by the translational motion of the target. This point should be the target's center of rotation. Thus, a prominent point, P_1, near the center of rotation can be used to removal the residual translational errors, as shown in Figure 5.4a.

To focus the image at this point, we need to estimate the residual phase function:

$$\Phi(t) = \frac{4\pi f}{c}\left[R_r + v_r t + \frac{1}{2}\gamma_r t^2\right] = \frac{4\pi f R_r}{c} + \frac{4\pi f v_r}{c} t + \frac{2\pi f \gamma_r}{c} t^2. \tag{5.17}$$

The polynomial phase function can be estimated by an exhausted searching for the polynomial coefficients. The first constant term can be set to zero because the translational motion has been removed. In most cases, the coefficient that relates to the jerk motion can be neglected. Thus, the polynomial coefficients can be expressed in the form:

$$p = [p(1), p(2), p(3), p(4)] = \left[\frac{1}{3!}p_2, \frac{1}{2!}p_1, p_0, 0\right], \tag{5.18}$$

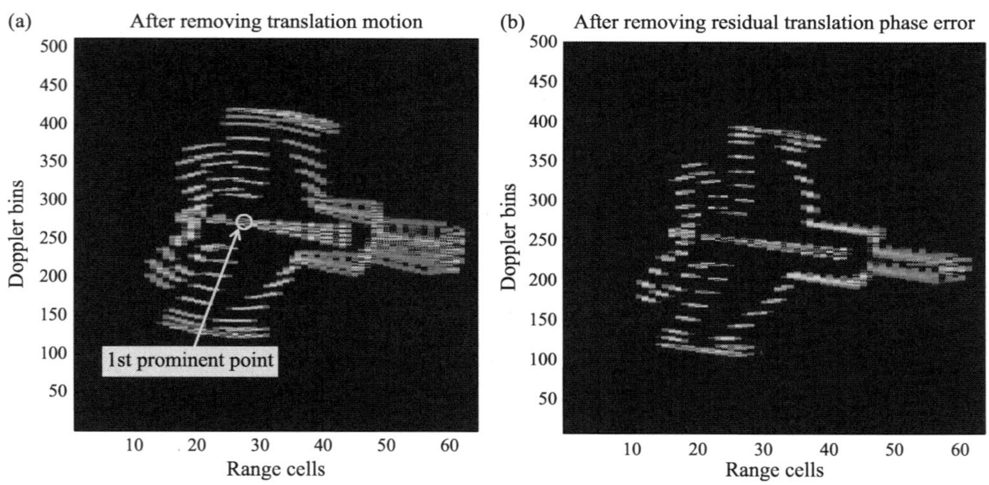

Figure 5.4 (a) After TMC selected 1st prominent point at the center of rotation; (b) ISAR image focused on the center of rotation after removing residual translational error.

and

$$\Phi(t) = p(4) + p(3)t + p(2)t^2 + p(1)t^3. \tag{5.19}$$

Usually, the velocity coefficient $p(3)$ is relatively large, and the acceleration $p(2)$ and the acceleration rate coefficients $p(1)$ are very small. After an exhausted searching process, polynomial coefficients and the optimal phase function can be found. The conjugate of the optimal phase function is used to compensate the residual translation phase errors. Then the ISAR image focused on the center of rotation is generated as shown in Figure 5.4b. The flowchart of removing residual translation and focusing on the rotation center is shown in Figure 5.5.

Figure 5.6 shows the corrected phase function and the removed phase errors. The Doppler spectral of the phase history at the range cell of the prominent point is shown in Figure 5.7. We can see that after the focus process the Doppler smearing in Figure 5.7a becomes focused in Figure 5.7b.

5.1.2 Removal of Rotational Phase Errors

After the removal of the residual translational phase errors, the remaining phase errors are rotational. The next step is to find a suitable second prominent point for estimating the rotation angle and compensating phase errors using the angular rotation.

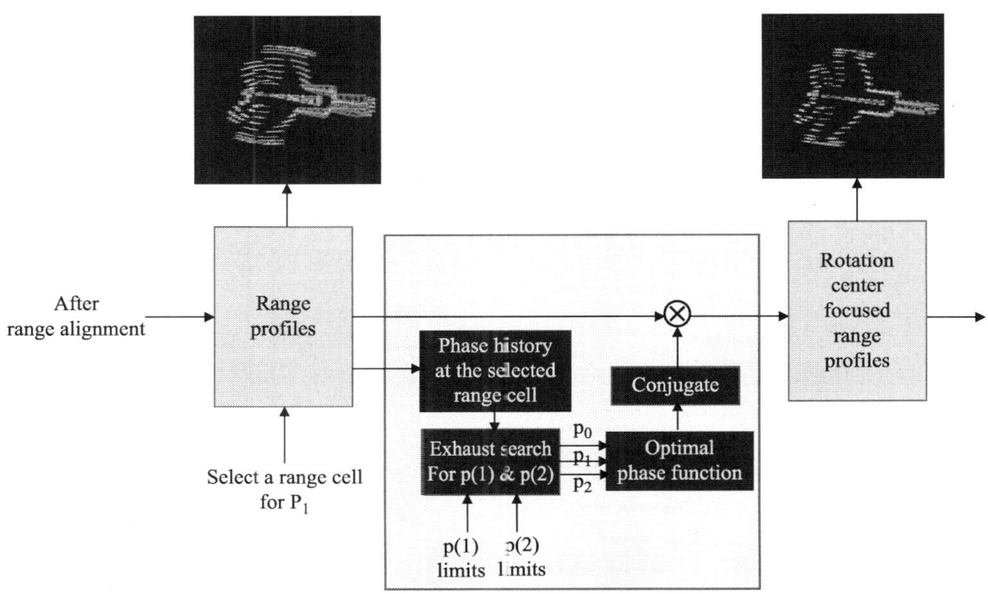

Figure 5.5 Flowchart of removing residual uncompensated translation and focused on the rotation center.

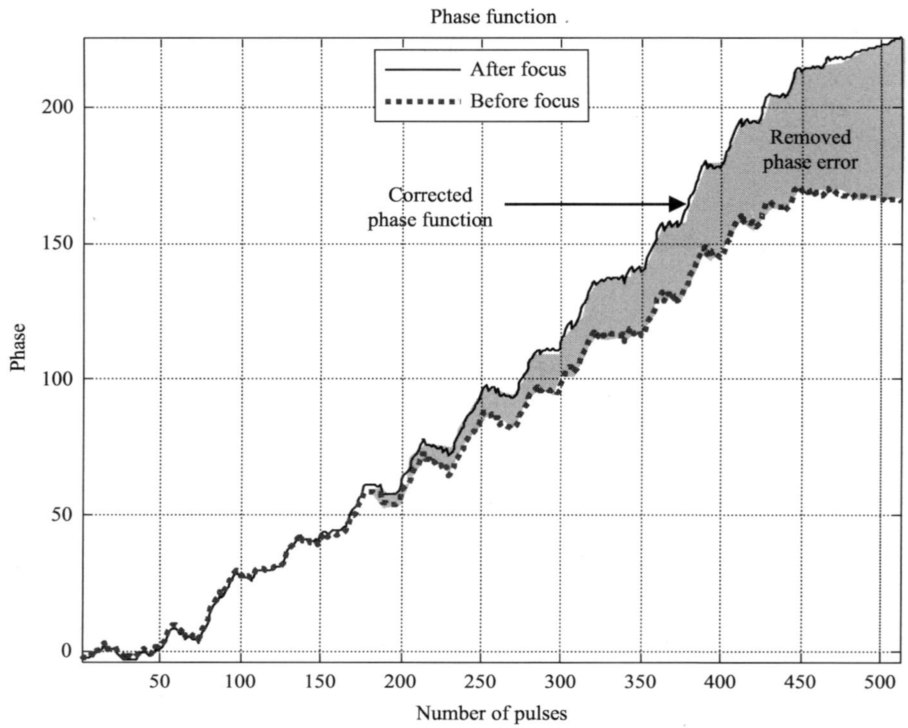

Figure 5.6 Phase functions before and after the focus process.

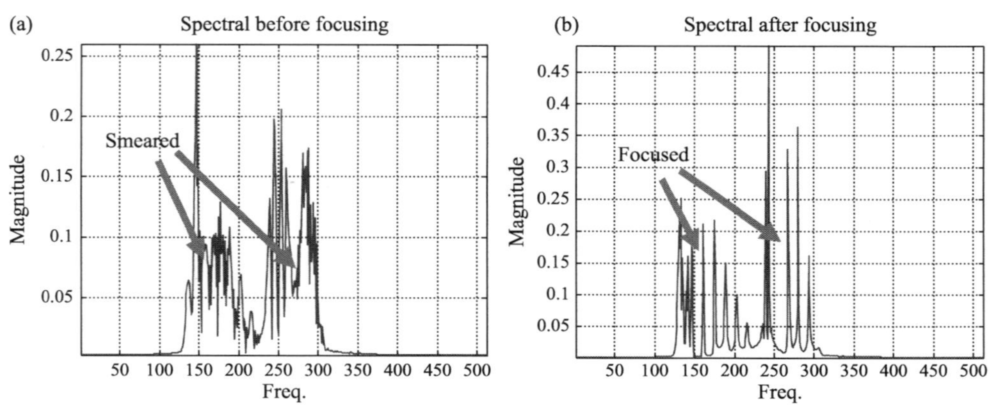

Figure 5.7 Doppler spectral at the selected range cell (a) before and (b) after the focus process.

Recalling the motion equations in (5.8), the angular velocity can be estimated by (5.12):

$$\Omega = (\ddot{R}_{0,P}/R_{0,P})^{1/2}, \tag{5.20}$$

where

$$\begin{cases} R_{0,P}(t) = r_P \cos\left(\theta_0 + \theta_P + \Omega t + \frac{1}{2}\gamma t^2\right) \\ \ddot{R}_{0,P}(t) = -\Omega^2\, r_P \cos\left(\theta_0 + \theta_P + \Omega t + \frac{1}{2}\gamma t^2\right). \end{cases} \tag{5.21}$$

To estimate rotation angle, we select a second prominent point that should be away from the first prominent point. We select the second point that has the same Doppler shift as that of the rotation center but is at a range cell away from the rotation center. Because the Doppler shift of the rotation center should be zero, we need to find a prominent point at the selected range cell that has zero Doppler shift to serve as the second point.

After calculating the range and acceleration from the motion equations, the estimated angular velocity can be estimated, and the ISAR PFA can finally be used. We apply the angular change during the CPI to the PFA, and the final corrected ISAR image of a rotating target can be obtained as shown in Figure 5.8.

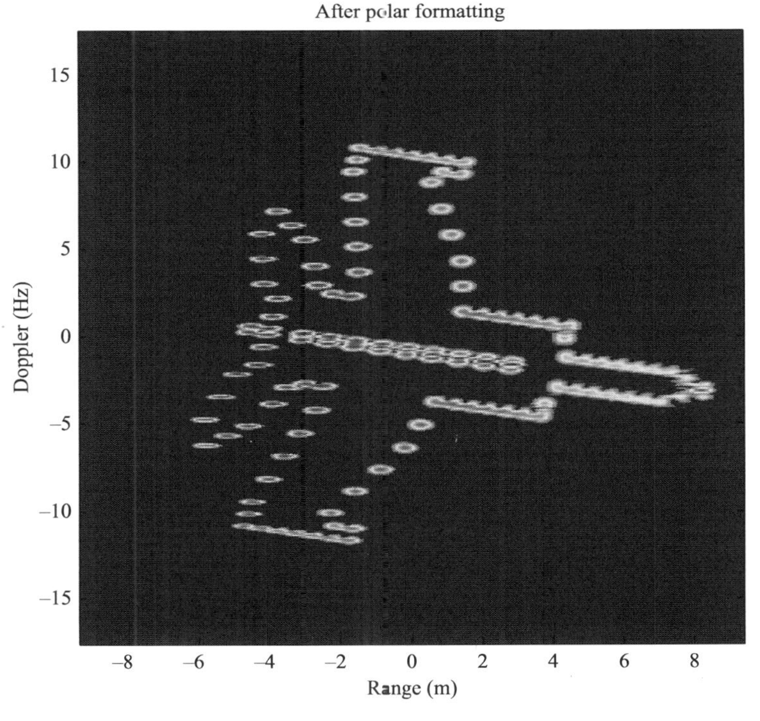

Figure 5.8 After applying the polar formatting algorithm, the rotational motion corrected ISAR image is generated.

5.2 Phase Gradient Autofocus Algorithm

The PGA, proposed in [8], has been widely used in SAR autofocus. It was developed to make a robust estimation of the gradient of the phase error in defocused SAR image data. If a complex target has no stable prominent scatterer point, the phase gradient (i.e., phase difference from pulse to pulse) can be estimated by measuring the pulse-to-pulse phase difference at each range cell and averaging them. Finally, the phase correction can be made iteratively.

The iterative PGA allows robust and nonparametric estimation and exploits the redundancy of the phase-error information contained in a degraded SAR image. Because the performance of the PGA is independent of the content in a SAR scene, there is no need to require isolated point-like reflections in the SAR scene like the PPP algorithm required.

To demonstrate the PGA, we use real SAR data. Figure 5.9 is the SAR scene before and after focusing using the PGA algorithm. In the scene, there are no stable prominent scatterers. Without applying the autofocus algorithm, the image of the scene is defocused (Figure 5.9a). By iteratively applying the PGA, a focused scene is shown in Figure 5.9b.

In general, the PGA can be applied to a phase-degraded complex SAR image no matter how the image was formed. Because a complex image and its phase history data are a Fourier transform pair, the range-compressed phase history data, called the range profiles, can be obtained by taking one-dimensional (1-D) Fourier transform in the azimuth direction. After performing range alignment in the range profiles, the phase function at range cell m usually consists of several terms:

$$\Phi(t_m) = \Phi_1(t_m) + \Phi_2(t_m) + \Phi_3(t_m) + \Phi_0, \tag{5.22}$$

where Φ_1 is the phase function related to scatterers' angular rotation during the CPI, Φ_2 is the residual translational phase error, Φ_3 is the phase errors induced by interfering scatterers from clutter and noise, Φ_0 is a constant initial phase of the m-th range cell, and t_m is the sampling time from pulse to pulse at the range cell m.

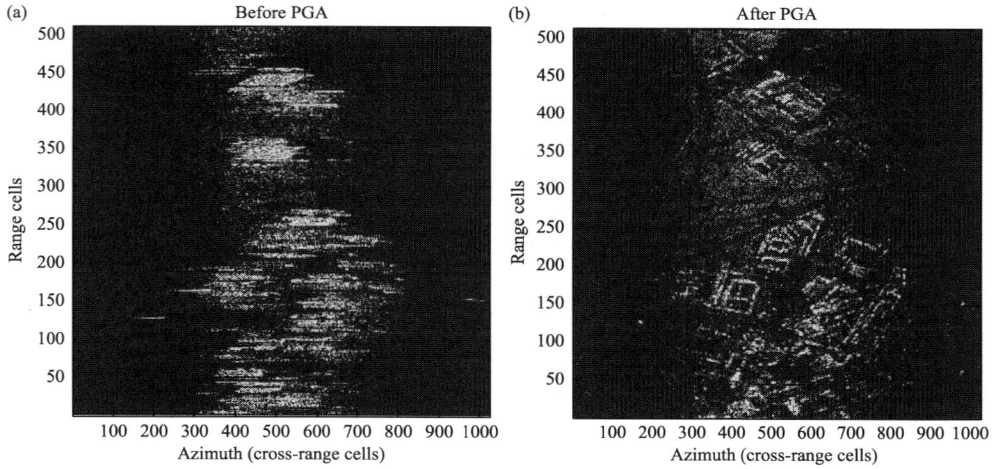

Figure 5.9 (a) Defocused SAR image of a scene; (b) focused SAR image after applying the PGA.

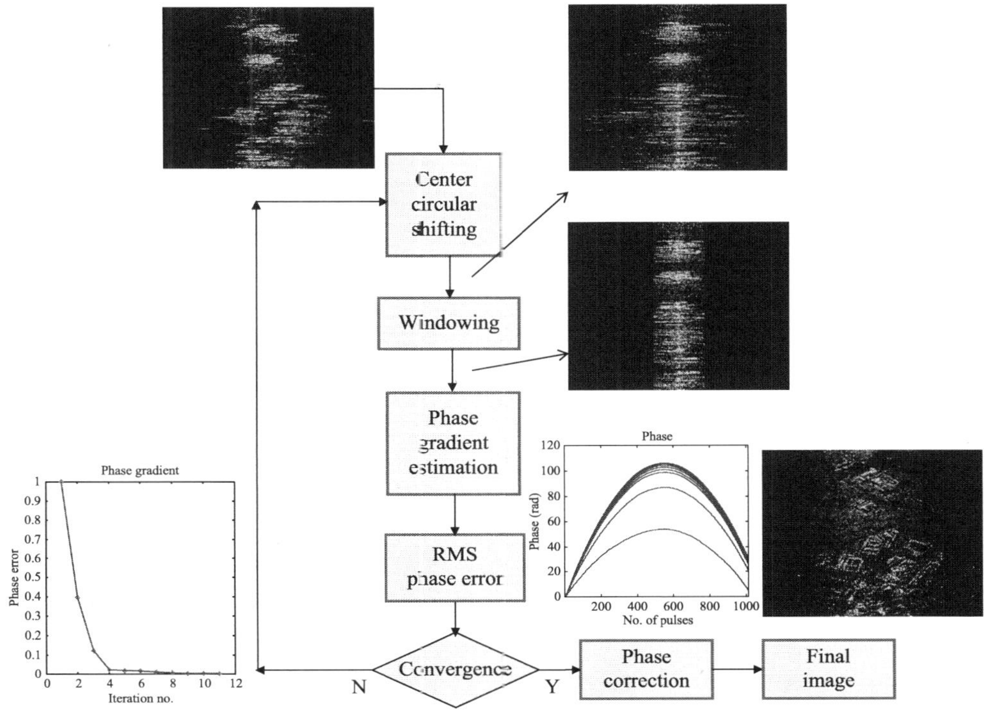

Figure 5.10 Block diagram of PGA algorithm.

To eliminate phase errors due to clutter, noise, and other residual errors, the PGA first takes the Fourier transform from pulse to pulse at each range cell to form M Doppler profiles and then circularly shifts the Doppler profiles such that the strongest scatterers in each range cell are relocated at the zero-Doppler cell (i.e., the center Doppler cell). Therefore, by low-pass filtering the centerized Doppler profiles, the effect of the interference, clutter, and noise Φ_3 can be easily reduced.

Next, we estimate the residual translation phase error, Φ_2. After taking the inverse Fourier transform, the residual translation phase error can be estimated from the phase gradient $\frac{d}{dt_m}\Phi(t_m)$. By integrating the phase gradient, the estimation of phase error, Φ_2, is obtained

$$\hat{\Phi}_2(t_n) = \int \left[\frac{d}{dt_n}\Phi(t_n)\right] dt_n, \tag{5.23}$$

which is used for correcting residual translational phase error.

The block diagram of the PGA is shown in Figure 5.10. In summary, there are four critical processing steps in the PGA:

1. Center (circular) shifting, in which a circular shift is performed on each Doppler profile of the image to place the strongest scatterer of the Doppler profile in the middle of the image (Figure 5.11a).

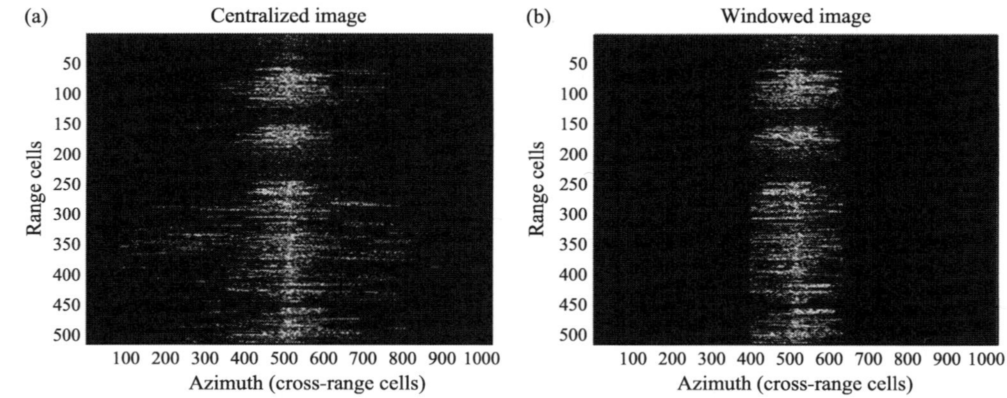

Figure 5.11 Image that is (a) centralized and (b) windowed.

2. Windowing, which consists of weighted windowing of each row of the image previously shifted. This operation allows the width of the central image points to be preserved and the others that do not contribute to the phase error estimation to be discarded (Figure 5.11b). Shifting and windowing together may achieve the highest signal-to-noise ratio (SNR) to maximize the accuracy of the phase error estimation.
3. Phase gradient estimation, which is a linear, unbiased minimum variance estimator. The phase error estimate is obtained by integrating the estimated phase gradient. Before operating the phase correction on each Doppler profile, it is necessary to remove the linear phase component from the estimated phase error to prevent any image shifting due to the phase correction.
4. Iterative phase correction, a procedure that is applied iteratively to the image until the root mean square phase error becomes small enough or convergence is reached. The iteration processing, the convergence, and the estimated phase correction function are shown in Figure 5.12. Finally, all the estimated phase errors are summed together to give a total phase error that is removed from the original image.

The PGA can also be used in ISAR image autofocus. Here, we introduce only the result of ISAR autofocus by using the PGA as shown in [9], where an ISAR PGA using simulated ISAR data was demonstrated. The simulation is based on a stepped-frequency modulated signal and operating at X-band with 600 MHz bandwidth and 10 MHz steps. The target is a small boat that consists of 20 point scatterers with dimensions of $5 \times 2 \times 5\ m^3$. The boat is located at a distance of 10 km and has translational motion with radial velocity of 5 m/s, acceleration of $-3\ m/s^2$, and jerk of $-2\ m/s^3$. Uncompensated ISAR image of the small boat is shown in Figure 5.13a, and the autofocused image after the first iteration of the PGA algorithm is shown in Figure 5.13b.

5.3 Image Contrast-Based Autofocus

The image contrast-based autofocus (ICBA) aims to form well-focused ISAR images by maximizing the image contrast (IC), which is an indicator of the image quality.

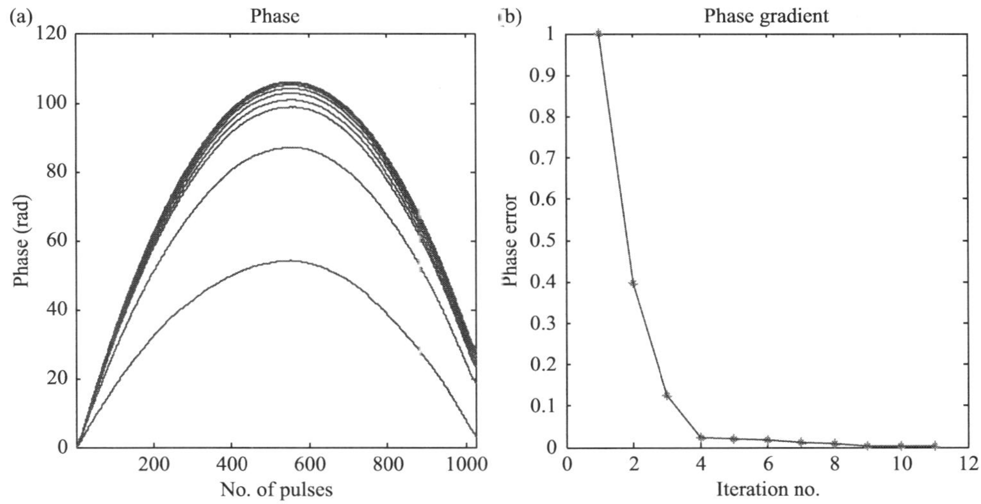

Figure 5.12 (a) Phase correction function is finally converged; (b) phase error is converged after a number of iterations.

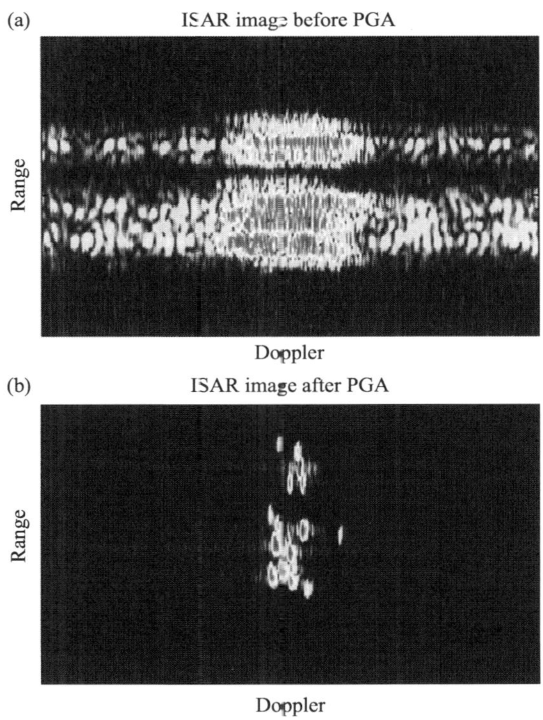

Figure 5.13 (a) Uncompensated ISAR image of the small boat; (b) autofocused image after the first iteration of the PGA algorithm.

This characteristic makes such an algorithm different from other techniques such as those described in Sections 5.1 and 5.2:

1. Parametric nature of the ICBA: the radial motion of a target's point is described by a parametric function (typically a Taylor polynomial)
2. Radial motion compensation is accomplished in one step, therefore avoiding the range alignment step

When the relative radar-target motion is regular, the distance from the origin of the target to the radar $R_0(t)$ can be approximated around the central time instant $t = 0$ by means of a Taylor's polynomial as follows:

$$R_0(t) \cong \tilde{R}_0(t) = \sum_{n=0}^{N} \frac{\alpha_n t^n}{n!}, \tag{5.24}$$

where $\alpha_n = d^n R_0(t)/dt^n$. Since the term $R_0(t)$ must be estimated and compensated for, the ISAR image autofocus problem reduces to the estimation of the α_n coefficients. Typically, a second- or third-order polynomial is sufficient to describe target's radial motions for short integration time intervals, which are usually good enough to form high-resolution ISAR images at C-band or higher frequencies.

Particular attention must be paid to the zero-order term, α_0. In fact, the phase term associated with the zero-order component is constant, being equal to $\exp(-j4\pi f\alpha_0/c)$. It is easy to demonstrate that such a term does not produce any image defocus effect as it is a constant phase term. This fact allows for the term α_0 to be neglected, therefore reducing the autofocus problem to the estimation of only the remaining coefficients—two for second-order and three for the third- order polynomials.

The ICBA technique is implemented in two steps: (1) preliminary estimation of the focusing parameters, which is accomplished with an initialization technique that uses the Radon transform (RT) and a semi-exhaustive search; and (2) fine estimation, which is obtained by solving an optimization problem where the function to be maximized is the image contrast (IC).

5.3.1 Initialization Technique

To simplify the details of the initialization technique, we formulate the solution for a second-order polynomial. Extension to any greater order polynomial will be discussed at the end of this section.

Estimation of α_1. Let $S_R(\tau, k\Delta T)$ be the range-compressed data collected during the k-th radar transmitted pulse, with τ representing the round-trip delay time and ΔT the pulse repetition interval. An example of real data range profile time history, $S_R(\tau, k\Delta T)$, is plotted in Figure 5.14a for several values of k. It is worth noting that the τ-axis is scaled by a factor $c/2$ to obtain the range coordinate, r. We can easily note that the stripes, due to the main scatterer's range migration, are almost linear. Each stripe represents the trace of the time history of a generic scatterer distance $R_s(k\Delta T)$. To estimate the value of α_1 we assume that:

1. To a first approximation, the distance $R_{si}(k\Delta T)$ relative to the i-th scatterer varies linearly with a slope equal to α_1, that is, $Rsi(k\Delta T) \approx R_{si}(0) + \alpha_1 k\Delta T$.
2. The focusing point distance $R_0(k\Delta T)$ has roughly the same quasi-linear behavior of each scatterer, that is, $R_0(k\Delta T) \approx R_0(0) + \alpha_1 k\Delta T$. It is worth noting that, in general, the focusing point does not need to be coincident with any real scatterer.

If conditions (1) and (2) are roughly satisfied, a preliminary estimation of α_1 can be obtained by calculating the mean slope of the scatterer distance traces. Let $\alpha_1 = \tan(\phi)$, where the angle φ, given by the scatterer's trace and the abscissa axis, can be estimated by means of the RT of $S_R(\tau, k\Delta T)$ as follows:

$$\hat{\phi} = arg\{\max_\phi[RT_{S_R}(r, \phi)]\} - \frac{\pi}{2}, \tag{5.25}$$

where $RT_{SR}(r, \phi)$ is the RT of $S_R(\tau, k\Delta T)$. Hence, the estimate $\hat{\alpha}_1^{(in)}$ is obtained by equating $\hat{\alpha}_1^{(in)} = \tan(\hat{\phi})$.

The RT of $S_R(\tau, k\Delta T)$ is shown in Figure 5.14b. In weak SNR conditions, it is convenient to mask the range profile time history of $S_R(\tau, k\Delta T)$ with a threshold and set all the values that are below it to zero. The result obtained after applying this mask is shown in Figure 5.14c and the relative RT in Figure 5.14d. In this case the threshold is equal to 80% of the peak value of $S_R(\tau, k\Delta T)$; hence the distance traces of the main dominant scatterers are selected.

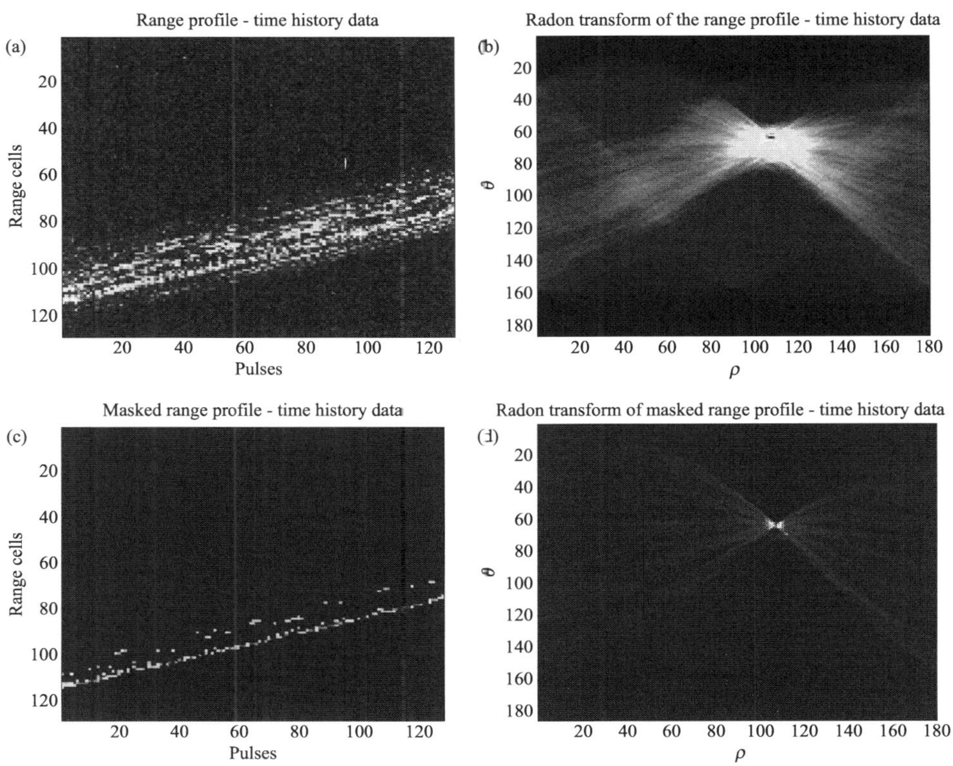

Figure 5.14 Radial velocity estimation by means of the Radon transform: (a) Range profile time history; (b) Radon transform of the range profile time history; (c) masked range profile time history; (d) Radon transform of masked range profile time history.

Estimation of α_2. Let $I(\tau, \nu; \tilde{\alpha}_1, \tilde{\alpha}_2)$ be the absolute value of the complex image obtained by compensating the received signal with two initial values $(\tilde{\alpha}_1, \tilde{\alpha}_2)$. The IC is defined as follows [3,10]:

$$IC(\tilde{\alpha}_1, \tilde{\alpha}_2) = \sqrt{A\left\{[I(\tau, \nu; \tilde{\alpha}_1, \tilde{\alpha}_2) - A\{I(\tau, \nu; \tilde{\alpha}_1, \tilde{\alpha}_2)\}]^2\right\}}/A\{I(\tau, \nu; \tilde{\alpha}_1, \tilde{\alpha}_2)\}, \quad (5.26)$$

where the operator $A\{\cdot\}$ represents the image spatial mean over the coordinates (τ, ν). The function $IC(\tilde{\alpha}_1, \tilde{\alpha}_2)$ represents the normalized effective power of the image intensity $I(\tau, \nu; \tilde{\alpha}_1, \tilde{\alpha}_2)$ and gives a measure of the image focusing. In fact, when the image is correctly focused, it is composed of several pronounced peaks (one for each scatterer) that enhance the contrast. When the image is defocused, the image intensity levels are concentrated around the mean value and the contrast is low. The final estimation of the focusing parameters α_1 and α_2 is obtained by maximizing the IC. Therefore, the following optimization problem must be solved:

$$(\hat{\alpha}_1, \hat{\alpha}_2) = \arg\left(\max_{\tilde{\alpha}_1, \tilde{\alpha}_2}[IC(\tilde{\alpha}_1, \tilde{\alpha}_2)]\right). \quad (5.27)$$

A preliminary estimate $\hat{\alpha}_1^{(in)}$ of α_2 is obtained by means of an exhaustive linear search, over the variable α_2, of the maximum of the image contrast $IC\left(\hat{\alpha}_1^{(in)}, \tilde{\alpha}_2\right)$ in a predefined interval $\left[\alpha_2^{(\min)}, \alpha_2^{(\max)}\right]$, where $\hat{\alpha}_1^{(in)}$ is obtained at the previous step, written as

$$\hat{\alpha}_2^{(in)} = \arg\left(\max_{\tilde{\alpha}_2}\left[IC(\hat{\alpha}_1^{(in)}, \tilde{\alpha}_2)\right]\right). \quad (5.28)$$

The initial guess for the iterative numerical search that solves the optimization problem is obtained by means of an exhaustive search within the preset interval $\left[\alpha_2^{(\min)}, \alpha_2^{(\max)}\right]$. If a strong acceleration of the target occurs and the value is found to be close to one of the boundaries, a new search interval is defined to investigate further values of α_2.

5.3.2 Estimation Refinement

A refinement of the preliminary estimates $\left(\hat{\alpha}_1^{(in)}, \hat{\alpha}_1^{(in)}\right)$ is obtained by maximizing the image contrast by using classic optimization algorithms. A numerical solution to the optimization problem may be based on deterministic approaches such as the Nelder-Mead algorithm [11] or random approaches using genetic algorithms [12]. The convergence of the algorithm to the global maximum depends on the initial guess, as the IC shows a good convexity close to the global maximum $(\hat{\alpha}_1, \hat{\alpha}_2)$ and a strong multimodal behavior away from it. An example of IC is provided in Figure 5.15a, and the sections along α_1 and α_2, corresponding to the global maximum, are shown in Figure 5.15b and Figure 5.15c, respectively.

In Figure 5.15, corresponding to a real data analysis, the IC shows a pronounced peak and quite regular behavior around it. A more detailed look at Figure 5.15b and Figure 5.15c shows a general multimodal characteristic along α_1 and quite regular behavior along α_2, within an interval around the global maximum position.

For the sake of clarity, a flowchart of the ICBA is shown in Figure 5.16.

5.3.3 Case Study

The ICBA initialization technique is first applied to a Boeing 737 and bulk loader data. The preliminary estimates $\left(\hat{\alpha}_1^{(in)}, \hat{\alpha}_1^{(in)}\right)$ are shown in Table 5.1.

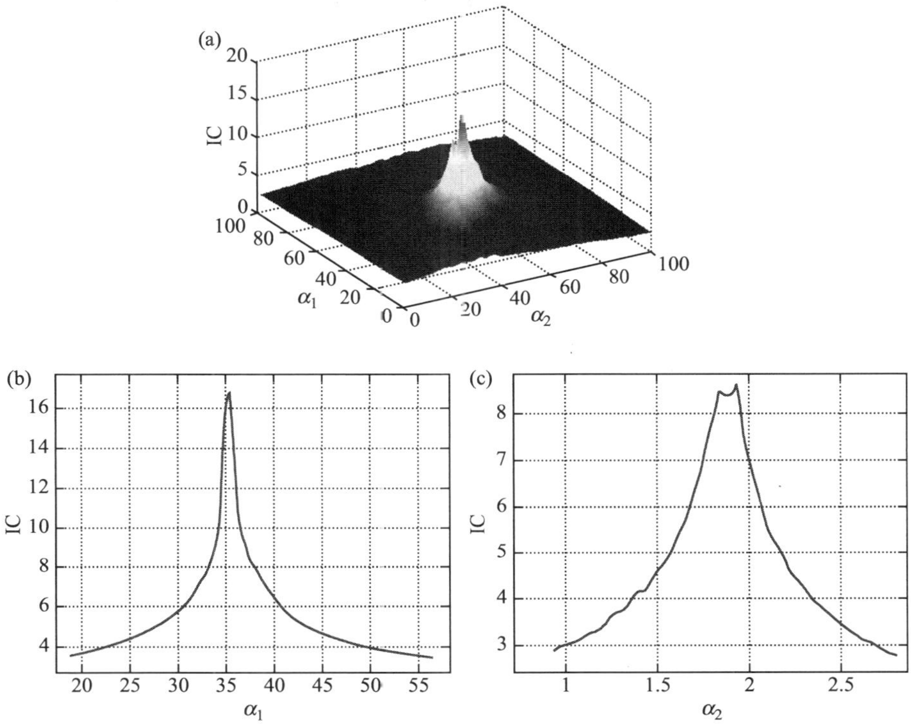

Figure 5.15 (a) Image contrast; (b) IC section along α_1; (c) IC section along α_2 in correspondence of the peak.

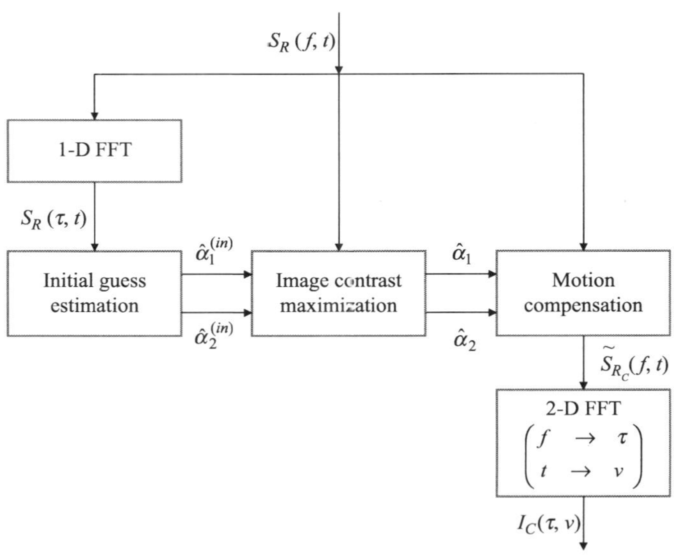

Figure 5.16 Flowchart of the ICBA technique.

Table 5.1 Preliminary Estimation of α_1 and α_2

Target	$\left(\hat{\alpha}_1^{(in)}, \hat{\alpha}_2^{(in)}\right)$ (m/s, m/s^2)
Bulk loader	(82.14, 0.03)
Boeing 737	(37.61, 1.74)

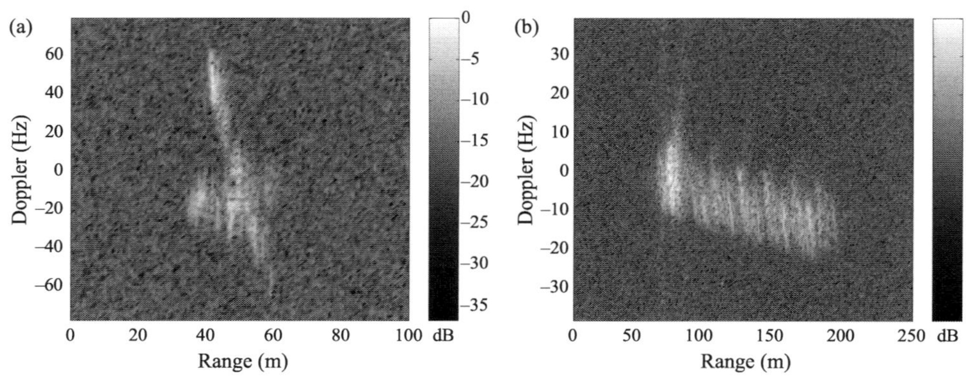

Figure 5.17 Real data ISAR images obtained using the rough estimates of the focusing parameters: (a) Boeing 737; (b) bulk loader.

The ISAR images obtained using $\left(\hat{\alpha}_1^{(in)}, \hat{\alpha}_1^{(in)}\right)$ are shown in Figure 5.17. It is clear that the images are not perfectly focused but the shapes of the targets are still recognizable. This means that the preliminary estimates are quite close to the global maximum point.

Results relative to the ICBA are presented in Figure 5.18 and Figure 5.19, respectively. The initial guesses used to initialize the ICBA are shown in Table 5.2. In the case shown in Figures 5.18 and 5.19, the numerical solution used to solve the optimization problem is based on the Nelder-Mead algorithm.

The ICBA provides an accurate estimate of the two focusing parameters, $\hat{\alpha}_1$ and $\hat{\alpha}_2$. It is important to note that these values have a physical meaning. In fact, they represent physical radial velocity and acceleration of the focusing point on the target. Even though the position of the focusing point on the target is not known, the estimation can be assumed to be close to the target center of mass radial velocity and acceleration. Table 5.2 gives the values of parameters $\hat{\alpha}_1$ and $\hat{\alpha}_2$ estimated by ICBA.

As a final consideration, it is worth pointing out that, in opposition to the PPP and PGA algorithms, the ICBA aims at focusing the entire image rather than only one or a few dominant scatterers and does not require the presence of stable dominant scatterers to ensure a good performance. Nevertheless, being a parametric technique, some assumptions about the target's motion should be satisfied, such as regular target motion during the integration time.

A similar method for measuring the image focus is represented by image entropy (IE). Next, we will introduce entropy minimization–based autofocus. The use of IE and IC is almost equivalent as they both represent a global measure of the image focus.

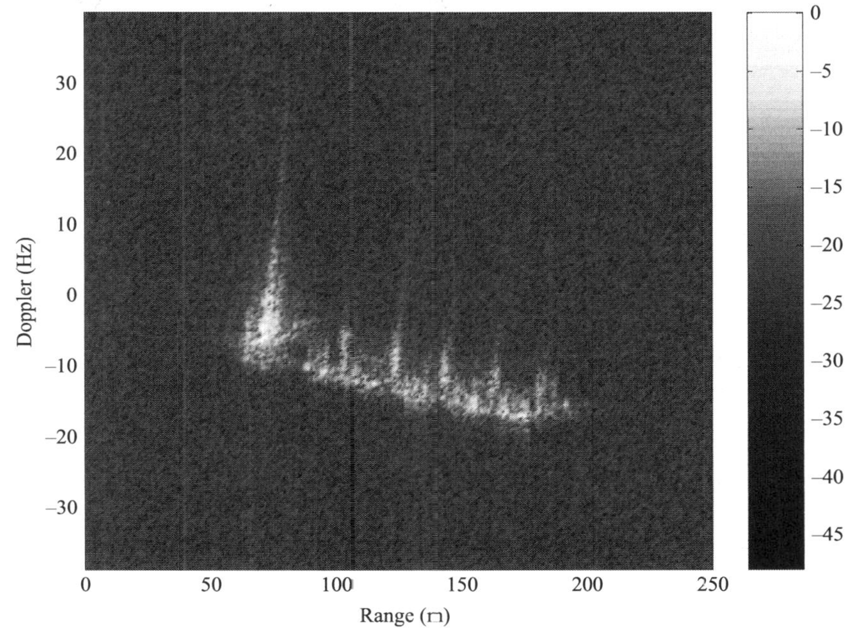

Figure 5.18 ISAR image of a bulk loader obtained using the ICBA.

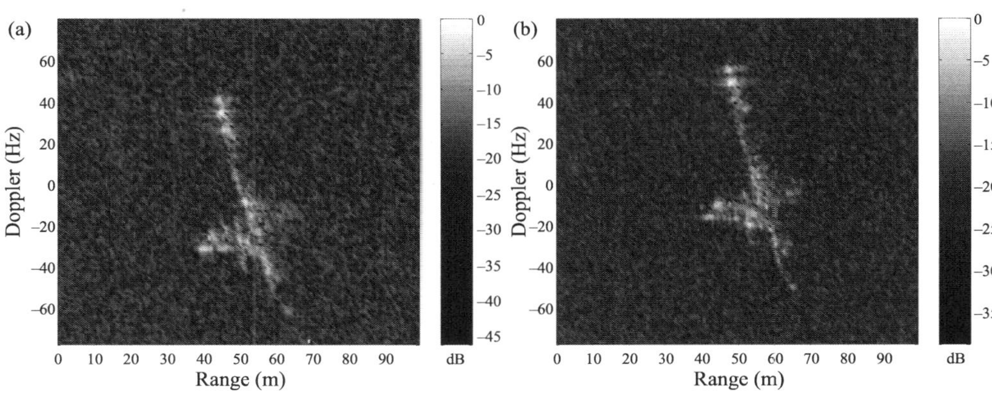

Figure 5.19 ISAR image sequence of a Boeing 737 obtained by means of the ICBA.

TABLE 5.2 Final Estimation of α_1 and α_2

	Bulk loader	**737 1:128**	**737 129:256**
α_1(m/s)	80.42	37.01	39.58
α_2(m/s^2)	−0.01	1.74	1.62

5.4 Entropy Minimization–Based Autofocus

In Section 4.13, we discussed entropy minimization for range alignment and phase adjustment. As introduced earlier, autofocus is a data-driven algorithm that automatically adjusts focusing parameters and correct phase errors based on data. Entropy minimization is one of the data-driven algorithms for ISAR autofocus.

Due to rotational phase errors and residual translation phase errors, ISAR images can be defocused. In many practical cases, dominant scatterers may not be well isolated; thus, it is difficult to precisely defined phase history for these scatterers. Therefore, autofocus techniques based on the assumption of well-isolated dominant scatterer may not work effectively [3,6,8]. In these cases, the autofocus algorithm based on minimization of the entropy cost function is helpful for ISAR autofocusing [4,13–15].

5.4.1 Phase Correction Based on 2-D Entropy Minimization

Entropy function can be used to indicate the quality of image focusing. 2-D entropy cost function for ISAR images was given in (4.12). Because the phase function in an ISAR image controls the focus of the image, we could use the entropy minimization method to correct the phase function.

The entropy minimization–based autofocus searches the optimal phase function $\hat{\mathbf{\Phi}}$ that can minimize the entropy cost function:

$$\hat{\mathbf{\Phi}} = \arg \min_{\mathbf{\Phi}}\{H(S_{mn})\}, \tag{5.29}$$

where S_{mn} ($m = 1, 2, \ldots, M$; $n = 1,2, \ldots, N$) is the $M \times N$ image matrix.

A flow chart of the entropy minimization autofocus is shown in Figure 5.20.

To effectively search for the optimal phase function, we should select a suitable model that represents the phase function (e.g., a polynomial function), searching parameters (no more than two), and the limit of searching range. In [14], a simplified model of two piece-wise linear phase function was used for searching parameters, where two parameters were used. For stepped-frequency signal waveform in a stepped-frequency continuous waveform (SFCW) radar, one parameter can be the index of the burst number and the other is the slope-angle of the piece-wise phase function. The following shows simulation results given in [14].

5.4.2 An Example of Entropy-Based Autofocus in ISAR Imaging

Entropy-based ISAR autofocus using a 3-D aircraft model (Figure 5.21) is demonstrated in [14]. The radar is a SFCW radar with 128 bursts and 128 frequency steps. The aircraft is at 20 km range and moving with a velocity of 300 m/s. Without applying autofocus, the ISAR image of a moving aircraft is shown in Figure 5.22a; the autofocused image using entropy-based minimization is shown in Figure 5.22b.

A 2-D surface plot of the converged entropy cost function is depicted in Figure 5.23. The minimum value of the cost function is $H_{\min} = 6.198$, where burst index is 58, and slope angle index is 46. The ISAR image with the minimum entropy reaches its best quality as seen in Figure 5.22b. Detailed algorithm can be found in [14].

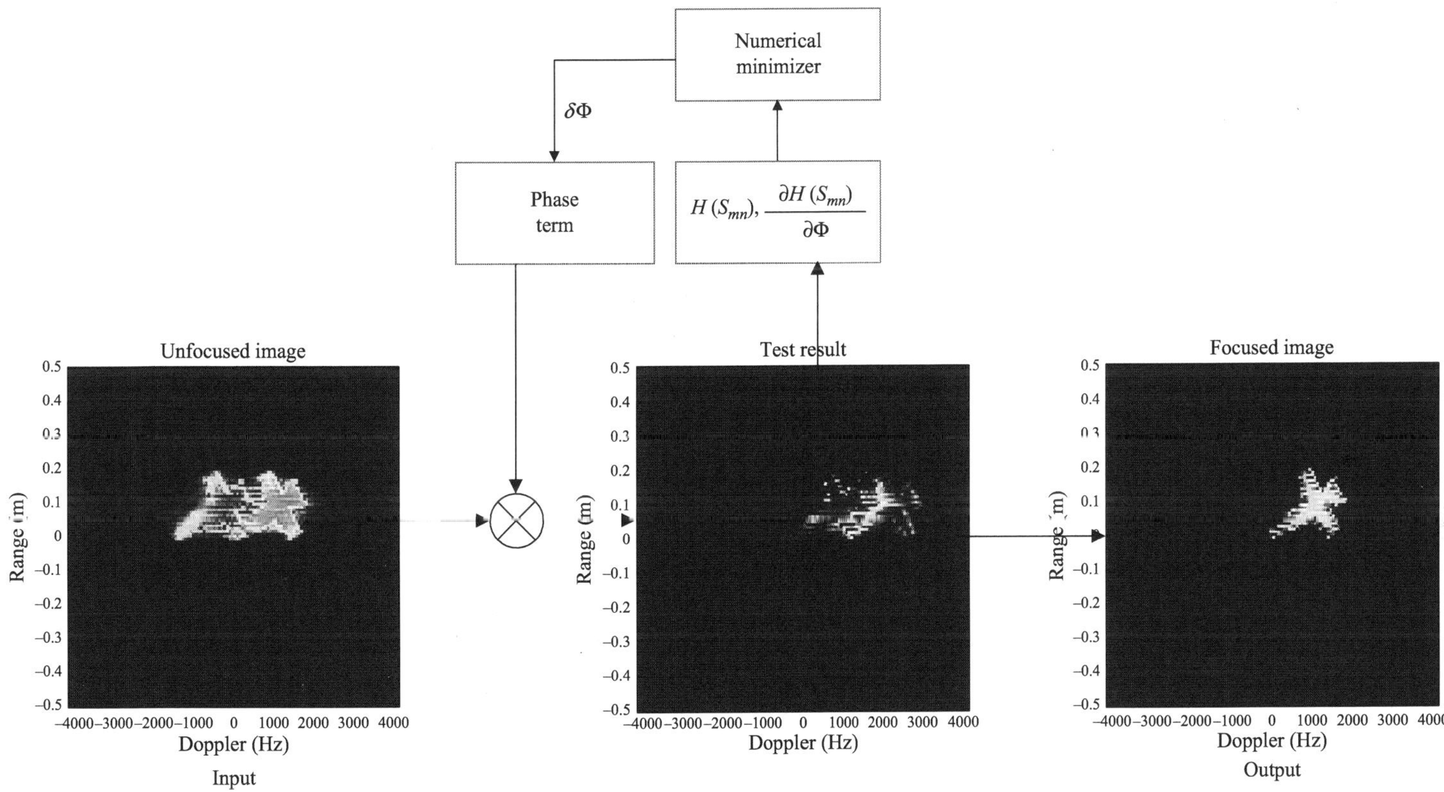

Figure 5.20 Block diagram of the entropy minimization autofocus algorithm.

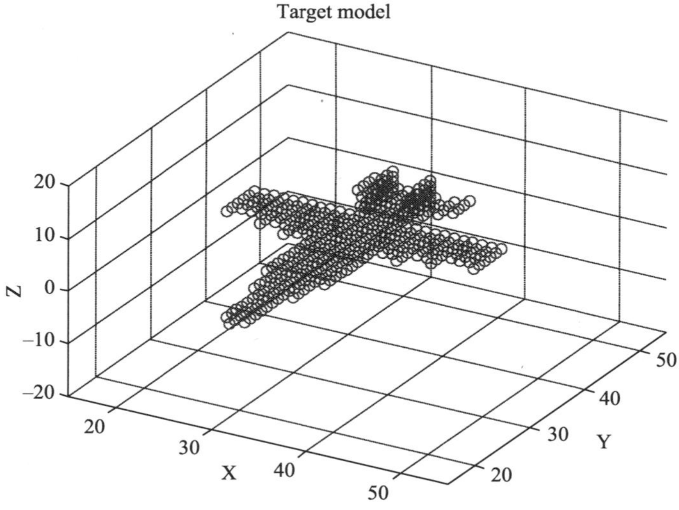

Figure 5.21 A 3-D model of an F-18 aircraft.

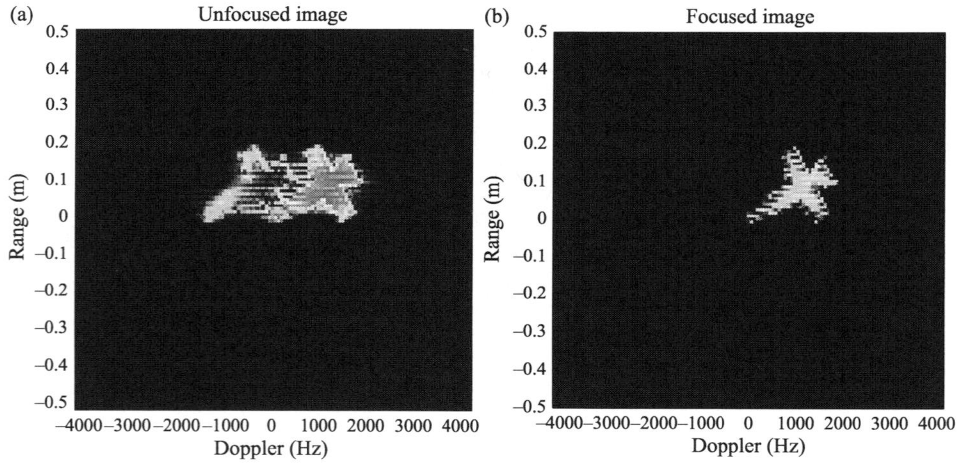

Figure 5.22 ISAR image of the aircraft (a) without autofocus and (b) after entropy-based autofocus.

5.5 Keystone Transform in ISAR

Keystone transform was originally called keystone remapping and was proposed in [16] for SAR imaging of moving targets. Due to target motion, range migration in SAR image causes image smearing. To focus the image, the keystone transform was used to rescale the time axis for each frequency by

$$t_k = \left(\frac{f_c}{f + f_c} \right) \tau, \tag{5.30}$$

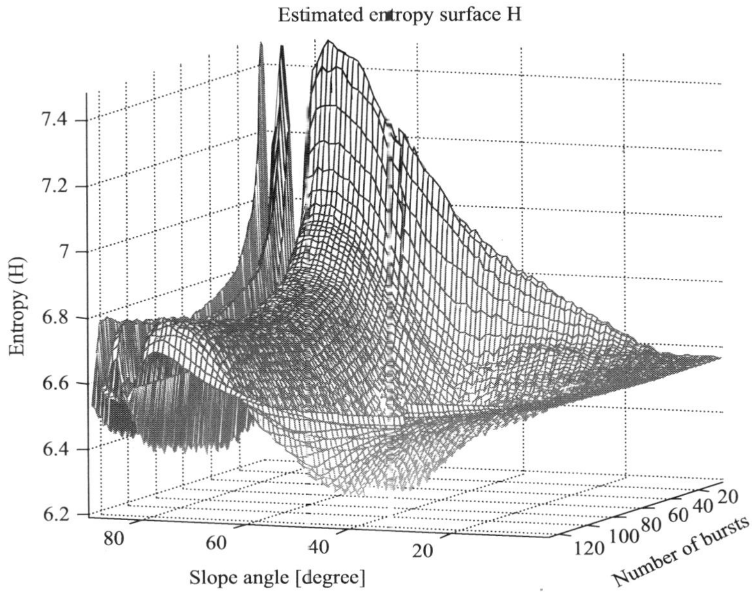

Figure 5.23 2-D surface of the converged entropy cost function showing the minimum entropy cost at $H_{min} = 6.198$, where burst index = 58, and slope angle index = 46.

where f_c is the carrier frequency, k is the index of the transmitted pulse, and τ is the rescaled time axis. The keystone transform actually remaps data from the (f, t_k) space to scaled data in the (f, τ) space. By applying the keystone transform, the range migration in the SAR image can be removed and the image of the moving target can be focused.

The keystone transform can also be used in ISAR autofocus for correcting range cell migration. Here, we introduce a keystone transform-based method for removing range cell migration in ISAR.

In ISAR imaging, after translational motion is compensated, the rotational motion can still lead to range cell migration and Doppler spectrum time varying. Thus, the keystone transform can be applied for removing the range cell migration [17].

We assume the radar transmits SFCW signal waveform with M bursts and N frequency steps. Within one burst, the signal is described by a sequence of N pulses with increased carrier frequencies: $f_n = f_c + n\Delta f$, $(n = 0, 1, 2, \ldots, N-1)$, where f_c is the carrier frequency, and Δf is the frequency step.

Assume a target is uniformly rotating as shown in Figure 5.24. The variation of the distance between the scatterer, P, located at (x_p, y_p) and the radar can be written as

$$r(t, \Omega_0) = d\sin\left(\theta_p + \Omega_0 t\right) = x_p \sin(\Omega_0 t) + y_p \cos(\Omega_0 t), \tag{5.31}$$

where t is the slow time, Ω_0 is the target rotation rate, $d = \sqrt{x_p^2 + y_p^2}$, and $\theta_p = \tan^{-1}\left(\frac{y_p}{x_p}\right)$. Here, Ω_0 can be considered constant if the CPI is not too long.

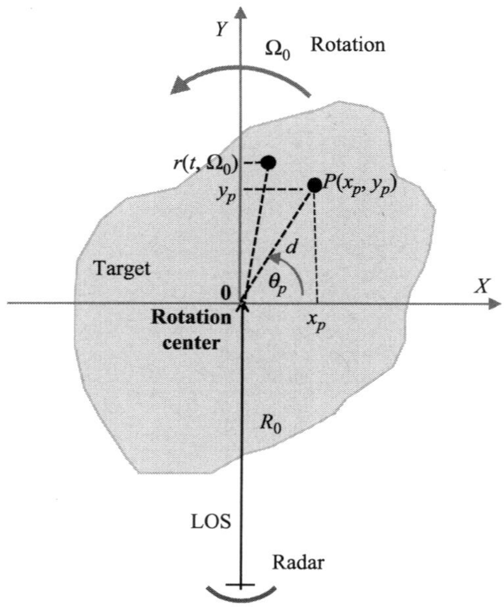

Figure 5.24 The geometry of a scatterer, P, in a rotating target.

The baseband signal returned from the scatterer, P, can be expressed as

$$s(t,n) = \rho \exp\left\{ -j\frac{4\pi}{c}(f_c + n\Delta f)\left[x_p \sin(\Omega_0 t) + y_p \cos(\Omega_0 t)\right]\right\}, \tag{5.32}$$

where n is the stepped frequency index, and ρ is proportional to the reflectivity of P. If the CPI is short, we have

$$\sin(\Omega_0 t) \approx \Omega_0 t, \tag{5.33}$$

and

$$\cos(\Omega_0 t) \approx 1 - \frac{1}{2}(\Omega_0 t)^2. \tag{5.34}$$

Therefore, we can rewrite the baseband signal as

$$s(t,n) = \rho \exp\left\{ -j\frac{4\pi}{c}(f_c + n\Delta f)\left[x_p \Omega_0 t + y_p - \frac{1}{2} y_p (\Omega_0 t)^2\right]\right\}. \tag{5.35}$$

Then, from the equation, the range profile can be obtained by taking the inverse Fourier transform with respect to the fast-time index n. However, from the range profile we can see that due to the rotation the range of the scatterer, P, migrates from its original position, y_p, to a new position $\left[x_p\Omega_0 t + y_p - \frac{1}{2}y_p(\Omega_0 t)^2\right]$. In other words, there is a range drift shift:

$$\Delta_r = x_p \Omega_0 t - \frac{1}{2} y_p (\Omega_0 t)^2. \tag{5.36}$$

If Δ_r is larger than the size of the range resolution cell, the energy of P will expand to more than one range cells and the image of the target becomes defocused. Thus, before the Doppler focus processing, it is necessary to compensate Δ_r along the range-direction, which is the range cell migration correction (RCMC).

For correcting the range cell migration, Δ_r, the linear term, $x_p\Omega_0 t$, is a dominant one and the quadratic term $-\frac{1}{2}y_p(\Omega_0 t)^2$ can be ignored. The variables x_p and y_p are comparable in size, but the rotational angle $\Omega_0 t$ is usually very small. For example, if $\Omega_0 = 2°/\text{sec}$ and $t = 1$ sec as in general ISAR imaging, then the rotational angle is about 0.03 rad, which is a small number. The quadratic term $-\frac{1}{2}y_p(\Omega_0 t)^2$ can be neglected compared with the size of the range resolution cell. For example, if $\Omega_0 = 2°/\text{sec}$, $t = 1$ sec, and $y_p = 40$ m, the quadratic term is less than 0.02 m. Compared with the normal size of the range resolution cell (e.g., 0.3 m), the quadratic term can certainly be ignored.

There are two main difficulties for the RCMC. First, the range shift depends on the unknown target rotational speed Ω_0, which cannot be known in advance. Second, the range shift depends on the cross-range position x_p, such that different cross-range position generates different size of the range shift. The advantage of using the keystone transform is that it can nicely handle both of the difficulties.

The keystone transform reformats ISAR data by using

$$\bar{t} = \frac{f_c + n\Delta f}{f_c} t. \tag{5.37}$$

Thus, we have

$$s(\bar{t}, n) = \rho \exp\left\{ -j\frac{4\pi}{c}(f_c + n\Delta f)y_p - j\frac{4\pi}{c} f_c x_p \Omega_0 \bar{t} + j\frac{2\pi}{c} y_p \Omega_0^2 \frac{f_c^2}{f_c + n\Delta f} \bar{t}^2 \right\}. \tag{5.38}$$

Because f_c is much larger than $n\Delta f$, we have

$$s(\bar{t}, n) \cong \rho \exp\left\{ -j\frac{4\pi}{c} n\Delta f \left(1 + \frac{1}{2}\Omega_0^2 \bar{t}^2\right) y_p - j\frac{4\pi}{c} f_c x_p \Omega_0 \bar{t} + j\frac{2\pi}{c} y_p \Omega_0^2 f_c \bar{t}^2 \right\}, \tag{5.39}$$

where ρ is a constant and independent of n and $\bar{t}$. We see that n and $\bar{t}$ have been decoupled in the second term of the phase function by applying the keystone transform. If we perform the inverse Fourier transform with respect to the fast-time index n, the peak of the range-profile will appear at the range $\left(1 + \frac{1}{2}\Omega_0^2 \bar{t}^2\right) y_p$, which means the instantaneous range shift is

$$\Delta_r' = -\frac{1}{2}\Omega_0^2 \bar{t}^2 y_p. \tag{5.40}$$

Therefore, the dominant range migration has been corrected by the keystone transform. As mentioned already, the residual range shift Δ_r' can usually be ignored.

After applying the inverse Fourier transform with respect to the fast-time index n, the exact range position, y_p, can be estimated from the peak of the range profile. Thus, we have the range-compressed signal at the range position, y_p:

$$s(\bar{t}) = \rho \exp\left\{ -j\frac{4\pi}{c} f_c x_p \Omega_0 \bar{t} + j\frac{2\pi}{c} y_p \Omega_0^2 f_c \bar{t}^2 \right\}. \tag{5.41}$$

Then, the Doppler profile can be obtained by any autofocus procedure.

In the following, we apply the keystone transform autofocus to the simulated Mig-25 data, which are generated by SFCW radar operating at 9 GHz carrier frequency with 64 frequency steps to form a bandwidth of 512 MHz, and the PRF of 15 KHz. Figure 5.25a shows the ISAR image without applying the keystone transform, where the defocusing is caused by the range cell migration. By applying the keystone transform autofocus, the ISAR image is shown in Figure 5.25b, where we can see the range cell migration has been corrected. After further applying Doppler autofocus, the completely focused ISAR image is shown in Figure 5.26, where fine range and cross-range resolutions are obtained.

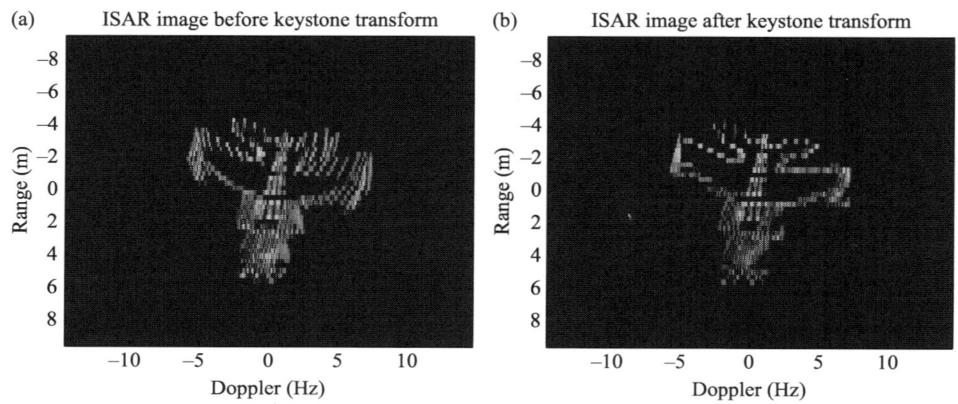

Figure 5.25 ISAR imaging (a) without the keystone transform autofocus; (b) after applying the keystone transform autofocus.

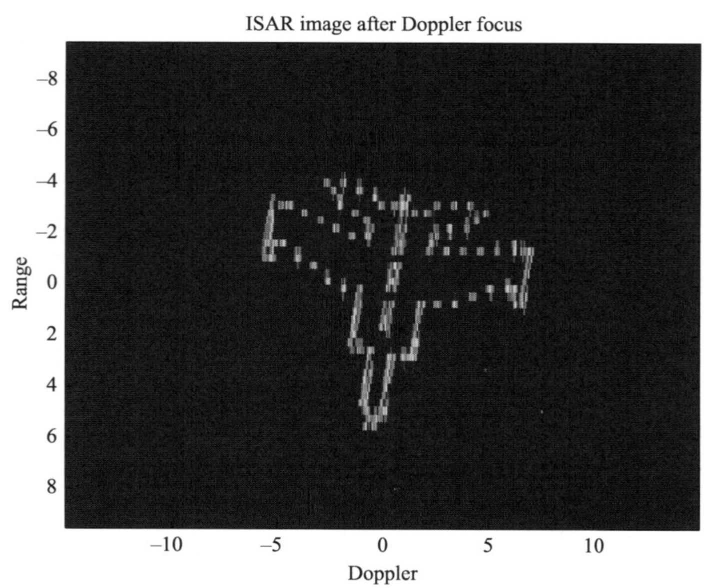

Figure 5.26 ISAR imaging after applying the keystone transform autofocus and Doppler autofocus.

References

[1] Carrara, W.G., R.S. Goodman, and R.M. Majewski, *Spotlight Synthetic Aperture Radar: Signal Processing Algorithms* (Chapter 6), Artech House, 1995.

[2] Jakowatz, C.V., D.E. Wahl, P.H. Eichel, D.C. Ghiglia, and P.A. Thompson, *Spotlight-Mode Synthetic Aperture Radar: A Signal Processing Approach*, Kluwer Academic, 1996.

[3] Berizzi, F., and G. Cosini, "Autofocusing of Inverse Synthetic Aperture Radar Images Using Contrast Optimization," *IEEE Trans.*, AES-32, no. 3, pp. 1185–1197, 1996.

[4] Li, X., G. Liu, and J. Ni, "Autofocusing of ISAR Images Based on Entropy Minimization," *IEEE Trans., AES-35*, no. 4, pp. 1240–1251, 1999.

[5] Werness, S., W. Carrara, L. Joyce, and D. Franczak, "Moving Target Imaging Algorithm for SAR Data," *IEEE Trans. on Aerospace and Electronic Systems*, vol. 26, no. 1, pp. 57–67, 1990.

[6] Wang, Y., H. Ling, and V.C. Chen, "ISAR Motion Compensation via Adaptive Joint Time-Frequency Technique," *IEEE Trans.*, AES-34, no. 2, pp. 670–677, 1998.

[7] Lipps, R., and D. Kerr, "Polar Reformatting for ISAR Imaging," *Proc. of 1998 IEEE Radar Conference*, pp. 275–280, 1998.

[8] Wahl, D.E., P.H. Eichel, D.C. Ghiglia, and C.V. Jakowatz, "Phase Gradient Autofocus—A Robust Tool for High Resolution SAR Phase Correction," *IEEE Trans.*, AES-30, no. 1 pp. 827–835, 1994.

[9] Nel, W., E. Giusti, M. Martorella, and M.Y. Abdul Gaffar, "A Time-Domain Based ISAR Autofocus Algorithm," Proceedings of 2011 IEEE CIE International Conference on Radar, Oct. 24–27, Chengdu, China, pp. 541–544, 2011.

[10] Martorella, M., F. Berizzi, and B. Haywood, "A Contrast Maximization Based Technique for 2D ISAR Autofocusing," *IEE Proc. on Radar Sonar and Navigation*, vol. 152, no. 4, pp. 253–262, 2005.

[11] Nelder, J.A., and R. Mead, "A Simplex Method for Function Minimisation," *Comput. J.*, vol. 7, pp. 308–313, 1965.

[12] Martorella, M., F. Berizzi, and S. Bruscoli, "Use of Genetic Algorithms for Contrast and Entropy Optimization in ISAR Autofocusing," *EURASIP Journal on Applied Signal Processing*, vol. 2006, p. 133, 2006.

[13] Zhu, D., L. Wang, Y. Yu, Q. Tao, and Z. Zhu, "Robust ISAR Range Alignment via Minimizing the Entropy of the Average Range Profiles," *IEEE Trans. GRS Letters*, vol. 6, no. 2, pp. 204–208, 2009.

[14] Lazarov, A.D., and Ch.N. Minchev, "ISAR Signal Modeling and Image Reconstruction with Entropy Minimization Autofocussing," Proceedings of IEEE/AIAA 25th Digital Avionics Systems Conference, Oct. 15–19, Portland, OR, pp. 1–11, 2006.

[15] Karakasiliotis, A.V., A.D. Lazarov, P.V. Frangos, G. Boultadakis, and G. Kalognomos, "Two-Dimensional ISAR Model and Image Reconstruction with Stepped Frequency-Modulated Signal," *IET Signal Processing*, vol. 2, no. 3, pp. 277–290, 2008.

[16] Perry, R., C. Dipietro, and R.L. Fante, "SAR Imaging of Moving Targets," *IEEE Transactions on Aerospace and Electronic Systems*, vol. 35, no. 1, pp. 188–199, 1999.

[17] Wang, J., and D. Kasilingam, "Global Range Alignment for ISAR," *IEEE Transactions on Aerospace and Electronic Systems*, vol. 39, no. 1, pp. 351–357, 2003.

CHAPTER 6

Signal Processing Issues in ISAR Imaging

In the previous chapters, we have dealt with the geometrical aspects of inverse synthetic aperture radar (ISAR) imaging, which have led to a theoretical approach of the problem of forming electromagnetic (EM) images of noncooperative targets using high-resolution radars. Nevertheless, real-world data are corrupted by noise, and clutter and targets usually undergo complex motion, which cannot easily be modeled or predicted. Moreover, other effects such as limited resolution or high sidelobe levels (SLLs) may strongly reduce the effectiveness of ISAR imaging in classification and recognition. In this chapter, we will introduce such problems and provide both classic and recent solutions to them.

6.1 ISAR Imaging in the Presence of Target's Complex Motion

The complex motion of a target may drastically complicate the issue on imaging of noncooperative targets via classic ISAR imaging techniques. As introduced in Chapters 1 and 2, a condition that guarantees the formation of a well-focused ISAR image is a time-invariant (or constant) rotation of the target. In the presence of complex motion, such as nonuniform pitching, rolling, and yawing or in the case of target's fast maneuvering, such a condition on time-invariant rotation is hardly met and the effect on the formation of ISAR images can be destructive. In the next subsection, we will show some examples of image degradation due to the presence of complex motion.

6.1.1 Effects of Target's Complex Motion on ISAR Imaging

To show what complex motion can cause defocusing of ISAR images, we will consider targets that undergo oscillating motion. A range-Doppler method will be used for reconstructing ISAR images. The magnitude and rate of oscillating motion play a role in the formation of the ISAR image relative to the coherent processing interval (CPI) and the pulse repetition frequency (PRF).

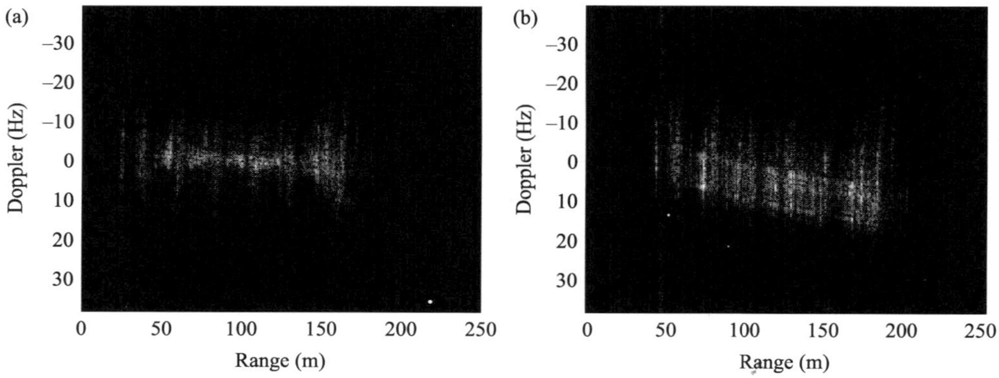

Figure 6.1 Two examples of defocused ISAR images of the same ship: (a) image frame no. 1; (b) image frame no. 2. (These images are obtained processing real data acquired by the Defence Science and Technology Organisation (DSTO) with a stepped-frequency radar; see Chapter 12 for more details.)

In the next example, two ISAR images are formed by processing two data subsets extracted from a long-time acquisition data set. The nonuniform roll, pitch, and yaw motion of a ship usually cause severe defocusing effect (Figure 6.1). Such complex motions induce rapid changes in the effective rotation vector and therefore in the image projection plane (IPP), so a constant or nearly constant effective rotation vector is a desirable characteristic that should be searched for in a data set to obtain well-focused and high-resolution ISAR images. Section 6.1.2 will show a method for searching data subsets with such a characteristic.

6.1.2 Optimal Time Windowing

When the target's effective rotation vector is constant, which typically occurs in the case of noncomplex target's motion, longer CPIs tend to produce ISAR images with a finer resolution. Therefore, one may attempt to increase the time window to obtain better images. Unfortunately, for a longer time interval the effective rotation vector does not often stay constant mainly because targets may change their motion at will or may get affected by a number of external forces, such as in the case of ships on rough sea surfaces or terrestrial vehicles on bumpy roads. As target's motion are not predictable in the majority of cases, it is practically impossible to foresee the exact optimal time window that produces a well-focused ISAR image with the finest possible resolution. For this reason, long time intervals of data are acquired and processed to extract one or more well-focused and high-resolution ISAR images. This procedure requires (1) the radar to dwell on the target for a enough long time to provide well-focused and high-resolution ISAR images with high probability; and (2) a postacquisition time window selection.

Whereas the first problem is a direct and unavoidable consequence of the unpredictability of target's motion, the second one can be addressed as the time window selection problem.

Since well-focused images have larger value of the image contrast (IC) than defocused images and high-resolution images also have larger IC values than low-resolution images,

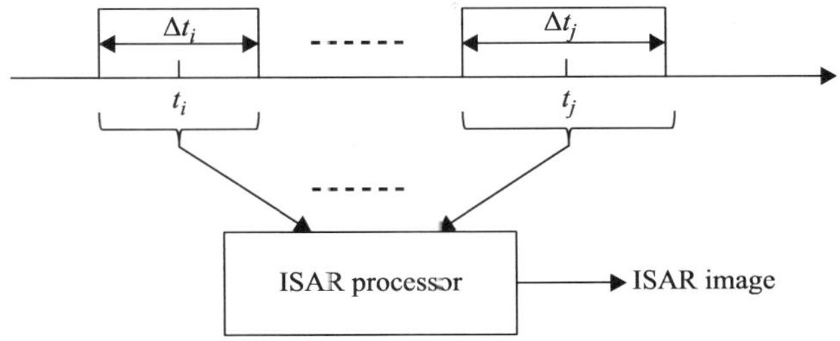

Figure 6.2 Time window selection.

the IC can be used as an indicator to measure how well an ISAR image is focused and the magnitude of its resolution.

The problem at hand can be formulated as the selection of two parameters, which uniquely identify a time window across the entire data set time length: the central time instant t, which indicates the time window position; and the window time length, Δt. For each parameter pair, a time window is identified and the corresponding data can be selected, which should fed to an ISAR processor to form an ISAR image. The IC within the selected window can be measured and compared to the IC obtained for any other time window. A pictorial view of this procedure can help us to understand the concept (Figure 6.2).

6.1.3 Maximum IC-Based Automatic Time Window Selection

We will now define the IC parameterized with respect to the time window parameters:

$$IC(t, \Delta t) = \sqrt{A\left\{\left[I(\tau, f_D;\ t, \Delta t) - A\{I(\tau, f_D;\ t, \Delta t)\}\right]^2\right\}} \Big/ A\{I(\tau, f_D;\ t, \Delta t)\}, \tag{6.1}$$

where $I(\tau, f_D;\ t, \Delta t)$ is the power of the image intensity, and $A\{\cdot\}$ is the mean operator over the variables (τ, f_D) which represent the time delay and the Doppler coordinate, respectively. As detailed in Chapter 5, the image contrast is a normalized standard deviation of the ISAR image intensity. The contrast gives a measure of the image focus. In fact, in a well-focused image, each peak associated with a scatterer is strongly pronounced and the overall image appears sharper.

It is also worth noting that the denominator represents the image mean power and is used to normalize the IC. Such a normalization is needed when images with different sizes must be compared in terms of their contrast and hence their focus.

The optimal time position and length of the time window are obtained by maximizing the IC with respect to the variable pair $(t, \Delta t)$. Therefore, the following optimization problem can be formulated:

$$\left(t_{opt}, \Delta t_{opt}\right) = \arg\max_{t,\ \Delta t}[IC(t, \Delta t)], \tag{6.2}$$

where $\Delta t \in [0, T_{obs}]$, $t \in \left[\frac{\Delta t}{2}, T_{obs} - \frac{\Delta t}{2}\right]$, and T_{obs} is the observation time interval.

Because the variables $(t, \Delta t)$ are discrete, (6.2) represents a discrete optimization problem. Specifically, such a problem can be classified as a nonlinear *knapsack problem* [1], which can be solved by taking a brute force approach. Nevertheless, to avoid an exhaustive

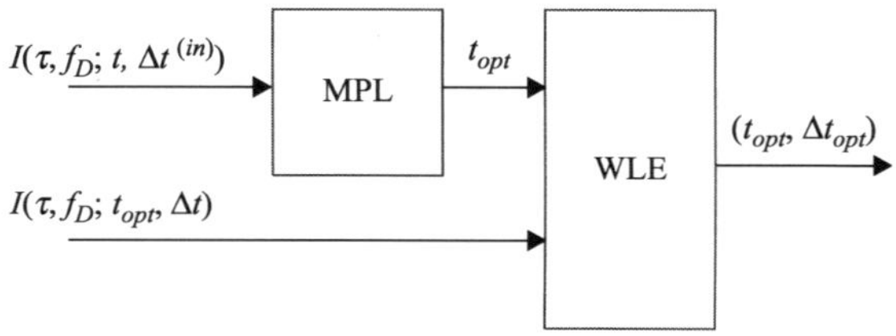

Figure 6.3 MC-ATWS flowchart.

search, which involves time resources that are usually not affordable, a heuristic but effective method can be applied.

A method for automatically selecting an optimal time window according to the problem detailed in (6.2), namely, the maximum contrast-based automatic time window selection (MC-ATWS), can be defined as a double linear search. This method can be briefly itemized as follows:

1. Maximization of the contrast with respect to t for a given guess $\Delta t = \Delta t^{(in)}$ (let t_{opt} be the solution)
2. Optimization with respect to Δt with $t = t_{opt}$

By referring to Figure 6.3, we indicate the two steps with maximum position locator (MPL) and window length estimator (WLE).

The MPL processes the ISAR images by calculating their IC and provides the time, t_{opt}, that produces the image with the highest contrast. The ISAR image sequence is reconstructed by using a sliding window with time length $\Delta t = \Delta t^{(in)}$. The frame by frame step can be one or more radar sweeps. Small steps provide high estimation accuracy but more expensive computational load. A trade-off between the accuracy of the MPL and the computational load can be defined according to the specific application. Also, the initial value $\Delta t^{(in)}$ can be set accordingly to the scenario that is considered. In particular, the radar PRF and the target motion velocity are two parameters that must taken into account when setting the initial value $\Delta t^{(in)}$. Two significant examples will be shown at the end of this section where the initial value of the time window length is set accordingly.

Once the t_{opt} is found, the value of Δt_{opt} must be searched using trial and error:

1. Initialize the window enlargement step with $n = k$ and set a variable $flag = 0$.
2. The window length is extended by adding $2n$ samples (n on the left-hand side and n on the right-hand side).
3. A new ISAR image is formed and its IC is calculated: if $IC_{i+1} \geq IC_i$ then go to step (1) and set $flag = 1$; otherwise substitute n with $n = n - 1$ and go to step (4).
4. If $n \geq 1$, then go to step (1); otherwise go to step (5).
5. If $flag = 1$ then stop; otherwise go to step (6).
6. Repeat from step (1) by substituting step (2) with "The window length is reduced by eliminating $2n$ samples (n on the left-hand side and n on the right-hand side)."

Another example provided here is to highlight the importance of selecting a suitable time window to form well-focused ISAR images. Figure 6.4 shows an ISAR image produced by integrating 0.41 sec of data. The image appears well focused but with limited resolution. Figure 6.5 displays an ISAR image formed by coherently processing 3.27 sec of data.

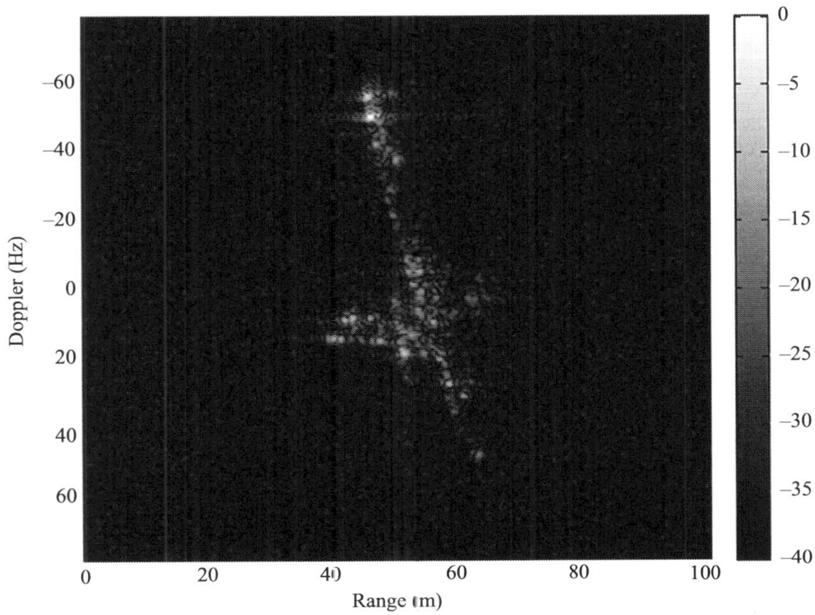

Figure 6.4 ISAR image with CPI = 0.41 sec.

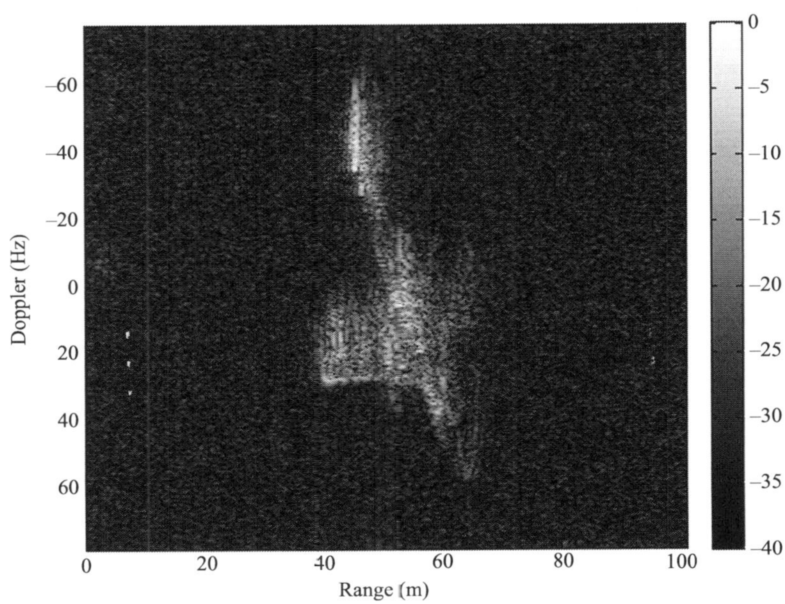

Figure 6.5 ISAR image with CPI = 3.27 sec.

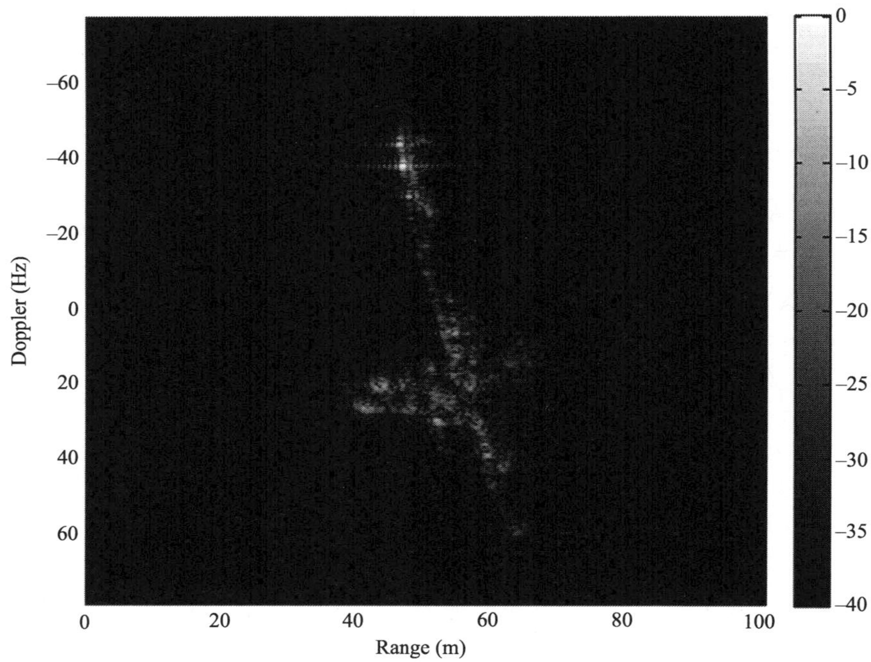

Figure 6.6 ISAR image with CPI = 0.89 sec.

The image appears defocused as the target's own motion changes the orientation of the target's effective rotation vector during the CPI. The third image in Figure 6.6 is obtained by running the automatic time window selector. The selected time window length is of 0.89 sec.

It is worth noting that Figure 6.6 shows an ISAR image with higher resolution than the one displayed in Figure 6.4 while remaining focused.

6.2 ISAR Imaging in the Presence of Strong Noise and Clutter

The presence of noise and clutter in the received data produces two major effects in ISAR imaging. The most destructive effect may prevent the ISAR system from forming a well-focused image when using conventional image formation techniques. In this case, the noise or clutter is either very strong or has characteristics that strongly affect the performance of some or all of the formation algorithm steps, such as the radial motion compensation. A less destructive effect in terms of image formation would be the presence of noise and clutter effects in the ISAR image in the form of strong artifacts. Although in the latter case a well-focused image of the target can still be formed, the information contained in the image may be so corrupted that any classifier's performance would be strongly penalized when making use of such an image.

6.2.1 ISAR Imaging in the Presence of Noise

Before analyzing the effects of noise in ISAR images we will provide the definition of the signal-to-noise ratio (SNR) for ISAR, which can be calculated either in the data domain (before image formation) or in the image domain (after image formation). The two SNR definitions lead to significantly different values of the SNR because they are calculated before and after coherent signal processing. Therefore, we define the SNR in the data domain as follows:

$$SNR_D = P_T/P_N, \tag{6.3}$$

where P_T is the power of the target's echo, and P_N is the noise power, as measured in the data domain. This is equivalent to the definition of SNR for each received signal sample.

The SNR in the image domain can be defined as follows:

$$SNR_I = \frac{E_T/N_T}{E_N/N_N}, \tag{6.4}$$

where E_T is the energy of the target's echo, E_N is the energy of the noise, as measured in the image domain, and N_T and N_N are the number of pixels covered by the target and noise, respectively.

A simple method for estimating the SNR_I when this is high enough is to use image segmentation where the target's echo is separated from the noise floor via a thresholding approach. A standard way of setting the threshold is

$$Th = mean(I) + K \cdot std(I), \tag{6.5}$$

where I is the image magnitude, the operator *mean* indicates the mean value, *std* is the standard deviation, and K is a parameter that allows shifting the threshold according with heuristic rules. Typically, the value of K is chosen around the unit. This approach is effective for estimating the value of the SNR_I only when this value is high enough. In fact, in this case, the target can be effectively separated from the noise by using a segmentation approach. This fact works reasonably well only if the target's intensity is significantly larger than that of noise.

An example of ISAR images that are gradually corrupted by random Gaussian noise is shown in Figure 6.7. The original ISAR image is represented in Figure 6.7a. An SNR_I roughly equal to 23.5 dB has been measured in the image domain by setting a threshold according to (6.5) and with $K = 1.3$. The ISAR images in Figure 6.7b, Figure 6.7c, and Figure 6.7e have been obtained by progressively adding white Gaussian noise to the data. Since the measured SNR in the data domain is quite high, we may neglect the presence of noise. This allows for noise to be introduced in the data so we can quickly calculate the SNR_D. Specifically, the ISAR image produced in Figure 6.7b is obtained with a valued of $SNR_D = 10$ dB, whereas the image in Figure 6.7c and Figure 6.7e are obtained with $SNR_D = -5$ dB and $SNR_D = -15$ dB, respectively. The ISAR image represented in Figure 6.7d is obtained by limiting the dynamic range of the ISAR image in Figure 6.7c between the range of –20 dB and 0 dB. This is to highlight the fact that the image in Figure 6.7c is still a well-focused image although the presence of noise generates artifacts of point-like scatterers randomly distributed across the whole image. When SNR_D is as low as –15 dB, it is unlikely that a well-focused image can be generated by directly applying ISAR imaging algorithms that are not suitably designed to handle a large amount of noise. Figure 6.7f shows a threshold version of Figure 6.7e to point out that the latter is not focused at all.

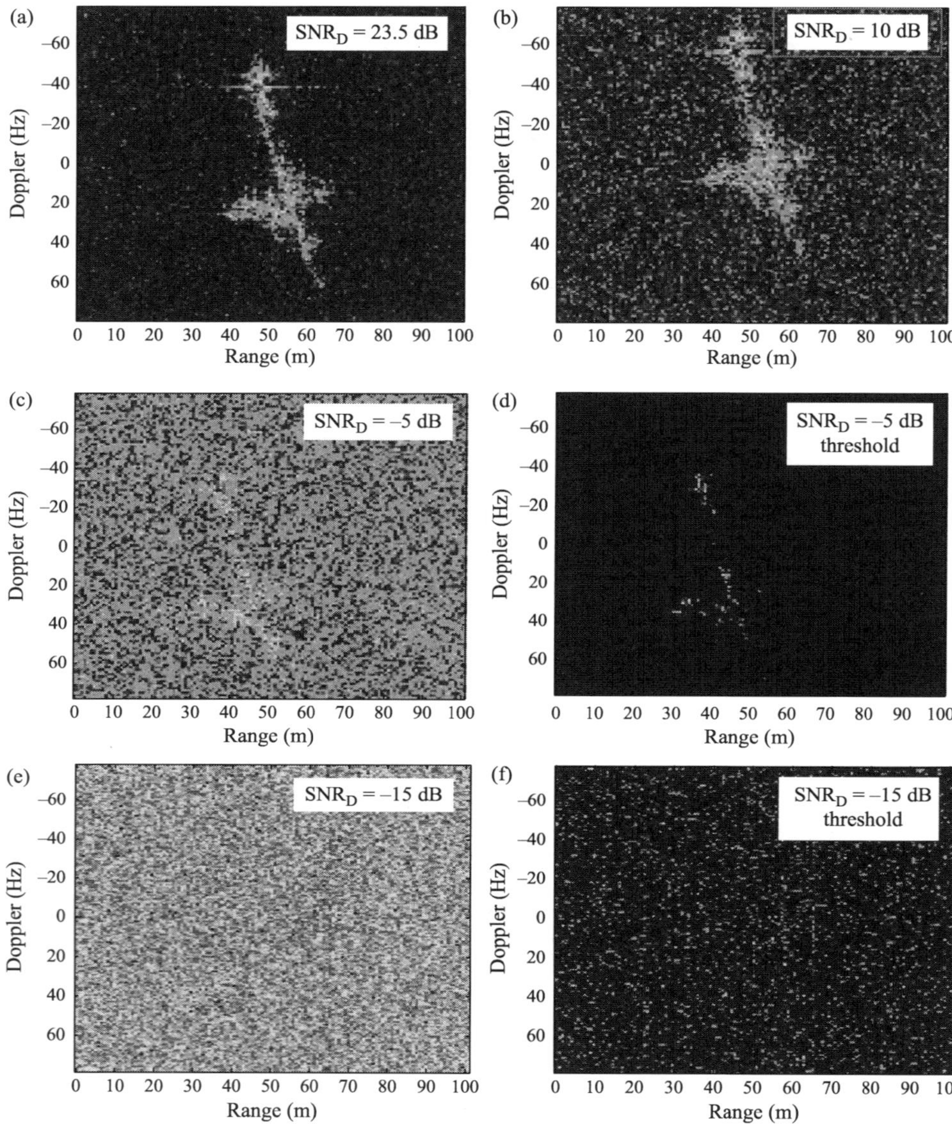

Figure 6.7 ISAR images obtained with different SNR.

Sometimes it is the presence of clutter the main disturbance rather than noise. An example of an ISAR image of a ground target with significant ground clutter is shown in Figure 6.8. Most of the clutter return is concentrated at a cross-range bin that corresponds to zero Doppler, due to the clutter stationarity. It should be pointed out that although the clutter can be modeled as an additive disturbance its power is proportional to the transmitted power; therefore, transmitting more power cannot defeat it.

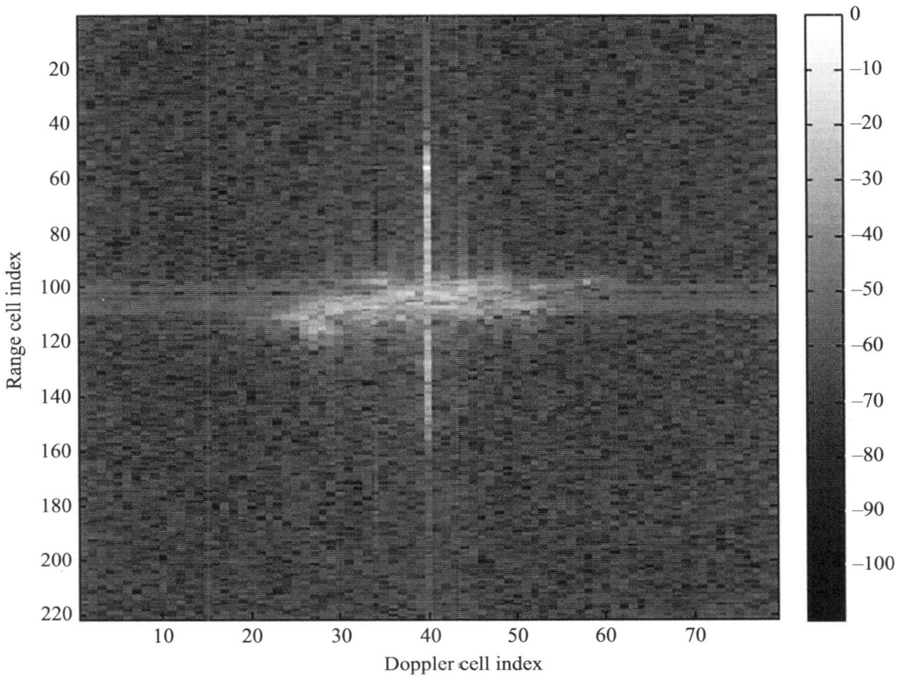

Figure 6.8 ISAR image of a ground target (T-72). A strong ground clutter can be noticed at the Doppler (cross-range cell) corresponding to zero Doppler.

6.2.2 Performance of ISAR Imaging Techniques in the Presence of Noise

The Fourier approach to the formation of ISAR images can be seen as the application of a matched filtering to the received signal when each target's component produces a stationary signal with a constant Doppler frequency. Therefore, it is to be considered as the optimal solution in terms of SNR in the output image. Linear time-frequency transforms (TFTs), such as the short-time Fourier transform (STFT), or bilinear TFTs (e.g., those defined by Cohen's class [2]), may be considered as an extension of matched filtering to nonstationary signals. These, in fact, are able to match the target's signal components in the extended time-frequency domain [3].

Problems may arise from the presence of strong noise when applying radial motion compensation algorithms, as they may be more or less sensitive to noise.

In the following analysis, four autofocus algorithms are tested against noise: the dominant scatterer autofocus (DSA) or hot spot (HS); the phase gradient autofocus (PGA); the image contrast-based autofocus (ICBA); and the image entropy-based autofocus (IEBA). The data are obtained from the airplane data set in Figure 6.4 by adding Gaussian white noise (GWN).

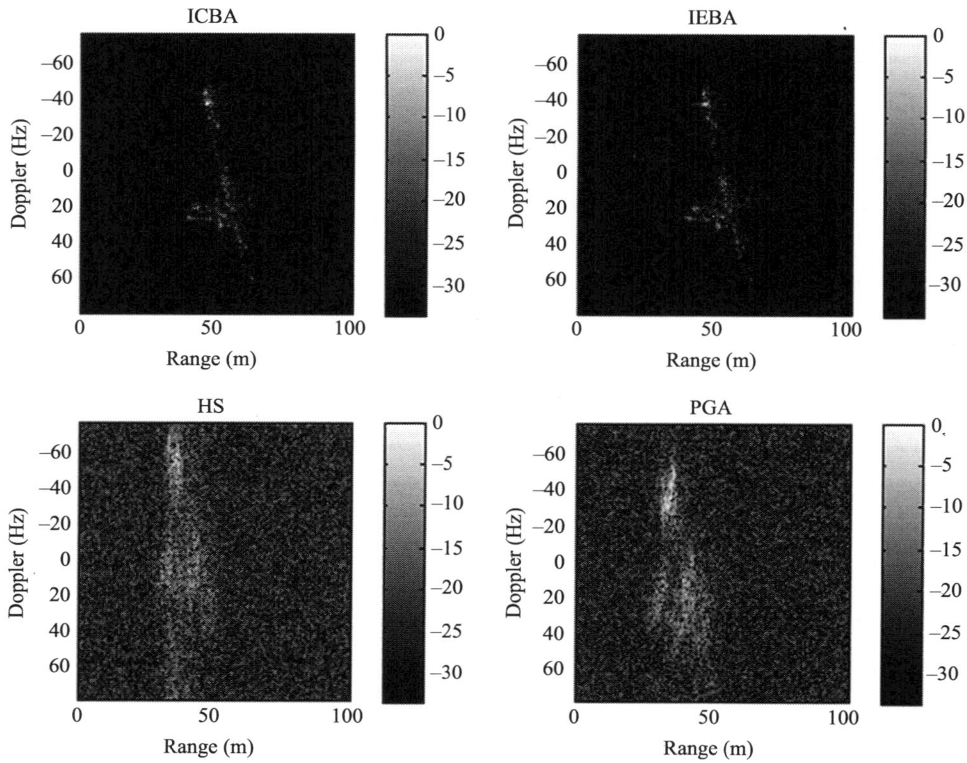

Figure 6.9 Autofocus technique comparison with $SNR_D = 0$ dB.

ISAR images in Figures 6.9 through 6.11 are formed by using the four algorithms in different SNR conditions: $SNR_D = [0, -5, -10]$ dB. The ISAR images formed using the ICBA and IEBA are focused correctly even when SNR = –10 dB, whereas the HS and PGA algorithms are not able to form focused images due to the strong presence of noise.

Plots of the IC, IE (image entropy) and IP (image peak) are also shown in Figures 6.12 through 6.14 to highlight the effect of the presence of noise onto image sharpness indicators. Specifically, Figure 6.12 illustrates the results relative to a comparison among the presented algorithms for several values of SNR (in the range [0,–10] dB with a step of 1 dB). Higher values of the IC indicate better performances. The ICBA and IEBA show a better robustness with respect to the presence of noise. Figure 6.13 shows the results in terms of IE. In this case, lower values of the IE indicate better performances. Similar to what was shown in terms of IC, the IE also indicates that the ICBA and IEBA are more robust to noise. Finally, Figure 6.14 shows results in terms of the IP. It should be noticed that the IP is not a reliable indicator of image quality unless it is used in conjunction with other indicators such as IC and IE. Nevertheless, also in this case the results show that the ICBA and IEBA outperform the PGA and HS in terms of robustness to noise.

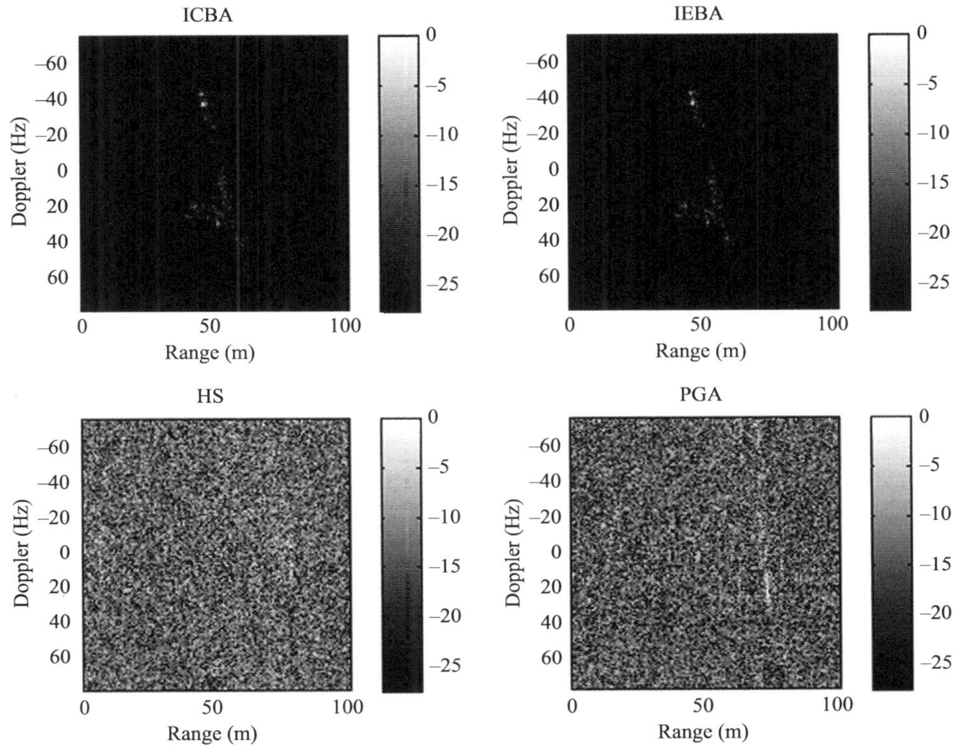

Figure 6.10 Autofocus technique comparison with $SNR_D = -5$ dB.

The results show that parametric techniques such as ICBA and IEBA are more robust with respect to noise as they tend to refocus the target, which is also in low SNR values. This result can be justified by simply recalling that parametric autofocus techniques aim at estimating a small number of parameters by using a large number of noisy samples. Conversely, nonparametric techniques aim at estimating a large number of parameters and therefore reduce the estimation accuracy.

6.2.3 Clutter Suppression in ISAR Imaging

The presence of strong clutter may seriously impede the formation of ISAR images or even the detection of the target itself. In this scenario, clutter reduction techniques must be applied to the received signal. We have to distinguish between stationary and moving platforms, as the characteristic of the measured clutter are quite different in the two cases and thus require substantially different technology and processing to remove the clutter before applying image formation. The problem of clutter removal in both the stationary and moving platform cases has been largely studied in the literature, and several solutions have been produced. Such solutions have been applied mostly to detections problems rather than imaging problems.

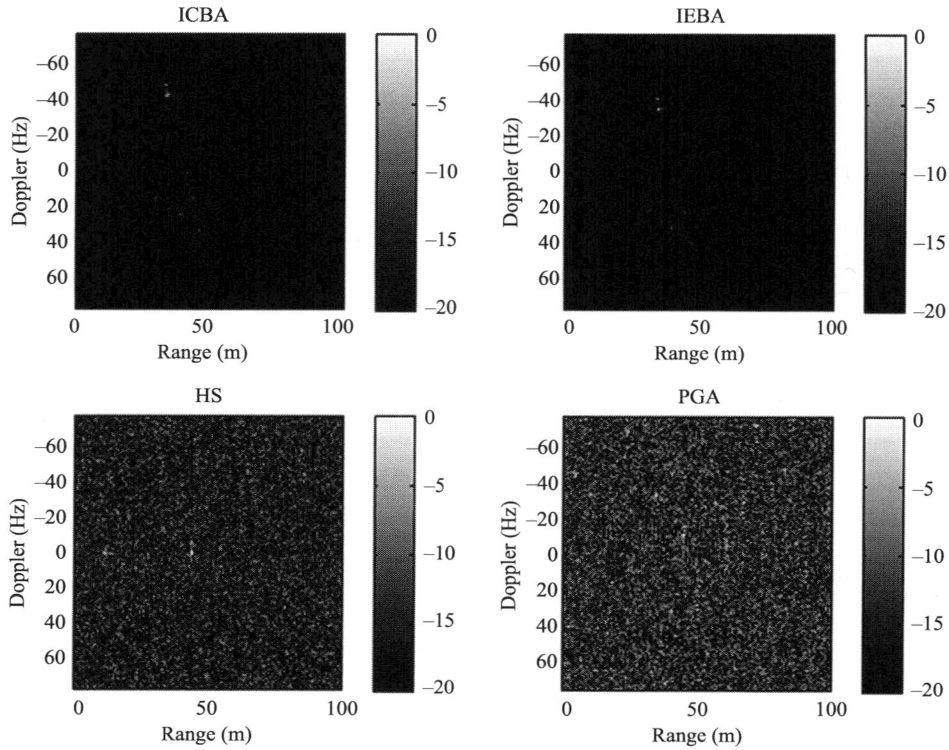

Figure 6.11 Autofocus technique comparison with $SNR_D = -10$ dB.

Specifically, in the stationary platform case, a number of algorithms have been derived to effectively remove clutter of different types, such as ground and sea clutter [4,5]. Conversely, in the case of moving platforms, multichannel systems have been employed as they provide the means for effectively canceling clutter, although at a greater expense. The proposed techniques for the latter case are generally addressed as ground moving target indicators (GMTI), the most common of which are the displaced phase center antenna (DPCA) and the space-time adaptive processing (STAP) [6,7]. As these topics have been largely covered by a number of books [4–7], we will not delve into the details of clutter cancellation. It is important, however, to state that such techniques can be used before any image formation algorithm is applied to the clutter-removed data. Such a step is important because clutter can seriously affect the image formation since it prevents algorithms from effectively apply motion compensation before the image formation. We will now present a simple example of how target removal can be applied before ISAR image formation.

6.2.4 Stationary Platform and Ground Clutter

In the case of stationary radars illuminating ground moving targets, some ad hoc techniques can be used, as demonstrated in [8]. Specifically, since the ground clutter produces a

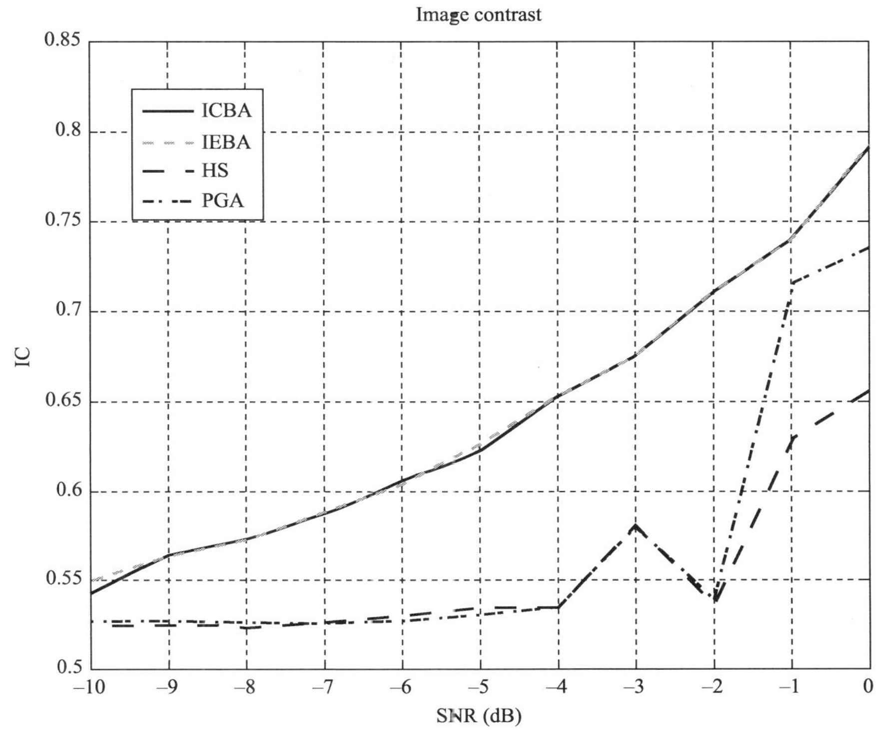

Figure 6.12 Autofocus technique comparison: image contrast.

stationary return whereas moving targets do not, a zero-Doppler clutter (ZDC) suppression can be obtained similar to the application of a moving target indicator (MTI) in coherent radar. The ZDC suppression can be implemented in the image domain either by subtracting an average value of clutter measurements from the formed ISAR image or by selecting image pixels where the ZDC has been detected and thereby forcing those pixel values to zero.

It is worth pointing out that ZDC suppression techniques are suitable for ISAR applications where the clutter is stationary and illuminated for a long enough time to obtain sufficient information for its estimation. A typical scenario where ZDC suppression techniques are applicable is that of a turntable experiment [8].

ZDC estimation and subtraction. ZDC can be estimated by averaging a sequence of ISAR images where both clutter and moving targets are present. Because the ground clutter is stationary and the moving target is not, the latter is filtered out by the averaging operation, and therefore a clutter estimation is obtained. Finally, the clutter estimate is subtracted from an ISAR image containing the moving target to produce a clutter-reduced ISAR image.

A functional block diagram is depicted in Figure 6.15 that describes the main ZDC estimation and subtraction (ZDC-ES) algorithm steps.

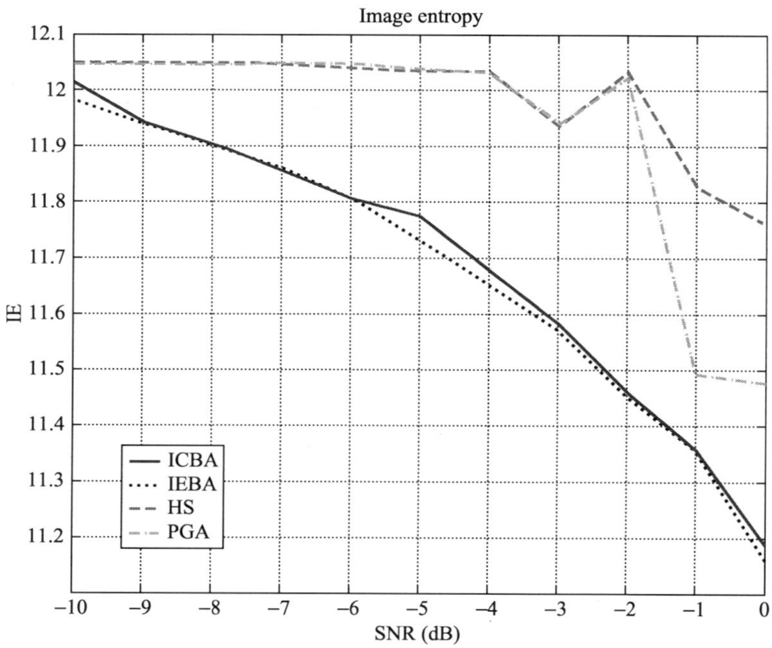

Figure 6.13 Autofocus technique comparison: image entropy.

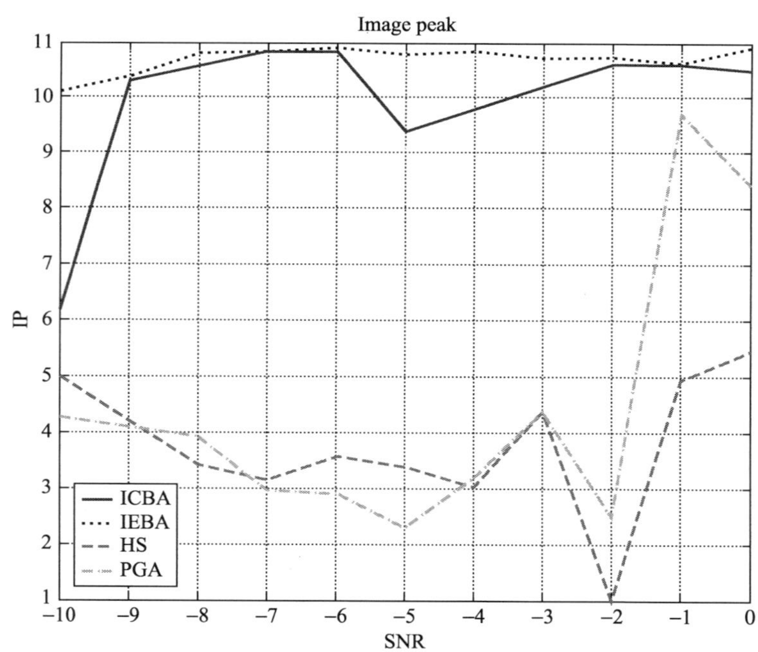

Figure 6.14 Autofocus technique comparison: image peak.

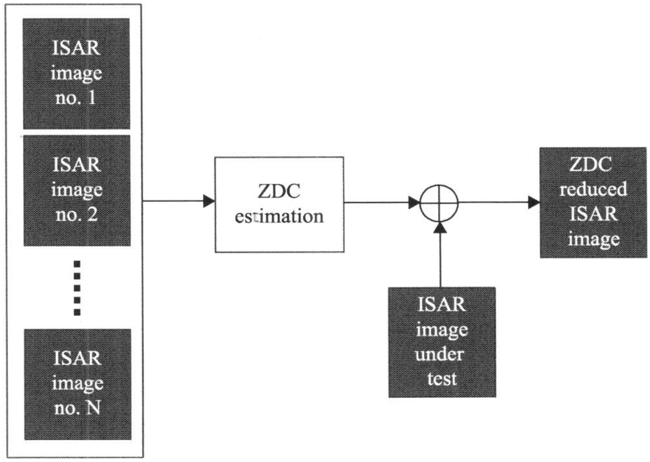

Figure 6.15 Zero-Doppler clutter reduction algorithm.

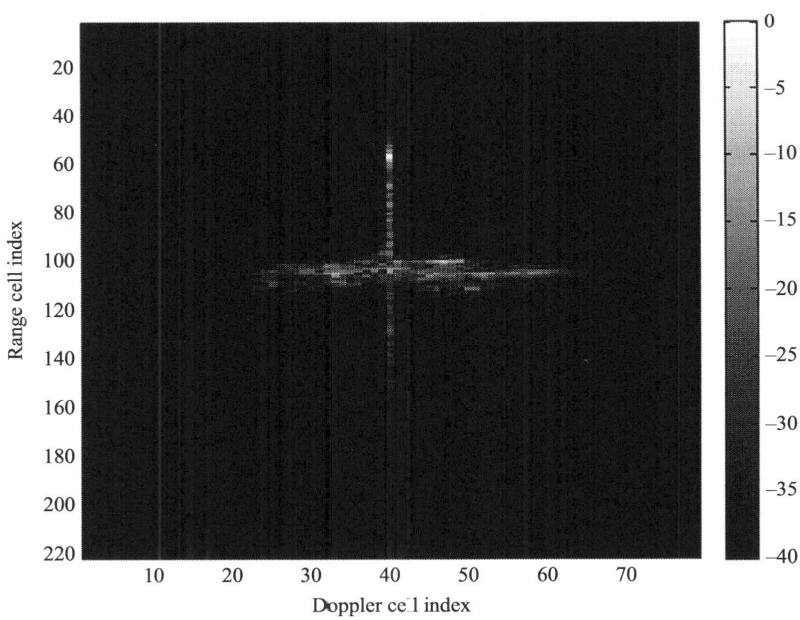

Figure 6.16 Original ISAR image with ZDC.

Figure 6.16 and Figure 6.17 show ISAR images before and after applying the ZDC-ES technique, respectively. The clutter reduction is evident when comparing the two figures.

ZDC detection and cancellation. Another way to proceed to suppress ZDC is to detect image pixels where the ZDC is likely to be present and create a mask to select and eliminate such pixels. This technique, known as ZDC detection and cancellation (ZDC-DC), was proposed

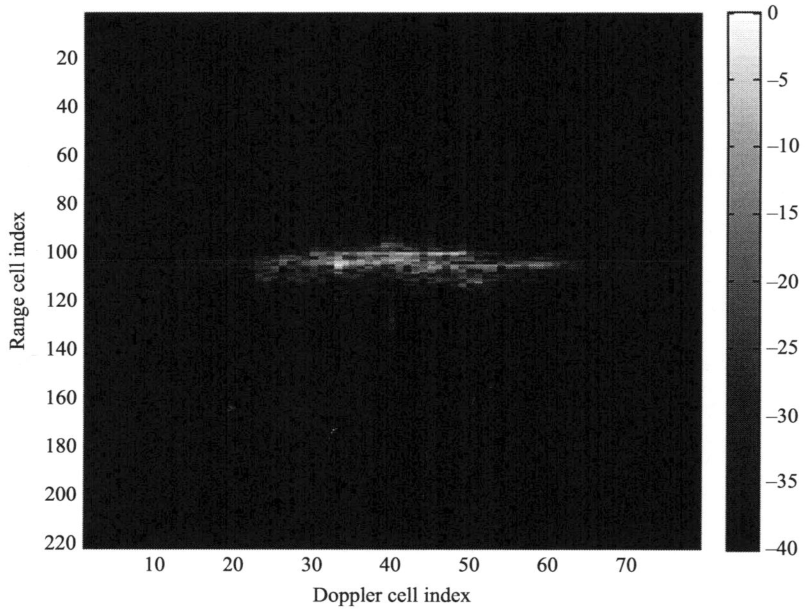

Figure 6.17 ISAR image with reduced ZDC.

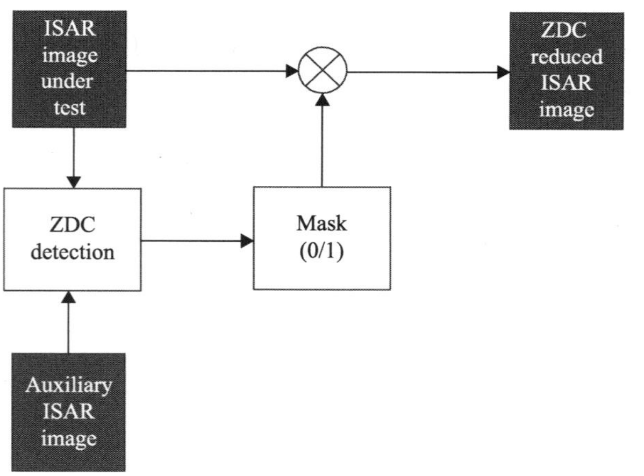

Figure 6.18 Zero-Doppler clutter suppression algorithm.

in [8]. A simple detection scheme is implemented as depicted in Figure 6.18. A detection threshold is applied to the magnitude of the complex difference of two ISAR images to detect pixels containing ZDC. Therefore a clutter-free ISAR image is produced after a pixel-by-pixel multiplication between the mask and the ISAR image under test. This technique produces significantly better results, although it is quite sensitive to the threshold selection, which makes it somehow less robust than the ZDC-ES technique.

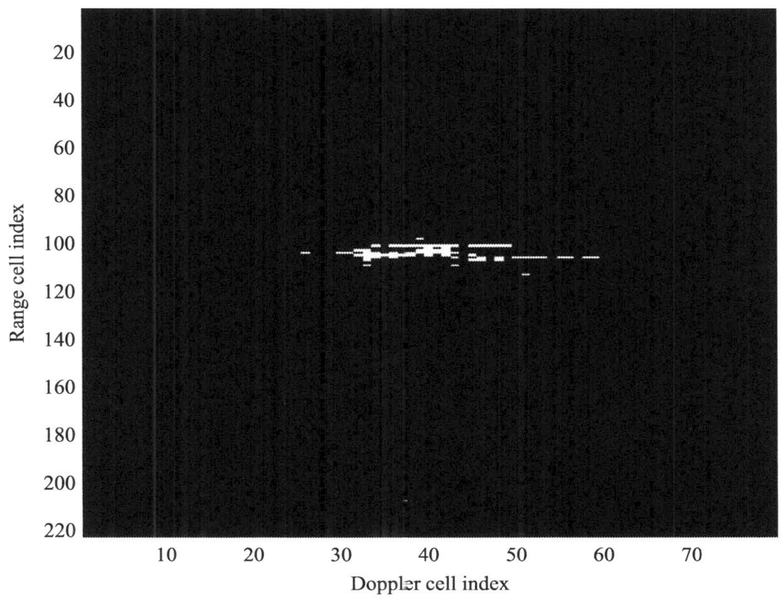

Figure 6.19 Zero-Doppler clutter detection mask.

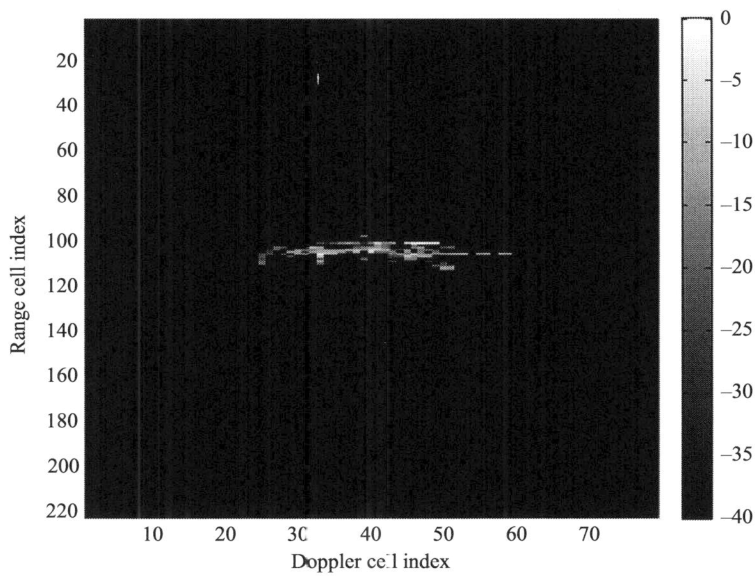

Figure 6.20 Zero-Doppler clutter canceled ISAR image.

The same ISAR image used in the previous example is tested after applying the ZDC-DC to it. The selection mask is shown in Figure 6.19, where the detected pixels containing ZDC are shown in blue and the remaining pixels in red. The image after ZDC cancellation is shown in Figure 6.20. The threshold used in this example was manually set to 5%.

6.3 Sidelobes and Their Reduction and Cancellation in ISAR Imaging

In Fourier-related transformations, such as those typically used in radar imaging, sidelobes are the effect of limited observations, which could be limited either in the spatial domain, the time domain, or both. Sidelobes typically are stronger close to the peak and decrease more or less monotonically to zero as the distance increases from the peak. The ratio between the peak and the highest sidelobe (typically the closest to the peak) is called sidelobe level (SLL) and is typically measured in decibels (dB). Rectangular time or spatial acquisition windows generate sidelobes of about 13 dB. Sidelobes create undesirable effects in radar images because they mask weaker signals that remain buried under stronger signal sidelobes.

The goal of this section is to present some techniques to reduce or even cancel sidelobes in ISAR imaging.

6.3.1 Weighting Functions

One reason for the occurrence of sidelobes is the abrupt way with which a signal transits from nonzero to zero values. During an acquisition, the received signal may vary from practically zero (except for noise) to high values, and the same may happen when an acquisition terminates (from high values to zero).

A widely used method for reducing sidelobes is via weighting functions, which aim to smooth such transitions. Typical windows used for this task are Hamming, Hanning, Kaiser, and Chebitchev. The trade-off when reducing sidelobes with weighting functions is a widening of the main lobe (peak), which is then associated with a resolution loss.

To visualize the effects of using weighting windows, we use some real ISAR ship data. An ISAR image of the ship is formed without using any weighting windows and is compared with an ISAR image obtained using a Kaiser weighting window with a shape parameter equal to 4, as shown in Figure 6.21. The results are visualized in Figure 6.22 and Figure 6.23. Specifically, Figure 6.22a shows the ISAR image of the whole ship, whereas Figure 6.22b

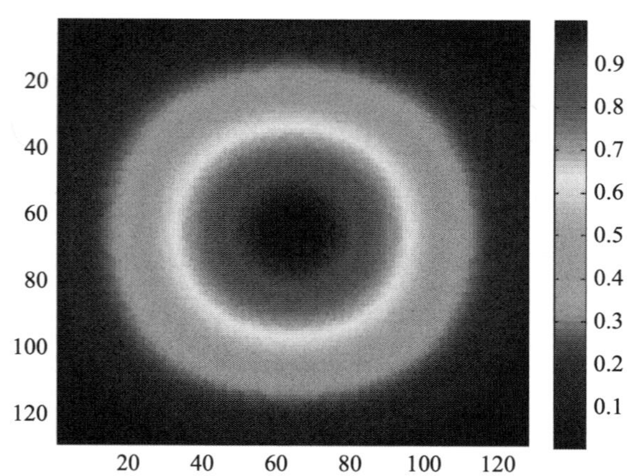

Figure 6.21 Kaiser window.

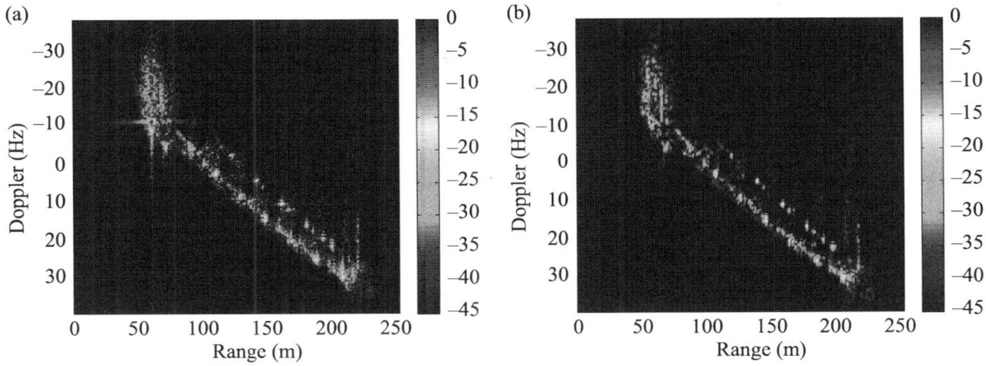

Figure 6.22 ISAR image (a) nonweighted and (b) weighted cases.

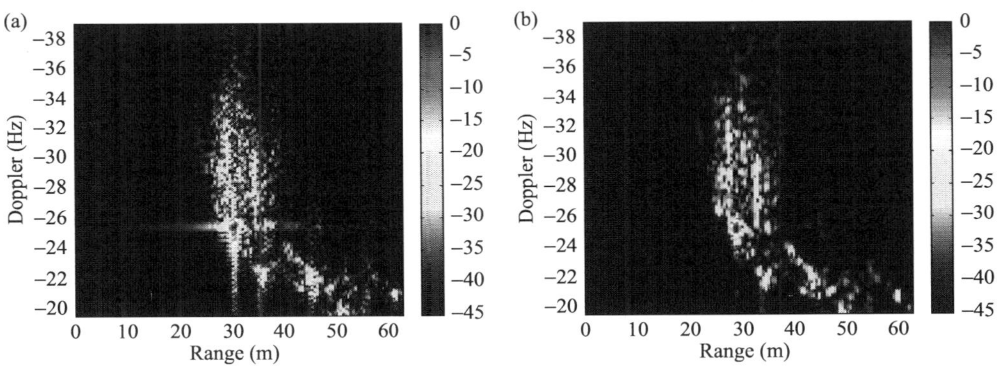

Figure 6.23 ISAR image zoom-in (a) nonweighted and (b) weighted cases.

shows the ISAR image of the same ship using the Kaiser window. The second image appears cleaner than the first as an effect of the sidelobe level reduction. A resolution loss is also appreciated when closely looking at the second image and comparing it with the first one. A close-up of the brighter scatterer is shown in Figure 6.23a for the nonweighted case and in Figure 6.23b for the weighted case. The sidelobe level reduction and resolution loss are even more clearly highlighted in this case.

6.3.2 Spatially Variant Apodization

A different way to perform sidelobe cancellation is by implementing apodization techniques initially imported from optical engineering [9]. Dual apodization (DA) and multiple apodization (MA) were introduced as nonlinear techniques to effectively suppress sidelobes. The basic way of implementing DA is to form two images: one with a flat weighting function and another with a given weighting function (e.g., Hamming, Hanning, Kaiser) and to obtain a third "apodized" image by selecting pixel by pixel the minimum value between two same pixels in the two images. Such an idea can then be extended to multiple images obtained by applying different weighting functions.

The spatially variant apodization (SVA) technique is a generalization of the apodization concept, where each pixel is weighted with a different apodization function, from which the term *spatially variant* is derived. The idea is based on the use of a *cosine on a pedestal* weighting function, which can be expressed mathematically as follows:

$$W(f) = 1 + 2k\cos(2\pi f/B), \tag{6.6}$$

where the weight k can range from 0 to 0.5. It is worth noting that $k = 0$ yields a flat weighting function (no effect), whereas $k = 0.5$ yields a Hanning window with all cosine and no pedestal.

The Nyquist sampled Fourier transform of (6.6) produces only three nonzero terms, as expressed by

$$w[n] = k\delta[n-1] + \delta[n] + k\delta[n+1], \tag{6.7}$$

where $\delta[n]$ is the Kronecker delta function, which yields 1 when $n = 0$ and 0 otherwise.

Therefore, the application of a cosine on a pedestal weighting function is equivalent to the convolution with the function described by the three Kronecker delta functions in (6.7). The generic element $x[n]$ is transformed into the following element:

$$x'[n] = kx[n-1] + x[n] + kx[n+1]. \tag{6.8}$$

The weight k is then chosen by minimizing the value of $|x'[n]|^2$ with the constraint $0 < k < 0.5$. A closed form solution for the unconstrained optimization problem exists and can be written as

$$k_u[n] = -Re\left\{\frac{x[n]}{x[n-1] + x[n+1]}\right\}. \tag{6.9}$$

Therefore, the constrained solution becomes

$$k[n] = \begin{cases} 0 & \text{if } k_0[n] < 0 \\ k_0[n] & \text{if } 0 < k_0[n] < 0.5\,. \\ 0.5 & \text{if } k_0[n] < 0 \end{cases} \tag{6.10}$$

ISAR images are typically 2-D images, and therefore a 2-D SVA must be employed to reduce the sidelobes along both the range and the cross-range coordinates. A simple way to obtain a 2-D apodization is to apply a 1-D SVA first along one coordinate and then along the second coordinate. Although this way to proceed is quite fast and simple, the second apodization operates on a nonNyquist sampled image, which makes the apodization technique work less effectively.

For this reason, a joint 2-D SVA can be defined as one that directly operates in a two-dimensional domain. A joint 2-D version has been proposed in [9] where an extension of the 1–D technique is defined. Specifically, a 2–D raised cosine is firstly defined followed by a minimization. Better results have been shown when applying a joint 2-D SVA instead of two 1–D SVA applied sequentially along the two coordinates.

An ISAR image and its apodized version are shown in Figure 6.24a and Figure 6.24b. A visually evident result is obtained as an effect of the application of the SVA.

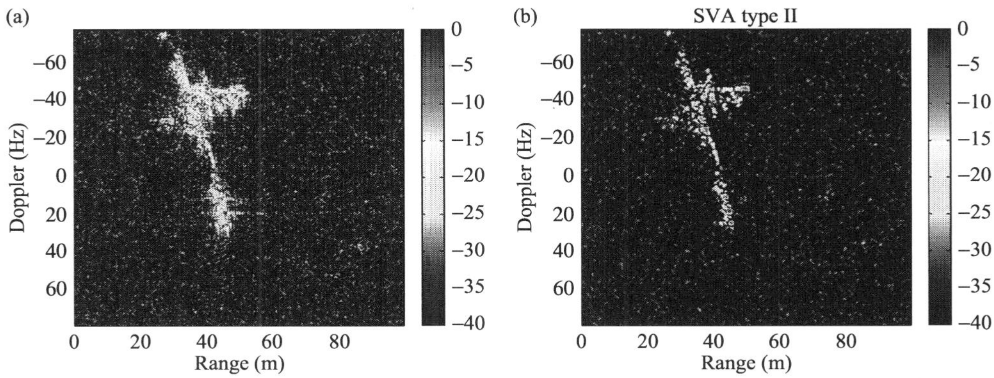

Figure 6.24 (a) Original ISAR image; (b) ISAR image after the application of SVA.

6.3.3 Extraction of Scattering Centers—Clean Technique

The CLEAN technique was initially introduced in radar array processing to suppress sidelobes [10]. For this reason, such a technique is often framed as a sidelobe suppressing technique. In the work of Choi et al. [11], the CLEAN technique was presented for the first time in a slightly different light, as it was applied to estimate position and amplitude of scattering centers present in a radar image. Following the work of Choi, some refinement and adaptation were necessary to apply the CLEAN technique to ISAR data because the target's noncooperation makes the problem of estimating the scatterer's position and amplitude harder to handle [12,13].

In the following subsection, the CLEAN technique is presented as ISAR signal processing. Such processing will be based on ISAR signal modeling to address the problem strictly from an ISAR point of view.

Finally a block diagram will be derived that implements the CLEAN technique for ISAR applications in an iterative manner. Two stop conditions will also be presented that can be used to stop the iterations and form the final CLEAN image.

Signal modeling. In defining the ISAR signal modeling that will be used to introduce the CLEAN technique, we will assume that the target is composed of several point-like scatterers. Under such an assumption and after performing radial motion compensation, the received signal can be written as follows [14]:

$$S_R(f,t) = W(f,t)\sum_{k=1}^{N}\rho(x,y)\exp\{-j\varphi(f,t)\}, \tag{6.11}$$

where $\varphi(f,t) = \frac{4\pi f}{c}\left[\boldsymbol{x}_{k_{LOS}}^{T_i}(t)\right]$, $W(f,t) = \mathrm{rect}(\frac{t}{T})\mathrm{rect}(\frac{f-f_0}{B})$, f_0 represents the carrier frequency, B is the transmitted signal bandwidth, T is the CPI, c is the speed of light in vacuum, $\boldsymbol{x} = (x,y,z)$ represents the vector that locates a generic point on the target with respect to a target-fixed system, and $\rho(\cdot)$ is the reflectivity function. Function rect(t) yields 1 when $|t| < \frac{1}{2}$, and 0 otherwise. It is worth pointing out that the x and y coordinates of vector $\boldsymbol{x}$ can be chosen arbitrarily. However, a convenient way is to relate them to the cross-range and range coordinates, respectively. When the target moves smoothly and the integration time is kept short enough, the target rotation vector can be assumed constant. Such an assumption will be addressed as almost constant rotation vector condition and is formalized as follows:

$$\Omega_{tot}(t) \cong \Omega_{tot}. \tag{6.12}$$

The phase associated with the signal component relatively to a generic scatterer, k, can be analytically expressed as

$$\varphi_k(f,t) = -\frac{4\pi f}{c}[\sin(\Omega_{eff}t)x_k + \cos(\Omega_{eff}t)y_k]\text{rect}\left(\frac{t}{T}\right), \tag{6.13}$$

where Ω_{eff} is the modulus of the effective rotation vector, which is defined as $\Omega_{eff} = \boldsymbol{i}_{LOS} \times (\Omega_{tot} \times \text{i}_{LOS})$, and $\boldsymbol{i}_{LOS}$ is the unit vector of the radar line of sight (LOS).

Moreover, if the observation time is such that $T \ll \frac{2\pi}{\Omega_{eff}}$, (6.13) can be approximated as

$$\varphi_k(f,t) = -\frac{4\pi f}{c}\left(y_k + x_k\Omega_{eff}t - \frac{1}{2}y_k\Omega_{eff}^2 t^2\right). \tag{6.14}$$

By substituting (6.14) in (6.11) and after motion compensation and range compression by 1-D fast Fourier transform (FFT), the complex range profile can be obtained as

$$\text{rngpro}^{(k)}(\tau,t) = B \cdot \text{sinc}\left[B\left(\tau - \frac{2}{c}y_k\right)\right]\exp\{j\varphi_k(f_0,t)\}\cdot\text{rect}\left(\frac{t}{T}\right), \tag{6.15}$$

where

$$\varphi_k(f_0,t) = -\frac{4\pi f_0}{c}\left(y_k + x_k\Omega_{eff}t - \frac{1}{2}y_k\Omega_{eff}^2 t^2\right). \tag{6.16}$$

Then the time history of each range cell is Fourier transformed to obtain an ISAR image, as analytically shown by

$$I_C(\tau, f_D) = BT_{obs}\sum_{k=1}^{k}\text{sinc}[B(\tau-\tau_k)] \cdot \text{sinc}[T(f_D - f_{D_k})] \otimes D(\tau, f_D), \tag{6.17}$$

where

$$\tau_k \stackrel{\Delta}{=} \frac{2}{c}y_k, \tag{6.18}$$

$$v_k = \frac{2f_0\Omega_{eff}}{c}x_k, \tag{6.19}$$

and

$$D(\tau, f_D) = FT_{t\to f_D}\left[\exp\{j\pi\mu_k t^2\}\right] \tag{6.20}$$

is a distortion term due to the quadratic signal phase, $\mu_k = \frac{2f_0}{c}y_k\Omega_{eff}^2$ and $\otimes$ is the convolution operator.

It is worth noting that the distortion introduced by $D(\tau, f_D)$ depends on the scatterer's position, which makes it a space-dependent distortion. The Fourier analysis is not suitable to handle such a distortion. For this reason, the polynomial Fourier transform will be used to solve this problem.

Signal component estimation and cancellation. The CLEAN technique can be implemented by iteratively estimating thepoint spread function (PSF) parameters for each scatterer and by removing its contribution via a coherent cancellation. A functional block scheme is depicted in Figure 6.25 that explain how these two operations are performed in an iterative fashion.

(1) CLEAN ALGORITHM

The algorithm of the CLEAN technique is as follows:

Step 1: The brightest scatterer is selected in the image domain.
Step 2: The parameters of the selected scatterer PSF are estimated and stored in a scatterer database.
Step 3: A partial CLEAN image is formed by coherently summing the PSF main lobes obtained with the extracted scatterer's parameters.
Step 4: The estimated PSF is coherently subtracted from the ISAR image.
Step 5: A check is performed on the residual ISAR image to determine if more scatterers are present that must be extracted. If the check is positive, steps 1–5 are repeated; if not, the algorithm ends and the latest updated CLEAN image becomes the final output.

PSF parameter estimation. The estimation of each scatterer's contribution is performed by minimizing the residual image energy after removing the estimated contribution. The idea is that of finding the PSF parameters that provide the best fit to the scatterer's image so that it may be coherently subtracted from the image. Although it is not strictly correct, one may argue that if the remaining energy is minimized the scatterer's image may be efficiently

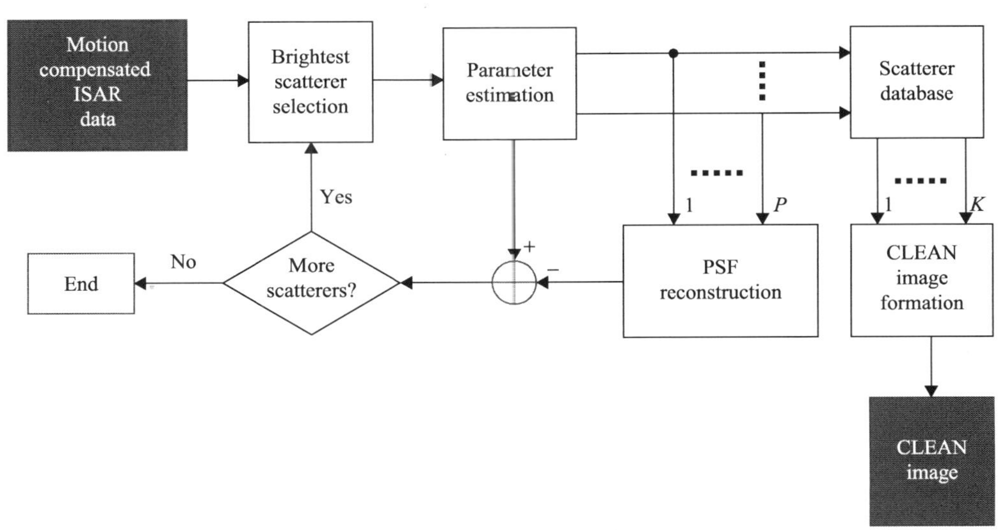

Figure 6.25 Coherent CLEAN technique functional diagram.

removed after performing a coherent subtraction. It should be mentioned that the subtraction operation may be performed either in the data or the image domain as the two domains are related by means of a Fourier transform and therefore a linear transform. Mathematically, this can be expressed as follows:

$$\hat{\mathbf{\Phi}}_k = \text{argmin}_{\mathbf{\Phi}_k} E(\mathbf{\Phi}_k) \text{ with } \mathbf{\Phi}_k = [\tau k, f_{D_k}, \mu_k, A_k], \tag{6.21}$$

where

$$E(\mathbf{\Phi}_k) = \iint\limits_{-\infty}^{\infty} \left| P_R(\tau, t) - \sum_{n=1}^{k} \hat{P}_R^{(n)}(\tau, t) \right|^2 d\tau dt \tag{6.22}$$

with

$$\hat{P}_R^{(k)}(\tau, t) = A_{Rk} \cdot \text{sinc}[B(\tau - \hat{\tau}_k)] \cdot \exp\left\{ j\varphi_k(t, \hat{f}_{D_k}, \hat{\mu}_k) \right\} \cdot \text{rect}\left(\frac{t}{T}\right), \tag{6.23}$$

$$\varphi_k\left(t, \hat{f}_{D_k}, \hat{\mu}_k\right) = -2\pi\left(\hat{f}_{Dk} t - \frac{\hat{\mu}_k}{2} t^2\right), \tag{6.24}$$

$$A_{Rk} = A_k \cdot B \cdot \exp\left\{ -j\frac{4\pi f_0}{c} y_k \right\}, \tag{6.25}$$

and $P_R(\tau, t)$, the matrix of range compressed profiles obtained from the received data.

It is worth pointing out that the residual energy (RE) in (6.22) can also be calculated in the data domain or in the image domain as the transformation from one domain to another is performed via a Fourier transform, which is an energy-preserving transformation. Mathematically, this can be expressed as follows:

$$E(\mathbf{\Phi}_k) = \iint\limits_{-\infty}^{\infty} \left| S_R(f, t) - \sum_{n=1}^{k} \hat{S}_R^{(n)}(f, t) \right|^2 df dt, \tag{6.26}$$

$$E(\mathbf{\Phi}_k) = \iint\limits_{-\infty}^{\infty} \left| I_C(\tau, \nu) - \sum_{n=1}^{k} \hat{I}_n(\tau, \nu) \right|^2 d\tau d\nu, \tag{6.27}$$

where $S_R(f, t)$ is defined in (6.11), $I_C(\tau, \nu) = \sum_{k=1}^{k} I_k(\tau, \nu)$, and

$$S_R^{(k)}(f, t) = A_k \exp\left\{ j\varphi_k'\left(t, \hat{f}_{D_k}, \hat{\mu}_k\right) \right\} W(f, t), \tag{6.28}$$

with

$$\varphi_k'\left(f, t, \hat{f}_{D_k}, \hat{\mu}_k\right) = -2\pi\left(\frac{2f}{c}\hat{y}_k \hat{f}_{Dk} t - \frac{\hat{\mu}_k}{2} t^2\right). \tag{6.29}$$

The image obtained from each scatterer is obtained by means of a 2-D Fourier transform of (6.28) and can therefore be expressed as follows:

$$I_k(\tau, f_D) = A_{Ik}\,\text{sinc}[B(\tau - \hat{\tau}_k)]\text{sinc}\left[T\left(f_D - \hat{f}_{D_k}\right)\right] \otimes D(\tau, f_D, \hat{\mu}_k), \tag{6.30}$$

where

$$D(\tau, f_D, \hat{\mu}_k) = FT_{t\rightarrow f_D}\left[\exp\{j\pi\hat{\mu}_k t^2\}\right], \tag{6.31}$$

$$A_{Ik} = A_k \cdot B \cdot T \cdot \exp\left\{-j\frac{4\pi f_0}{c} y_k\right\}. \tag{6.32}$$

It is worth noting that the image model of the k-th scatterer carries the same distortion of the ISAR image of the same scatterer, provided that a correct estimation of the chirp rate, $\hat{\mu}_k$, is performed. This allows for a correct cancellation of the scatterer contribution to be obtained, which ensures that the following scatterer extraction ($k + 1$) is not affected by previous erroneous estimations (error propagation).

Stop condition. The algorithm stops iterating when there are no more scatterers to be extracted. The stop condition can be implemented either by considering the signal energy [12] or its statistics [13]. The former can be applied by:

1. Calculating the initial signal (or image) energy. This can be done by either applying (6.26) or (6.27) when $n = 0$
2. Calculating the signal energy at each iteration
3. Calculating the ratio between the energy at iteration n and the initial energy ($n = 0$)
4. Comparing the ratio calculated at step (3) with a preset threshold (e.g., 10%)
5. Stopping if the value of the ratio is below the preset threshold

The energy-based stop condition is easy to implement although it can easily lead to either an extraction of false scatterers or scatterer's miss detection, even if the threshold is chosen adaptively. When background statistics can be exploited, a statistically based approach can be followed, similar to the case of classic detection problems. In the case of Gaussian background, the procedure to be followed consists of the following:

1. Separating the target from the background (segmentation), which can be reasonably done in the image domain with the application of an adaptive threshold
2. Verifying that the background statistics match a Gaussian distribution
3. For each range cell where the target is present, a statistical analysis is performed to check whether the signal vector has a Gaussian distribution or not
4. If the received signal in a range cell does not have a Gaussian distribution, a scatterer extraction is performed via the CLEAN technique, which proceeds with the elimination of the scatterer's contribution
5. If the received signal in a range cell has a Gaussian distribution, no scatterers are extracted and a test is run on another range cell until all the remaining range cells have a Gaussian distribution

Details of this techniques can be found in [13].

6.3.4 Examples

The examples shown in this section are obtained using real data from data set A and data set B. A comparison can be made by looking at the CLEAN ISAR images obtained using the energy-based and statistically based stop conditions in Figure 6.26 and Figure 6.27.

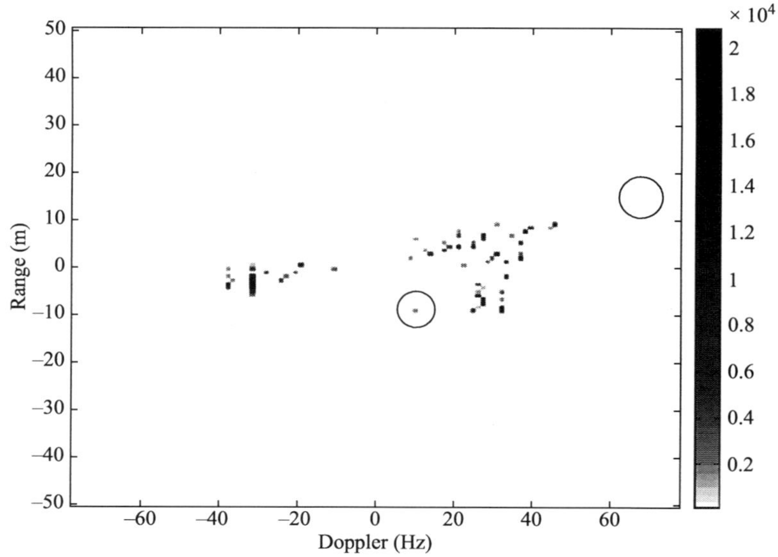

Figure 6.26 ISAR image, CLEAN, ground-based radar, Boeing 737, the nose of the airplane is not detected (circle on the right) and a false alarm is produced (circle on the left).

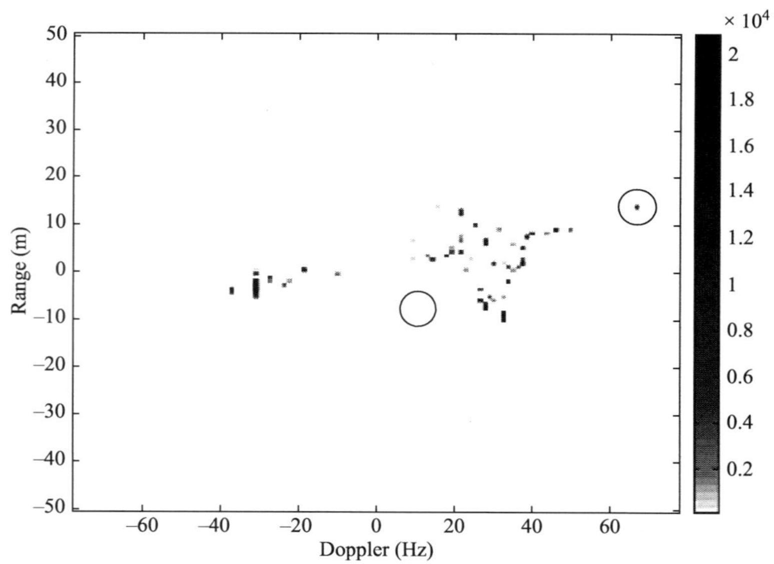

Figure 6.27 ISAR image, S-CLEAN, ground-based radar, Boeing 737, the nose of the airplane is correctly detected (circle on the right) and no false alarm is produced (circle on the left).

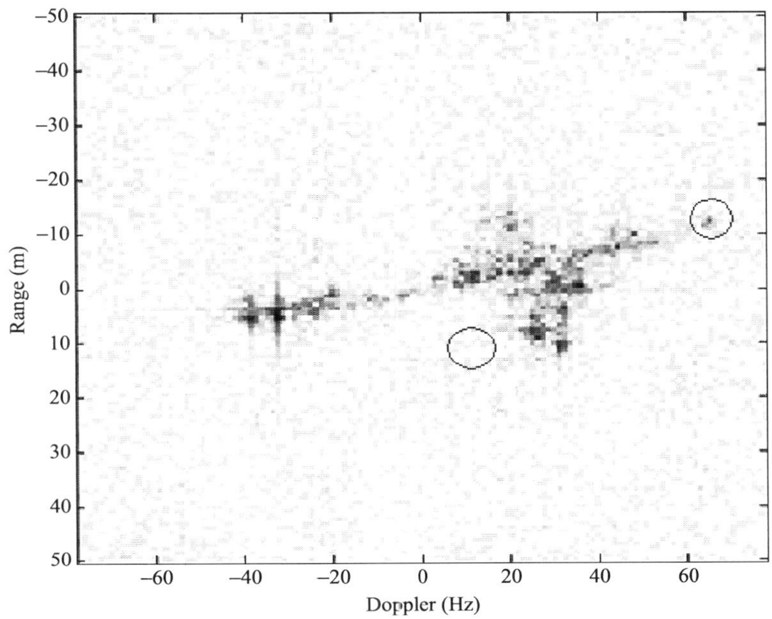

Figure 6.28 ISAR image, FFT, ground-based radar, Boeing 737, the circles highlight image regions of interest.

The comparison is run by extracting the same number of scatterers. The original RD ISAR image is shown in Figure 6.28 with the aim of highlighting the results obtained by applying the CLEAN technique in both cases.

References

[1] Parker, R.G., and R.L. Rardin, *Discrete Optimization*, Academic Press, 1988.

[2] Cohen, L., "Time-Frequency Distributions—A Review," *Proc. IEEE*, vol. 77, pp. 941–980, 1989.

[3] Xia, X., G. Wang, and V.C. Chen, "Quantitative SNR Analysis for ISAR Imaging Using Joint Time-Frequency Analysis—Short Time Frequency Transform," *IEEE Trans. on AES*, vol. 38, no. 2, pp. 649–659, 2002.

[4] Billingsley, J.B., *Low-Angle Radar Land Clutter: Measurements and Empirical Models*, IET, 2002.

[5] Ward, K., R. Tough, and S. Watts, *Sea Clutter: Scattering, the K-Distribution and Radar Performance*, 2d ed., IET, 2013.

[6] Gini, F., and N. Sidiropoulos, *Academic Press Library in Signal Processing: Volume 2, Communications and Radar Signal Processing*, Elsevier, 2013.

[7] Klemm, R., *Principles of Space Time Adaptive Processing*, 3d ed., IET, 2006.

[8] Showman, G.A., M.A. Richards, and K.J. Sangston, "Comparison of Two Algorithms for Correcting Zero-Doppler Clutter in Turntable ISAR Imagery," *Asilomar Conference on Signals, Systems & Computers*, vol. 1, pp. 411–415, 1998.

[9] Stankwitz, H.C., R.J. Dallaire, and J.R. Fienup, "Nonlinear Apodization for Sidelobe Control in SAR Imagery," *IEEE Tran. AES*, vol. 31, no. 1, pp. 267–279, 2005.

[10] Tsao, J., and B.D. Steinberg, "Reduction of Sidelobe and Speckle Artifacts in Microwave Imaging: The CLEAN Technique," *IEEE Trans. Antennas and Propagation*, vol. 36, no. 4, pp. 543–556, Apr. 1988.

[11] Choi, I.S., and H.T. Kim, "Two-Dimensional Evolutionary Programming-Based CLEAN," *IEEE Trans. AES*, vol. 39, no. 1, pp. 373–382, Jan. 2003.

[12] Sun, Y., and P. Lin, "An improved method of ISAR image processing," *Circuits and Systems, 1992., Proceedings of the 35th Midwest Symposium on*, vol. 2, pp. 983,986, 9–12 Aug 1992. doi: 10.1109/MWSCAS.1992.271131.

[13] Martorella, M., N. Acito, and F. Berizzi, "Statistical CLEAN Technique for ISAR Imaging," *IEEE Trans. GRS*, vol. 45, no. 11, Part 1, pp. 3552–3560, 2007.

[14] Martorella, M., and F. Berizzi, "Time Windowing for Highly Focused ISAR Image Reconstruction," *IEEE Trans. AES*, vol. 41, no. 3, pp. 992–1007, 2005.

CHAPTER 7

ISAR Target Feature Extraction

Automatic extraction of features from imagery has been the subject of extensive research for decades. Development of robust machine feature extraction is a challenging task. Because of the difference between inverse synthetic aperture radar (ISAR) and optical images as mentioned in Chapter 1, Section 1.5, feature extraction from two-dimensional (2-D) radar imagery is more complicated than that in computer vision. When different features present similar characteristics, these especially create confusion in performing target classification.

In contrast to machine feature extraction, humans have the ability to scan a large area and group simple features into a meaningful structure to recognize expected object. Humans can also perceive shapes and motions in noise and clutter environment and adapt to varying conditions and can interpret imagery very well by evaluating a variety of cues [1].

In computer vision, model-based methods are commonly used for recognizing an object. Model-based method often applies constraints and relationships (e.g., shape, size, orientation, curvature, and symmetry) to define an object [2]. Mathematical morphology has been used for automated feature extraction [3–5]. Hough transform is a suitable tool to detect linear features, curves, and shaped templates. In computer vision, template matching is often used to describe general characteristics of features and then applies a machine learning approach to search for specific features [2,6]. The aforementioned model-based method, template matching, and machine learning can also applied to 2-D ISAR images for feature detection, extraction, and target recognition.

ISAR imagery is the projection of a three-dimensional (3-D) target onto a 2-D range-Doppler (or cross-range) plane as discussed in Section 2.6. It does not look like the photo image of the target because it is the result of a completely different projection with respect to the optical image. As illustrated in Figure 1.12, the ISAR image of a Boeing 737 is quite different from its optical image. However, in the ISAR image, some important features such as nose, wingtips, and fuselage of the Boeing 737 can still be recognized. Thus, to effectively extract features from 2-D ISAR range-Doppler imagery is an important issue. For example, dimension of a target can be estimated from ISAR image extents in the range and cross-range.

Figure 7.1 demonstrates three ISAR images. Figure 7.1a is a normal ISAR image of an airplane shown in Figure 1.12, where we can estimate the size and orientation of the airplane from its range and cross-range extents, and the airplane's rotation parameters can also be determined based on the estimated target dimension [7], which we will introduce in

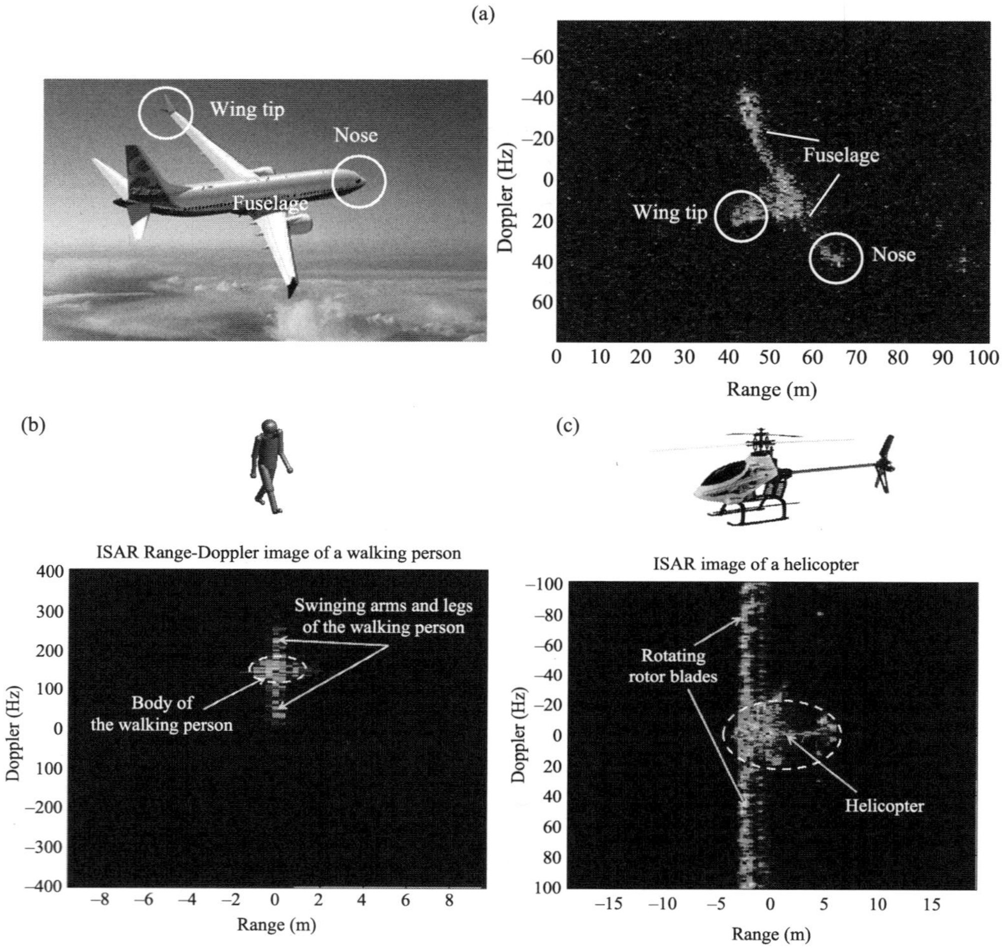

Figure 7.1 Features appeared in a 2-D ISAR imagery.

Section 7.1. Figure 7.1b is an ISAR image of a walking person with swinging arms and legs, where the dotted-line feature in the Doppler axis makes the ISAR image smearing in the Doppler domain. The dotted-line feature is caused by the kinematic motion of swinging arms and legs. Figure 7.1c shows an ISAR image of a helicopter with rotating rotor blades, where we can also see the dotted-line features due to rotating rotor blades. Features associated with target's rotation, vibration, or articulation often make 2-D ISAR image smearing in the Doppler axis.

Therefore, another distinct issue on ISAR target feature extraction is to extract target motion features showing in ISAR images. These motion-related features can be extracted from ISAR range profiles instead of 2-D ISAR imagery [8]. Certain types of motion—oscillatory displacements of scattering sources with multiple periods during the coherent processing interval (CPI)—provide deterministic Doppler signatures via the phase term of the radar returns that produce predicable sidebands around the carrier frequency. These deterministic Doppler sidebands are what we call *micro-Doppler* signatures: a frequency distribution caused by periodic micro-motion as seen in Figure 7.1b and Figure 7.1c. Although the exploitation of micro-Doppler signatures does not a priori require ISAR image formation, but a 2-D ISAR image is useful in revealing the range location of the target or target structural component that produces it. Awareness of this information allows the selection and extraction of more relevant subsets of the 2-D range profiles.

Examples of oscillatory motion include mechanical displacements or vibrations of target surfaces, rotational motion of components such as rotating antennas, rolling wheels, rotating rotor blades, even human locomotion such as the swinging of arms and legs. Any periodic motion with multiple cycles during the CPI modulates the Doppler frequency shift produced by relative translational target motion and manifests itself as time-dependent Doppler frequency sidebands. We refer to such frequency modulation as the micro-Doppler effect [9,10]. The time-varying micro-Doppler signatures provide information that reveals target features, characteristics, or disposition. Analysis of these signatures allows the estimation of various kinematic features, such as antenna rotation rates, vibrations frequencies, and human gait strides. In Section 7.2, we will introduce methods for exploiting, extracting, and analyzing these micro-Doppler signatures from ISAR data.

7.1 Feature Extraction from 2-D ISAR Imagery

ISAR image is a 2-D range-Doppler image. The range information is given by the signal's time delay from the target, and Doppler information is obtained from the relative rotation between the radar and the target. The relative motion between radar and targets is the most important condition for ISAR imaging. However, motion also brings defocusing and smearing in ISAR range-Doppler images. Thus, a variety of algorithms for producing well-focused range-Doppler images and the conditions for generating finely resolved images have been proposed as discussed in Chapters 3–5.

ISAR images are formed by processing sequences of radar returns from targets of interest. The range-compressed radar signals are usually arranged as a 2-D matrix of range profiles. The information about targets and clutter captured from reflected electromagnetic waves is embedded in the range profiles.

7.1.1 Rotation Estimation and ISAR Image Cross-Range Scaling

ISAR images are typically scaled in the range and Doppler coordinates. This hybrid representation of the ISAR image in the range-Doppler domain is a direct consequence of not knowing the radar-target geometry. Specifically, the target's effective rotation vector, which is the projection of the target's rotation vector onto the plane orthogonal to the radar line of sight (LOS), generates the inverse synthetic aperture and therefore defines the relationship between the Doppler frequency and the cross-range coordinate, as detailed in (2.67), and shown here for the sake of convenience:

$$f_D(t) = \frac{2f_c}{c}\Omega_{\text{eff}}(t)r_{cr}, \tag{7.1}$$

where Ω_{eff} is the effective rotation vector, f_c is the radar central frequency, and r_{cr} is the cross-range displacement. By assuming that the effective rotation vector is constant within a given observation time, the cross-range coordinate can be obtained by inverting (7.1):

$$r_{cr} = \frac{c}{2f_c\,\Omega_{\text{eff}}} f_D. \tag{7.2}$$

This equation is used to convert range-Doppler images into range-cross range images. Representing an ISAR image in homogeneous spatial coordinates allows distances between points belonging to the target to be measured and therefore allows important geometrical features, such as target's size, shape, and position of dominant scatterers to be extracted. Unfortunately, the value of Ω_{eff} is not known prior to the radar, and it must be estimated somehow. Unless external measurements or a priori information are available, such an estimation has to be carried out by simply using the received radar data.

In this subsection, we will detail a method for estimating the modulus of the effective rotation vector, namely, Ω_{eff}, to be able to use (7.2) to rescale images from the range-Doppler domain into the range-cross range domain.

7.1.2 Chirp Rate Method

This algorithm is based on the assumption of a quasi-constant effective rotation vector $\Omega_{\text{eff}}(t) = \Omega_{\text{eff}}$. When the target rotation vector can be assumed as constant within the coherent integration time, the chirp rate produced by the scattering centers can be related to the modulus of the effective rotation vector by means of a simple analytical expression. By considering (1.18–1.20) and by assuming that the CPI is not too long, the received signal phase term can be approximated by a second-order polynomial, as mathematically described in (6.14) and shown here for the sake of convenience:

$$\varphi_k(f,t) = -\frac{4\pi f}{c}\left(y_k + x_k\Omega_{\text{eff}}t - \frac{1}{2}y_k\Omega_{\text{eff}}^2 t^2\right), \tag{7.3}$$

where $\varphi_k(f, t)$ represents the phase time-frequency history relative to a generic k-th scatterer of the target with the target-fixed coordinates (x_k, y_k, z_k). As put forward in (7.3), each scattering center carries information about the modulus of the target rotation vector through its chirp rate:

$$\mu_k = \frac{2f}{c} y_k \Omega_{\text{eff}}^2. \tag{7.4}$$

Such a relationship linearly relates the range coordinate of a scattering center with its chirp rate. The coefficient of such a linear relationship contains the modulus of the effective rotation vector:

$$K = \frac{2f}{c}\Omega_{\text{eff}}^2. \tag{7.5}$$

By estimating the range position of each scatterer and the chirp rate that it generates when the target rotates with a given Ω_{eff}, a number of duplets $(\hat{y}_k, \hat{\mu}_k)$ with $k = 1,2,\ldots, N$

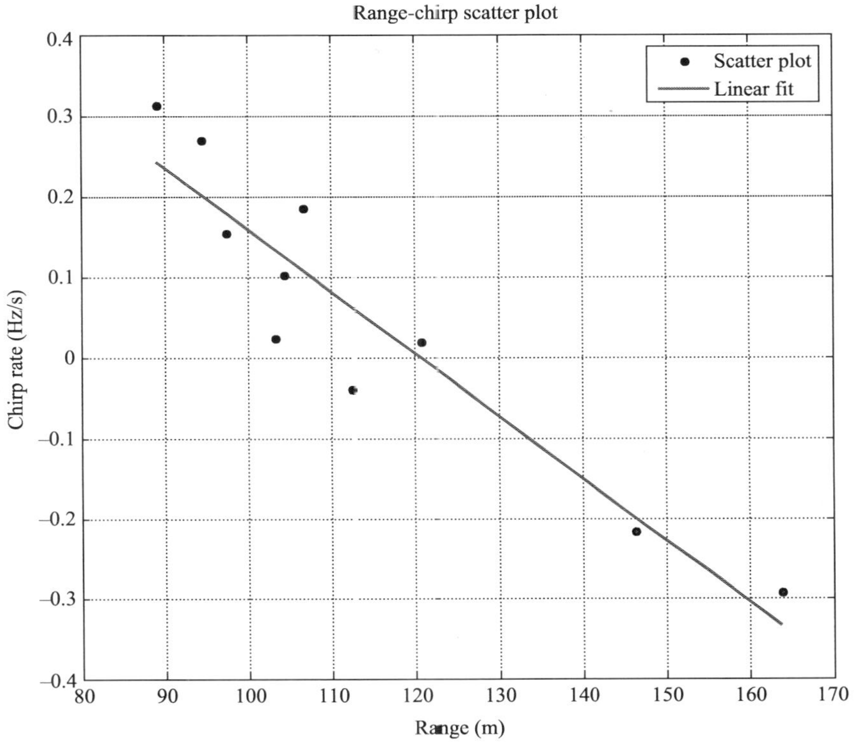

Figure 7.2 Range-chirp scatter plot.

can be generated. A linear regression line can then be calculated that produces an estimate of the factor $\hat{K}$ and therefore an estimate of the effective rotation vector:

$$\hat{\Omega}_{\text{eff}} = \sqrt{\frac{c\hat{K}}{2f}}. \tag{7.6}$$

An example of the scatter plot formed by a number of duplets is shown in Figure 7.2.

A crucial step of this method is to obtain accurate and robust estimates of the chirp rate associated with scatterers. Usually it is not easy to separate scattering centers from each other in ISAR data. An effective way was proposed in [11], where an ISAR image segmentation-based approach was used in conjunction with an ISAR image contrast optimization. The procedure of the cross-range scaling algorithm is recalled here and the flowchart shown in Figure 7.3.

1. The ISAR image is segmented to extract N subimages $I_T^{(n)}(r, \nu)$ relative to N dominant scattering centers. Their relative scattering center location, C_n, is also calculated.
2. Each subimage is inversely Fourier transformed from the Doppler domain into the time domain (i.e., azimuth decompression) to obtain N range-time signals $S_{MC}^{(n)}(r, t)$, with $n = 1, \ldots, N$.

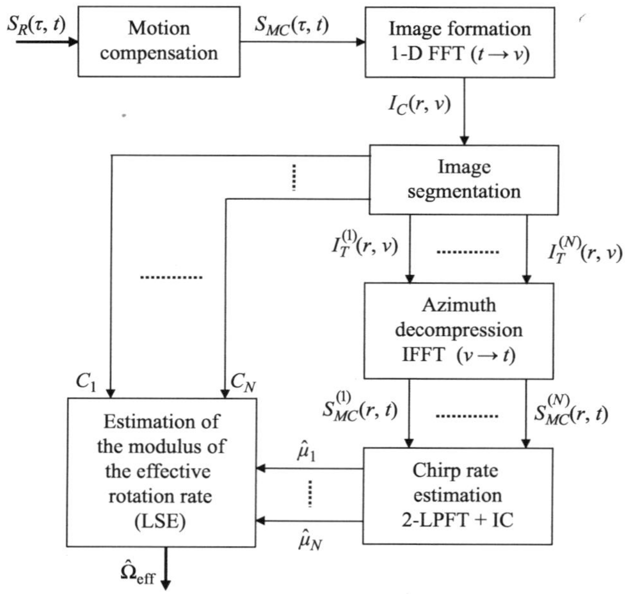

Figure 7.3 Cross-range scaling algorithm.

3. The N signals $S_{MC}^{(n)}(r, t)$ are then used to estimate each scatterer's chirp rate by maximizing the image contrast (IC) with respect to the chirp rate parameter. Specifically the chirp rate is estimated by maximizing the IC value of the image obtained by applying a second-order local polynomial Fourier transform (LPFT) to $S_{MC}^{(n)}(r, t)$.
4. The chirp rate, $\hat{\mu}_n$, and the scatterer's location, $C_n(n = 1, \ldots, N)$, obtained in the previous steps are used to estimate the modulus of the effective rotation vector, $\hat{\Omega}_{\text{eff}}$, via a least square error (LSE) approach.

All mathematical details are left out and can be found in [11].

7.1.3 Example of Using Chirp Rate Method

As an example, we will consider the real data of a Boeing 737. The ISAR image of such a target is shown in Figure 7.4a. Five dominant scatterers are then selected, and their range position and chirp rate are estimated and plotted in Figure 7.4b. A regression line is then calculated to allow for the estimation of the linear factor, K, and therefore the modulus of the effective rotation vector, Ω_{eff}. Once the effective rotation vector is estimated, the cross-range scaling operation can be performed using (7.2). The final cross-range scaled image is shown in Figure 7.5, where target's features such as length and wingspan can be measured.

7.2 Extraction of Micro-Doppler Features from ISAR Data

ISAR imagery is generated from the range profiles and displayed in the range and Doppler (cross-range) dimensions. ISAR image is represented by distributed scatterers with different

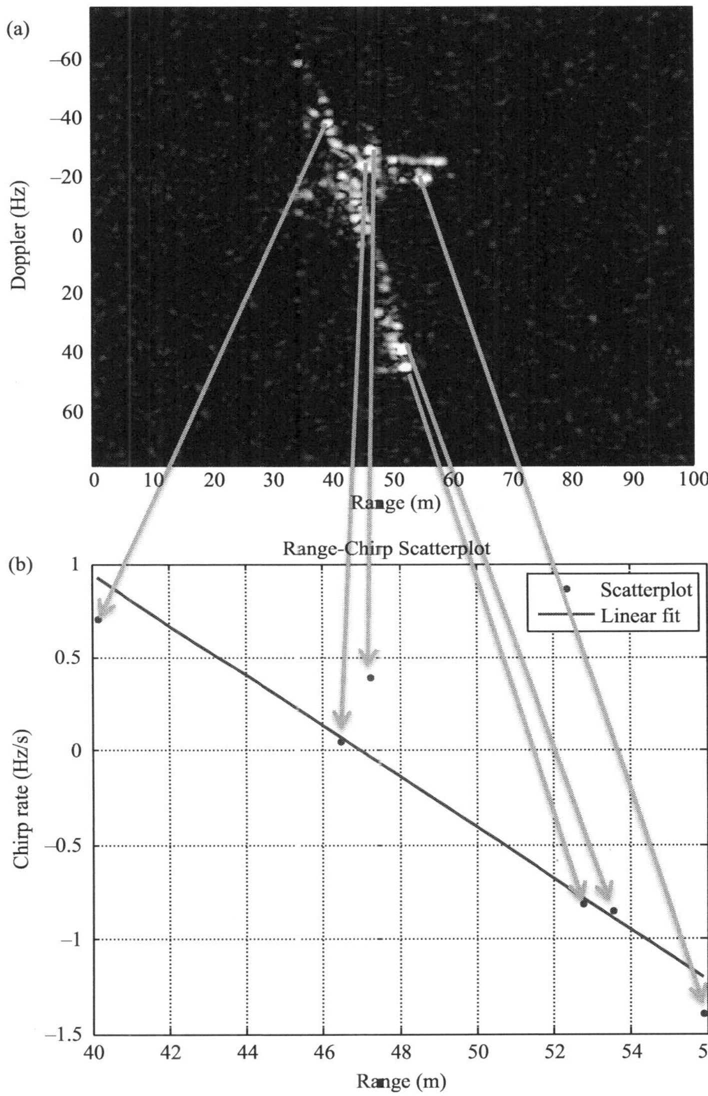

Figure 7.4 Effective rotation vector estimation example.

intensities corresponding to different reflectivity centers on the target. The Doppler frequency shift of the target geometric center indicates the radial velocity of the target translation. Each scatterer on the target has its own range and cross-range locations. Thus, from ISAR image of a target, we can identify different parts of the target along with their range and radial velocity information.

Figure 7.6a is two ISAR images of a walking person taken at two time instants. The radar is operating at X-band with 800 MHz bandwidth. Thus, the radar has 7.5-in high range

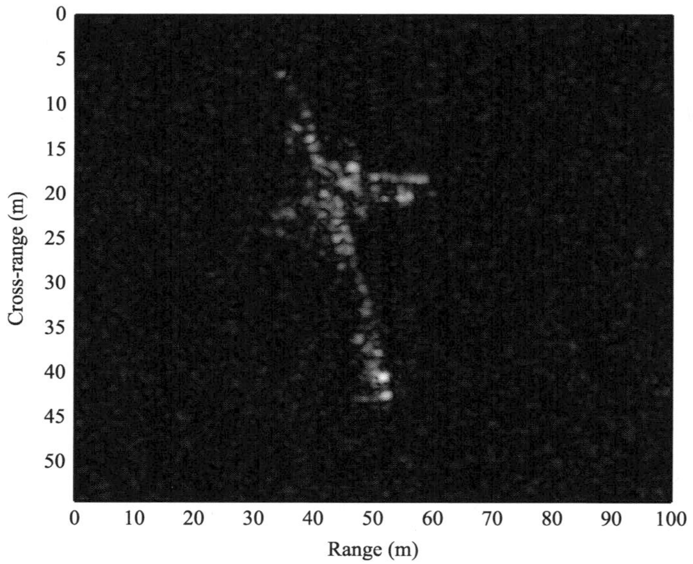

Figure 7.5 Cross-range scaled ISAR image.

resolution. The ISAR images are provided by ONERA, French Aerospace Laboratory from its X-band HYCAM radar data collection [12,13]. During the data collection, the HYCAM radar was at 17.7 m height in a building and looking down at 44.5° depression angle for a walking person as illustrated in Figure 7.6b. The image frame time is 0.1 sec.

The upper image of the walking person can be interpreted as in the double support phase, where both feet are on the ground as marked in the image. The lower image can be interpreted as in the balancing phase, where one foot is on the ground and the other is swinging. From the ISAR image in the balancing phase, the swinging leg is smeared in the velocity axis. Motion kinematic features of the walking person can be extracted from range profiles. Figure 7.6c is the micro-Doppler signature of the walking person, which can be used to extract kinematic features of the person.

7.2.1 Extraction of Target Identification Features from Micro-Doppler Signatures

Micro-Doppler signature is a signature of target's motion characteristic. The Doppler frequency shift is related to the radial motion velocity. If kinematic features of a target can be estimated, it is possible to reconstruct the target's movement from micro-Doppler signatures.

An example of showing how to extract features from micro-Doppler signature is helicopter rotor blades [10]. The geometry of the radar and a helicopter is depicted in Figure 7.7a. An ISAR image of the helicopter is shown in Figure 7.7b, where we can see that the fuselage and hub of the helicopter occupy a relatively small Doppler extent around the

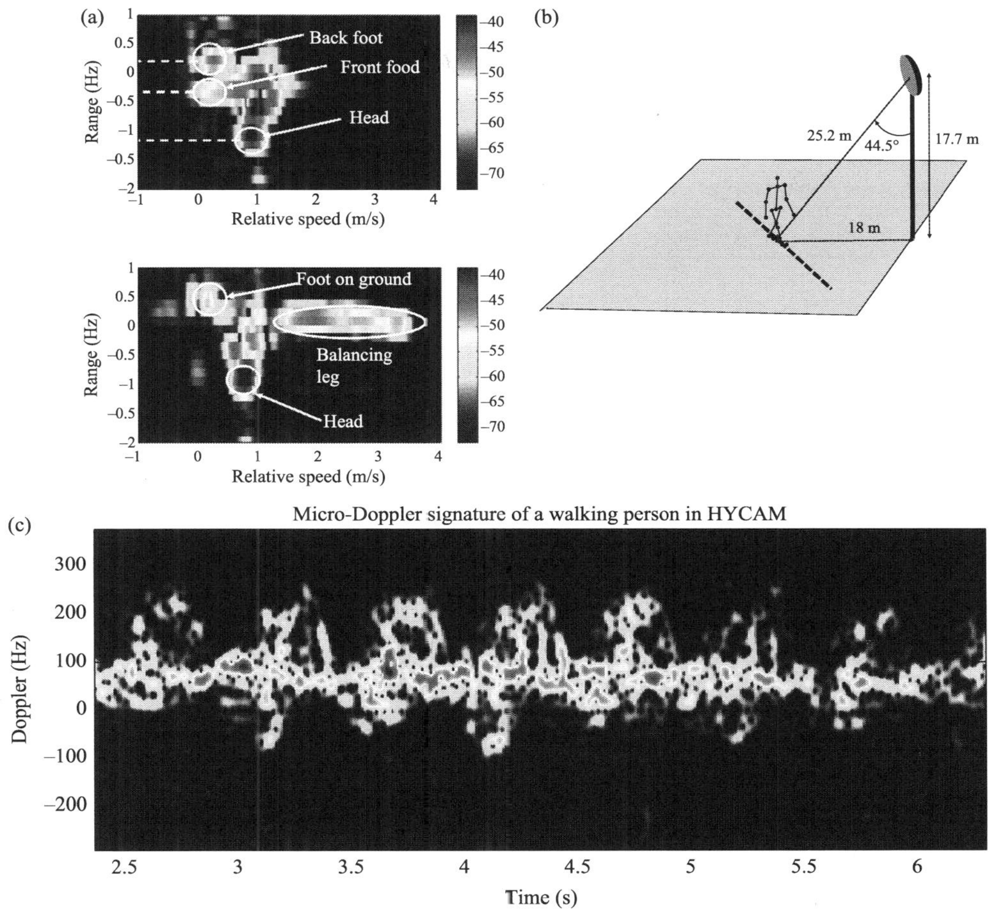

Figure 7.6 (a) ISAR images of a walking person; (b) radar and target geometry; (c) extracted micro-Doppler signature.

center of the Doppler axis while the fast-rotating rotor blades exhibit strong dotted line running across the Doppler dimension and overlapping with the body of the helicopter. Micro-Doppler signature of the dotted line uniquely represents the characteristics of the rotor blades. Rotation feature of rotor blades is an important feature for identifying helicopters of interest. Micro-Doppler signature generated by the rotating rotor blades is regarded as such a unique signature of helicopters.

Figure 7.8 shows the micro-Doppler signature of a two-blade rotor on a scale model helicopter measured by an X-band frequency modulated continuous wave (FMCW) radar with the wavelength $\lambda = 0.03$ m. Rotation rate of the rotor is about $\Omega = 2.33$ revolution/sec (r/s) or $\Omega = 2.33 \times 2\pi$ (rad/sec), and the blade length is $L = 0.2$ m. Thus, the tip velocity is $V_{tip} = 2\pi L\Omega = 2.93$ m/s, and the maximum Doppler shift is $\{f_D\}_{\max} = 195$ Hz. From the

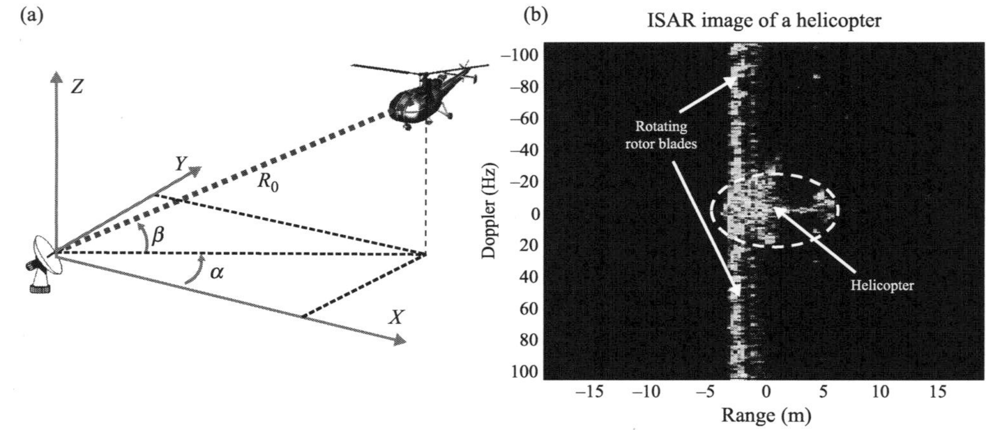

Figure 7.7 (a) Geometry of radar and a helicopter and (b) ISAR image of the helicopter.

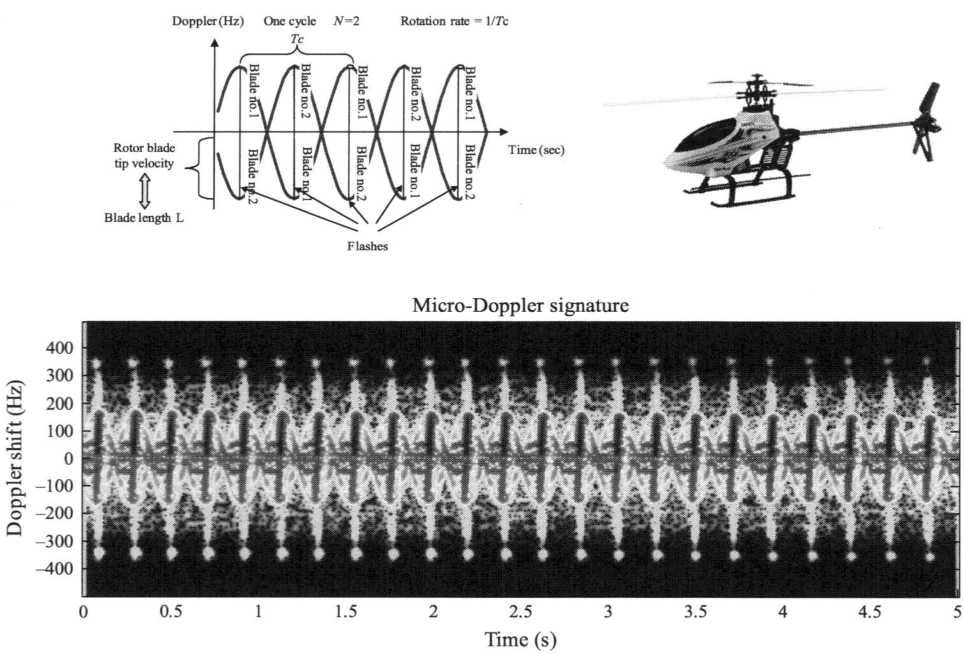

Figure 7.8 Micro-Doppler signature of a two-blade rotor on a scale model helicopter.

micro-Doppler signature, the number of blades, the length of the blade, and the rotation rate of the rotor can be estimated.

Micro-Doppler features of rotating rotor blades with two blades and three blades are depicted in Figure 7.9a and Figure 7.9b, respectively. It is obvious that the Doppler patterns

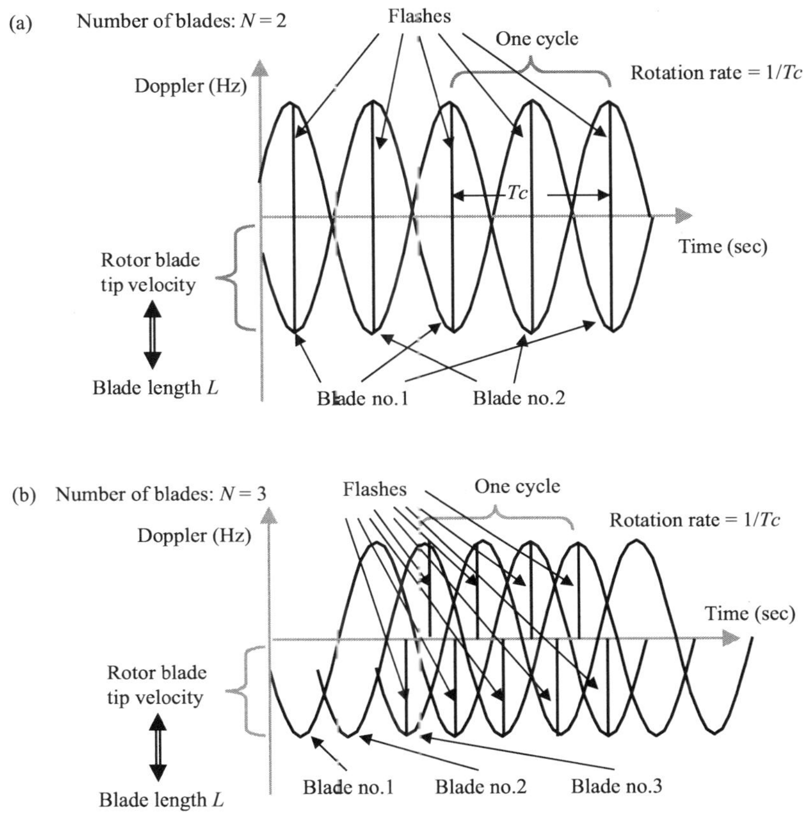

Figure 7.9 Micro-Doppler features of rotating rotor blades with (a) two blades and (b) three blades.

of an even number of blades and an odd number of blades are different. An even number of blades generates a symmetric Doppler pattern around the mean Doppler frequency, but an odd number of blades generates an asymmetric pattern around it. From the micro-Doppler signature of the rotor blades represented in the joint time-frequency domain, the number of blades, the length of blade, the rotation rate of the blades, and the speed of the tip can be estimated. These features are important for identification of an unknown helicopter.

7.2.2 Extraction of Motion Parameters from Micro-Doppler Signatures

Another example of showing how to extract motion parameters from micro-Doppler signature is a human walking model [10,14]. The most often motion performed by humans is walking. The extraction of motion features from a walking human is a basic method for the study of human motion.

Figure 7.10 shows a micro-Doppler signature of a walking person. Micro-Doppler components of the feet, tibias, clavicles, and torso are marked in the figure. The mean value of the Doppler frequency shift of the torso is about 133 Hz.

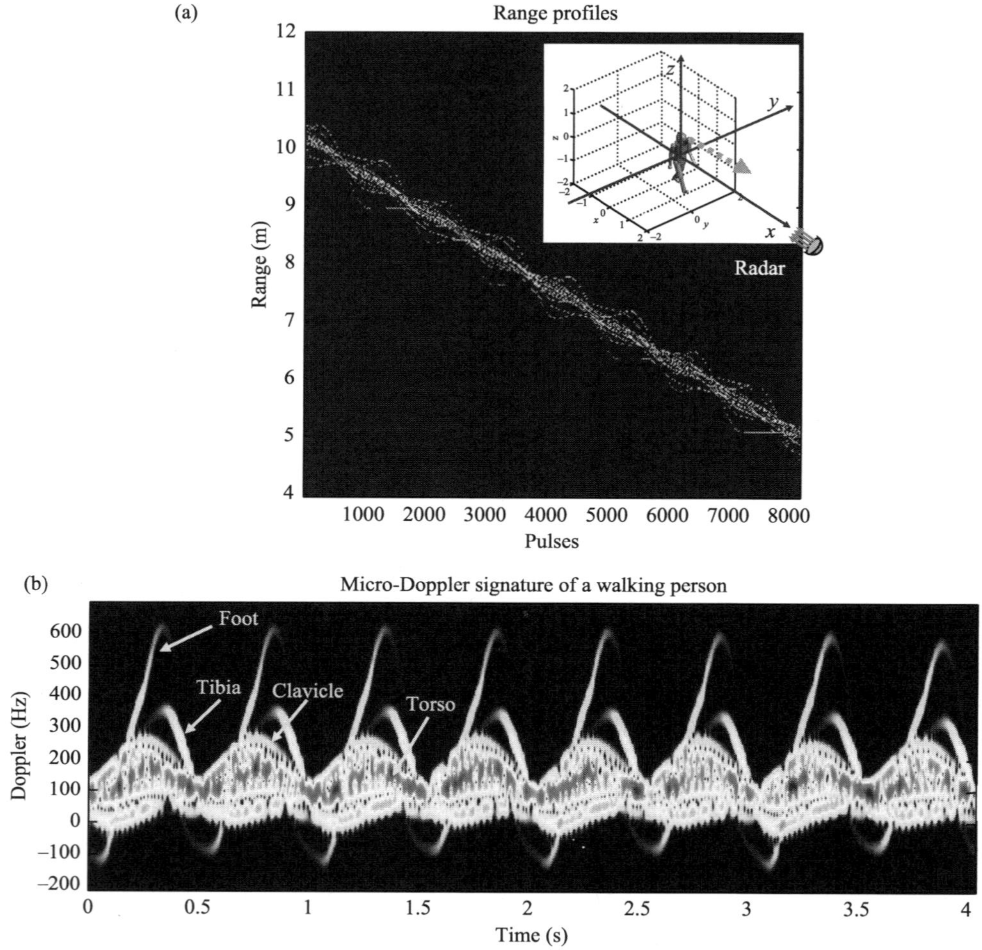

Figure 7.10 (a) Geometry of a human walking model and the range profiles; (b) micro-Doppler signature of the walking person.

Micro-Doppler signature of human movement shows strong reflections from the human torso due to its larger radar cross section (RCS). The decomposed micro-Doppler component of the torso can be used to measure the average torso velocity, the cycle of torso oscillation, and the amplitude of the torso Doppler oscillation. From the simulation result with a Ku-band radar operating at 15 GHz, by estimating peaks of the torso oscillation, the torso motion velocity is oscillating from 1.0 m/s and going up to 1.67 m/s (Figure 7.11). The mean value of the torso velocity is $V_{torso} = f_D\, \lambda/2 = 133 \times 0.02/2 = 1.33$ m/s. The oscillating period of the torso is 0.5 sec, or the cycle of the torso oscillation is 2 Hz.

The corresponding lower leg (tibia) motion parameters are shown in Figure 7.12. The average tibia velocity is $V_{tibia} = 129 \times 0.02/2 = 1.29$ m/s, the cycle of the tibia oscillation is 1 Hz, and the maximum amplitude of the tibia oscillation velocity is about 3.2 m/s.

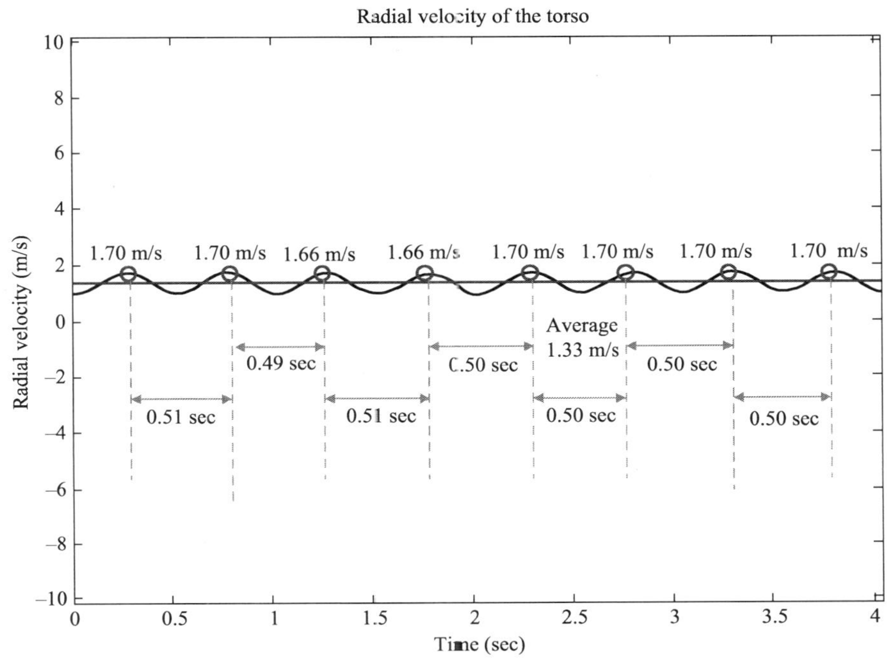

Figure 7.11 Torso velocity of a walking person.

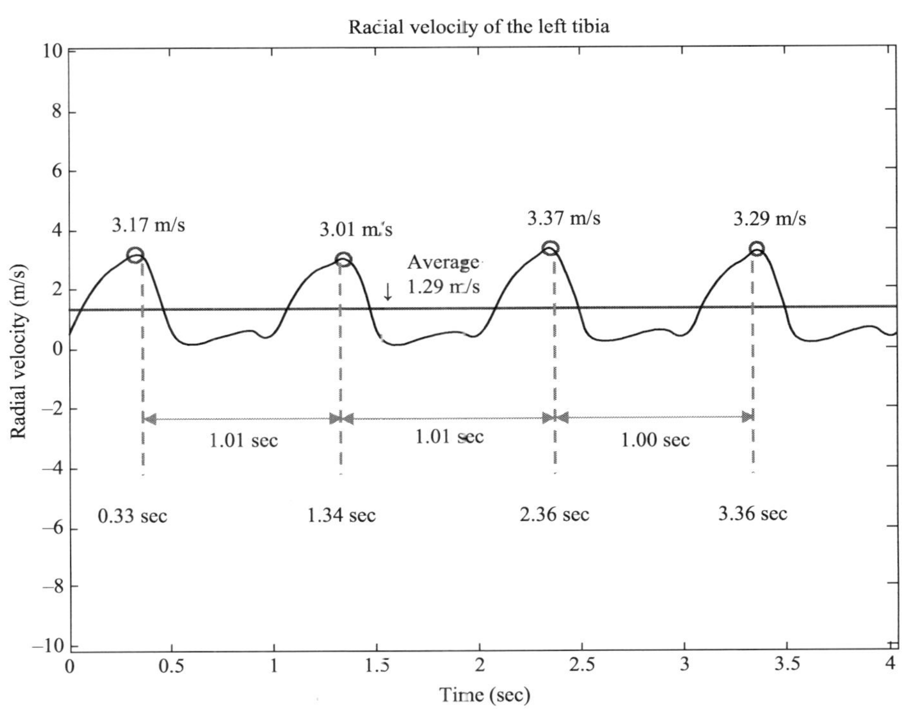

Figure 7.12 Tibia velocity of a walking person.

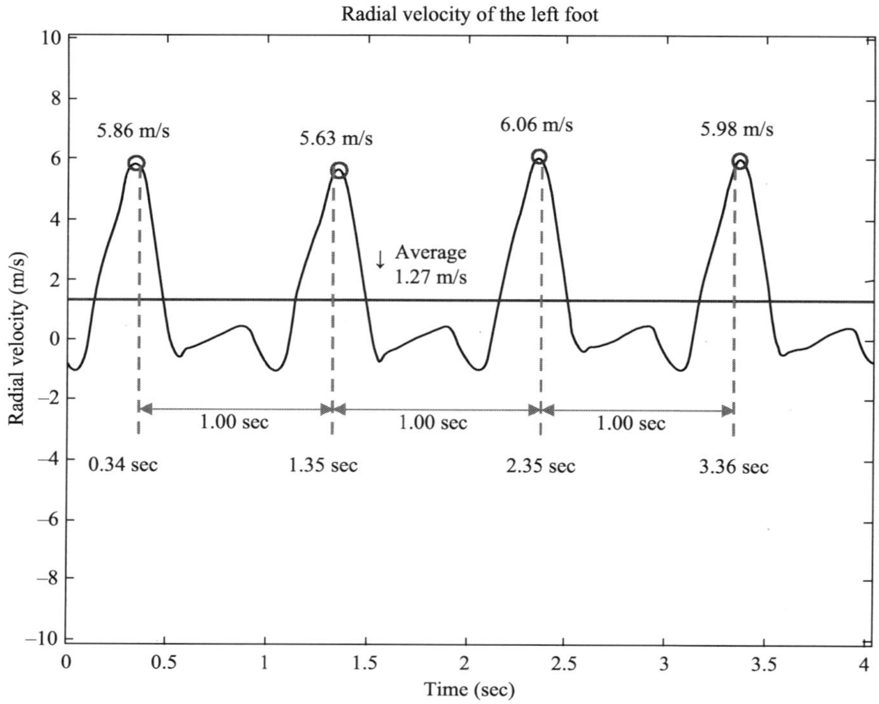

Figure 7.13 Left foot velocity of a walking person.

The corresponding foot motion parameters are shown in Figure 7.13. The average foot velocity $V_{foot} = 127 \times 0.02/2 = 1.27$ m/s, the cycle of the foot oscillation is 1 Hz, and the maximum amplitude of the foot oscillation velocity is about 5.9 m/s. Half the foot oscillation cycle is the foot-forward motion, and the other half cycle is the foot in contact with the ground. This makes the average velocity lower. The foot motion has the highest velocity, which is approximately four to five times the average foot velocity. The oscillation frequency of the torso is two times the tibia or the foot oscillation frequency because the torso accelerates while either foot swings.

7.3 Summary

ISAR imagery is quite different from optical images. Thus, feature extraction from 2-D radar imagery is more complicated. In this chapter, we introduced two distinct instances of feature extraction in ISAR: 2-D feature extraction from a normal ISAR range-Doppler (or cross-range) image; and target motion feature extraction from micro-Doppler signatures.

For 2-D feature extraction, image feature extraction in computer vision (e.g., the model-based method, morphology method, template matching, and machine learning) may be used. In Section 7.1, we introduced how to estimate the size and orientation of a target from its range and cross-range extents and how to estimate the target rotation parameters based on the estimated target dimension.

In Section 7.2, we introduced how to extract target motion features showing in ISAR images, which are extracted from ISAR range profiles instead of 2-D ISAR imagery. Time-varying micro-Doppler signatures provide information that allows the estimation of various kinematic features. We demonstrated these with helicopter rotor blades and simulated walking human data.

References

[1] Gundon, B., "The Development of Perceptual Grouping Operators to Aid in Extraction of Planimetric Features from High Resolution Satellite Images," *Proc. of ASPRS Annual Conference*, Portland, OR, pp. 257–260, 1999.

[2] Suetens, P., P. Fua, and A.J. Hanson, "Computational Strategies for Object Recognition," *ACM Computing Surveys*, vol. 24, no. 1, pp 5–61, 1992.

[3] Serra, J., "Introduction to Mathematical Morphology," *Computer Vision, Graphics and Image Processing*, vol. 35, no. 3, pp. 283–305, 1986.

[4] Destival, I., "Mathematical Morphology Applied to Remote Sensing," *Acta Astronautica*, vol. 13, no. 6–7, pp. 371–385, 1986.

[5] Dong, P., "Implementation of Mathematical Morphological Operations for Spatial Data Processing," *Computer and Geosciences*, vol. 23, no. 1, pp. 103–107, 1997.

[6] Karnielli, A., A. Meisels, L. Fisher, and Y. Arkin, "Automatic Extraction and Evaluation of Geological Linear Features from Digital Remote Sensing Data Using Hough Transform," *Photogrammetric Engineering & Remote Sensing*, vol. 62, no. 5, pp. 525–531, 1996.

[7] Hsu, S.M., " Extracting Target Features from Angle-Angle and Range-Doppler Images," *Lincoln Laboratory Journal*, vol. 6, no. 1, pp. 213–227, 1993.

[8] Chen, V.C., W.J. Miceli, and B. Himed, "Micro-Doppler Analysis in ISAR—Review and Perspectives," IEEE 2009 International Radar Conference, Oct. 12–16, Bordeaux, France, 2009.

[9] Chen, V.C., F. Li, S. Ho, and H. Wechsler, "Micro-Doppler Effect in Radar—Phenomenon, Model, and Simulation Study," *IEEE Trans. AES*, vol. 42, no.1, pp. 2–21, 2002.

[10] Chen, V.C., *Micro-Doppler Effect in Radar*, Artech House, February 2011.

[11] Martorella, M., "A Novel Approach for ISAR Image Cross-Range Scaling," *IEEE Trans. AES*, vol. 44, no. 1, pp. 281–294, 2008.

[12] Vignaud, L., A. Ghaleb, J. Le Kernec, and J.-M. Nicolas, "Radar High Resolution Range and Micro-Doppler Analysis of Human Motions," *Proc. of 2009 IEEE International Radar Conference*, 2009.

[13] Ghaleb, A., L. Vignaud, and J.M. Nicolas, "Micro-Doppler Analysis of Wheel and Pedestrians in ISAR Imaging," *IET Signal Processing*, vol. 2, no. 3, pp. 301–311, 2008.

[14] Vaughan, C.L., B.L. Davis, and J.C. O'Connor, *Dynamics of Human Gait*, 2d ed., Kiboho Publishers, 1999.

CHAPTER 8

ISAR Imaging for Refocusing Moving Targets in SAR Images

In this chapter we will introduce the problem of refocusing synthetic aperture radar (SAR) images of moving targets by treating the problem as an ISAR one.

SAR processors are designed to form radar images at very high resolution. Nevertheless, they are based on the assumption that the illuminated area is static during the synthetic aperture formation [1]. As a consequence of such an assumption, the existing SAR image formation algorithms are unable to focus moving targets and leading to blurred and displaced images of an object that is not static during the synthetic aperture formation. In the existing SAR literature, many attempts have been made to compensate for the target's motion and therefore to form focused images even in the presence of moving targets [2,3]. Nevertheless, such attempts are based on some assumptions on the target's motion that limits the effectiveness of such algorithms to some extent.

On the other hand, as detailed in the first chapters of this book, inverse synthetic aperture radar (ISAR) techniques do not base their functioning on the assumption that the target is static during the synthetic aperture formation. Instead, they exploit, at least partly, the target's own motions to form the synthetic aperture. Although ISAR techniques do not make use of a priori information about the target's motion, some other constraints apply to the ISAR image formation. Such constraints may include the image size, the achievable cross-range resolution, and the fact that the imaging system performance is not entirely predictable. Nevertheless, despite such constraints, ISAR imaging provides a more robust and flexible solution to cases of targets undergoing complex motions, such as pitching, rolling, and yawing ships. A functional block scheme is represented in Figure 8.1 that aims at describing a detection and moving target refocusing system [4].

As shown in the system depicted in Figure 8.1, targets are detected directly in the SAR image domain. A sub-image cut around the target is then selected and used to form a refocused image of the target using ISAR processing.

A number of techniques have been proposed in the literature to detect targets in SAR images [5–9]. Two scenarios should be treated differently if effective target detection must be attained and are relative to the detection of maritime and ground targets. They should be considered separately because of the intensity of sea or ground (e.g., soil, vegetation, urban areas) clutter present in the SAR image. Ground clutter is typically one or more order of

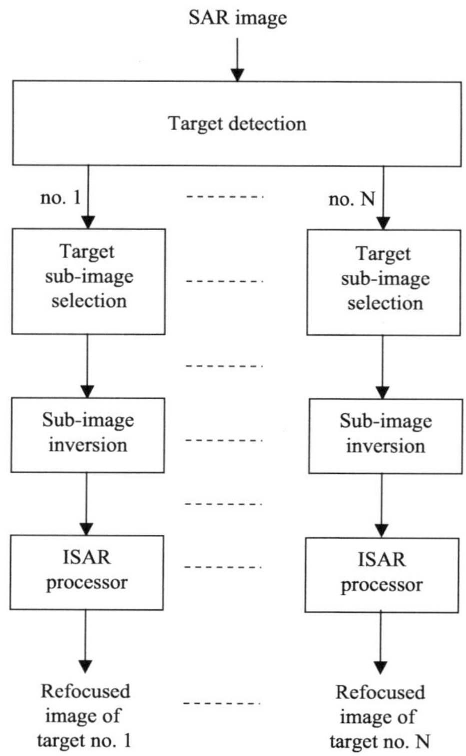

Figure 8.1 Moving target refocus scheme.

magnitude stronger than sea clutter, depending on the type of ground clutter and on the sea conditions/roughness. Another distinction should be made in terms of stationary or moving target detection. In the ground target case, stationary targets are typically detected by using change detection techniques [9–13], whereas moving target may be detected using the ground moving target indication (GMTI) [14–16]. In both cases, multiple passes or multiple channels should be used, both of which increase the system complexity or cost. In the maritime target case, stationary targets can be detected in single-channel SAR images, provided that a reasonable signal-to-clutter ratio (SCR) is present. In the case of maritime moving targets, the smearing effect due to the target motion may inhibit the detection performance, although a sufficiently high SCR may be retained.

In this chapter, we will skip details related to the detection step, which has been amply discussed in several papers and books [5–16], and we will aim to provide technical details about the refocusing process. Specifically, four types of image inversion mapping algorithms are explored: Fourier transform (FT), polar formatting (PF), chirp scaling (CS), and the Ω-k (OK) algorithms. This is not an exhaustive list of all available SAR image formation algorithms. Other equally important algorithms, such as back-projection [17] and the range migration algorithm [18], have been left out because a review of SAR image formation algorithms is not in the scope of this chapter. Nevertheless, in principle, all SAR image formation algorithms may be inverted exactly or not exactly via ad hoc methods, such as those devised in this chapter. An ISAR processor is then used to form well-focused images of the moving target. In particular, the autofocus, the image formation, the time window selection, and the cross-range scaling problems will be considered as integral parts of the ISAR image refocusing problem.

Before detailing the moving target refocusing technique, some commonly used SAR processing algorithms will be reviewed. This will be important to define data inversion techniques that will be producing filtered data in the wavenumber domain to be used as input to the ISAR processor.

8.1 Review of Spotlight SAR Algorithms

In this section we will recall some of the most commonly used spotlight SAR image formation algorithms, and we will introduce the notation that will be used to define the inversion algorithms. The choice of spotlight SAR algorithms is due to the interest in obtaining very high-resolution images. In addition, spotlight SAR algorithms may be used to form high-resolution side-looking SAR images, whereas low-resolution SAR imaging algorithms cannot be used to produce well-focused SAR images in the spotlight configuration. Specifically, the Ω-k, the PF, and the CS algorithms will be recalled. A common SAR geometry is represented in Figure 8.2 and is used as a reference for the SAR formation algorithms covered here. In Figure 8.2, the radar moves along the cross-range coordinate x with velocity v, and r represents the slant-range coordinate, $\theta_c(t)$ and $\theta_n(t)$ are the aspect angles of the center of the scene and of the n-th scatterer, respectively. L denotes the synthetic aperture length.

8.1.1 Ω-k Algorithm

SAR image reconstruction algorithms typically make use of a plane-wave approximation. However, in some applications the wavefront curvature must be taken into account. The Ω-k

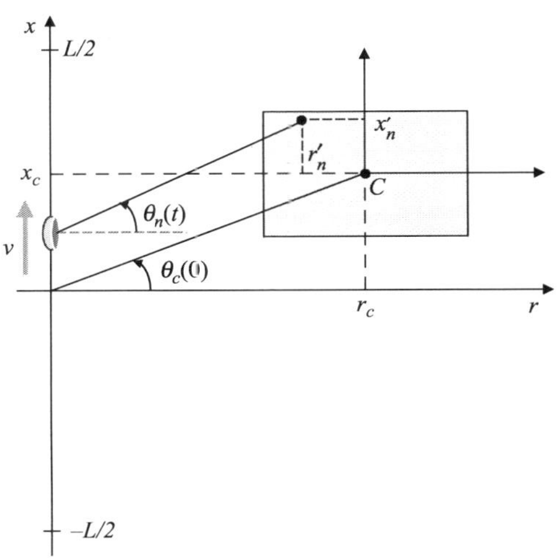

Figure 8.2 Spotlight SAR geometry.

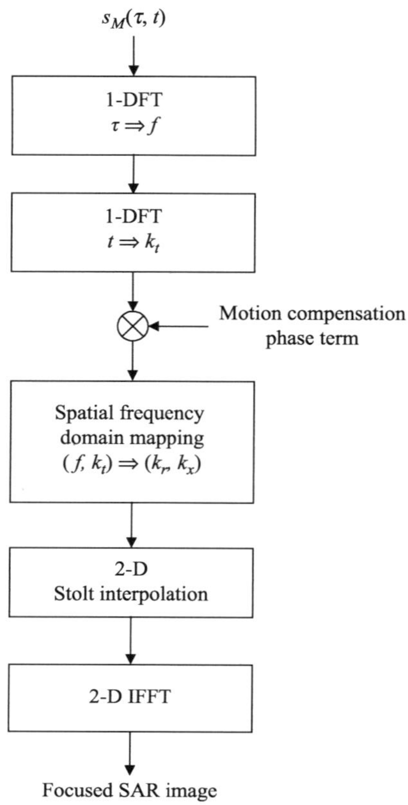

Figure 8.3 Flowchart of the Ω-k algorithm.

algorithm, also known as the wavenumber algorithm, improves the resolution by accurately modeling the actual spherical wave. Depending on the imaging parameters, a very accurate interpolation must be performed to avoid an artifact in the resulting image [1]. The flowchart of the Ω-k algorithm is shown in Figure 8.3, where $s_M(\tau, t)$ is the received signal at the output of the matched filter, τ is the fast time variable, t is the slow time variable, and $k = 2\pi f/c$ is the wavenumber.

Since the spatial frequency domain defined by the variable pair (k_r, k_x) is irregularly sampled, a two-dimensional (2-D) interpolation is needed prior to applying a 2-D inverse fast Fourier transform (IFFT).

Assuming that the target can be considered to be composed of a superposition of N_S independent and ideal scatterers, the received signal can be written as follows:

$$\begin{aligned} s_R(\tau, t) &= \sum_{n=1}^{N_s} \sigma_n p(\tau - \tau_n) \mathrm{rect}\left(\frac{t}{T_{obs}}\right) \\ &= \sum_{n=1}^{N_s} \sigma_n p\left[\tau - \frac{2}{c}\sqrt{(r'_n + r_c)^2 + (x'_n + x_c - vt)^2}\right] \mathrm{rect}\left(\frac{t}{T_{obs}}\right), \end{aligned} \tag{8.1}$$

where (x_c, r_c) are the cross-range and slant-range coordinates of the center of the scene, (x'_n, r'_n) are the cross-range and slant-range coordinates of a generic scatterer in the body-fixed coordinate system with its origin at the center of the scene, $p(t)$ is the transmitted pulse, T_{obs} is the observation time, and σ_n is the complex reflectivity of the n-th scatterer. It should be noted that in (8.1) the expression is written by considering the slow time as a continuous variable t. In the case of pulsed radar, such a domain is not continuous, and the variable t should be replaced with a discrete variable mT. Anyway, to simplify the following mathematical passages, a continuous form is kept. It should also be added that, when no Doppler ambiguity is present or the pulse repetition frequency (PRF) is sufficiently high, the use of the continuous slow-time notation does not lead to any difference with respect to the discrete notation.

The Ω-k algorithm can be summarized by means of the following steps:

1. Fourier transform with respect to the fast time variable τ after matched filtering the received signal:

$$S_{MF}(f,t) = |P(f)|^2 \sum_{n=1}^{N_s} \sigma_n \exp\left\{-j\frac{4f}{c}\sqrt{(r'_n + r_c)^2 + (x'_n + x_c - vt)^2}\right\} \text{rect}\left(\frac{t}{T_{obs}}\right). \quad (8.2)$$

2. Fourier transform with respect to the slow-time variable t, by applying the stationary phase method (SPM) [19–21]:

$$\begin{aligned} S_{MF}(f,k_t) &= FT\{S_{MF}(f,t)\} \\ &= |P(f)|^2 \sum_{n=1}^{Ns} \sigma_n \exp\left\{-j\left[\sqrt{\frac{16f^2}{c} - k_t^2} \cdot (r'_n - r_c) + k_t(x'_n - x_c)\right]\right\} I_n(f,k_t), \end{aligned} \quad (8.3)$$

where

$$I_n(f,k_t) = \begin{cases} 1 & k_t \in \left\{-\frac{4f}{c}\sin\left[\vartheta_n\left(-\frac{L}{2}\right)\right], \frac{4f}{c}\sin\left[\theta_n\left(-\frac{L}{2}\right)\right]\right\} \\ 0 & \text{otherwise} \end{cases}. \quad (8.4)$$

3. Motion compensation of the radar platform. This step is performed by multiplying $S_{MF}(f, k_t)$ with the focusing function

$$S_{MF}(f,k_t) = \exp\left\{-j\left(\sqrt{\frac{16f^2}{c} - k_t^2} \cdot r_c + k_t x_c\right)\right\}. \quad (8.5)$$

4. Coordinates transformation

$$\begin{cases} k_r = \sqrt{\frac{16f^2}{c} - k_t^2}. \\ k_x = k_t. \end{cases} \quad (8.6)$$

5. Stolt interpolation to obtain a regularly sampled rectangular grid.
6. 2-D Fourier transform to obtain an estimate of the reflectivity function in a spatial grid.

8.1.2 Polar Formatting Algorithm

The polar formatting algorithm (PFA) is well-known for processing stripmap and spotlight mode SAR data. This algorithm involves the assumption that the spherical wavefront of the radar signal can be approximated by planar wavefront around a central reference point in the imaged scene. The surface on which data are focused is also assumed planar when compensating for platform motion. These assumptions are quite correct for small scenes. Conversely, for large scenes, conversely, such approximations lead to errors in parts of the scene far away from the central reference point. Many algorithms have been proposed in the existing literature to compensate for the phase errors due to the wavefront curvature, such as that proposed in [22]. The plane wave assumption, typically used when formulating the PFA, limits not only the scene size but also the squint angle. However, in this chapter only traditional polar formatting is considered. It must be remarked, though, that individual targets occupy a very small area compared with the entire SAR image scene area. Therefore, errors introduced by the use of the PFA do not affect the target image when a local image refocus is carried out using ISAR imaging techniques.

Under the assumption that the scene size is smaller than the distance between the radar and the scene center, the PFA can be correctly used. With reference to the flowchart in Figure 8.4, the algorithm steps are summarized as follows:

1. Fourier transform (8.1) with respect to the fast time variable, τ, and, therefore, obtaining the result in (8.2).

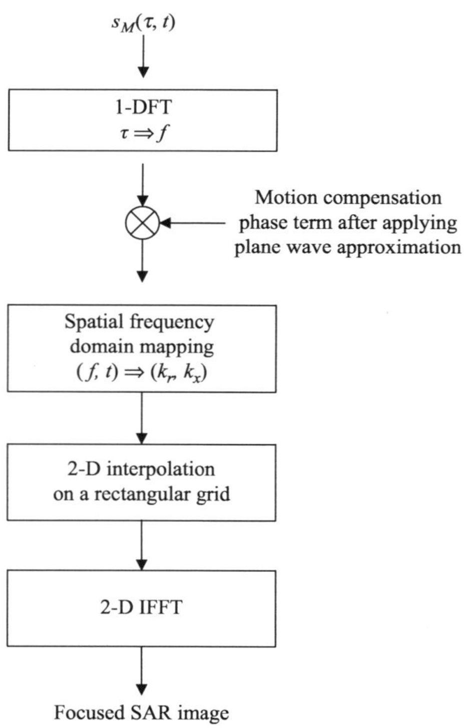

Figure 8.4 Polar format algorithm flowchart.

2. Radar platform motion compensation. The motion compensated signal can be obtained by multiplying $S_{MF}(f, k_t)$ with the focusing function

$$S_c(f,t) = \exp\left\{j\frac{4f}{c}\sqrt{r_c^2 + (x_c - vt)^2}\right\}. \tag{8.7}$$

3. Coordinates transformation. Under the straight is orange approximation the coordinates transformation is defined as follows, where $\theta_c(t)$ is the aspect angle with respect to the center of the scene:

$$\begin{cases} k_r = \dfrac{4f}{c}\cos\theta(t) \\ k_x = \dfrac{4f}{c}\sin\theta(t) \end{cases}. \tag{8.8}$$

4. 2-D interpolation to obtain a regularly sampled rectangular grid (to allow for 2-D FFT to be used to calculate a 2-D FT).
5. 2-D FFT to obtain the estimate of the reflectivity function.

8.1.3 Chirp Scaling Algorithm

The chirp scaling algorithm (CSA) has been proposed for high-quality SAR processing. This algorithm avoids any interpolation in the SAR processing chain and has been found suitable for the high-quality processing of several spaceborne SAR systems (e.g., SEASAT, ERS-1, RADARSAT). The CSA basically consists of multiplying the SAR data in the range-Doppler domain by a quadratic phase function (or called the chirp scaling) to equalize the range cell migration for a reference range, followed by an azimuth and range compression in the wavenumber domain. After transforming the signal back to the range-Doppler domain, a residual phase correction is carried out. Finally, azimuth IFFT is performed to generate a focused image. The chirp scaling multiplier or range perturbation function (RPF) is exactly a linear frequency modulation (FM; i.e., a quadratic function of range) when the following conditions are satisfied:

1. The radar pulse is linear frequency modulated.
2. The azimuth FM rate parameter is range invariant.

It should be remarked that in some cases the second assumption is not satisfied, such as in the cases of high squint angle and large swath widths.

Furthermore, the azimuth frequency variable should be centered on the Doppler centroid frequency at each range cell. The classical implementation of the CSA algorithm relies on the assumption that the Doppler centroid does not change from range cell to range cell. However, when the look angle variation is large (large swaths), the Doppler centroid can vary significantly from near to far range, and therefore it should be accounted for. An extended chirp scaling algorithm (ECSA) was developed to account for strong motion errors and variable Doppler centroids in range and also in azimuth [23–25]. For the sake of simplicity, only the classical CS algorithm is considered in this chapter.

Let the transmitted signal, $s(t)$, be a chirp signal (see Chapter 2, Section 2.2.1), the received signal after demodulation can be written as follows:

$$s_R(\tau,t)=\sum_{n=1}^{N_s}\sigma_n\exp\left[-j\frac{4f}{c}r_n(t)\right]\exp\left[j\pi\mu\left(\tau-\frac{2r_n(t)}{c}\right)^2\right]\cdot w_r\left[\tau-\frac{2r_n(t)}{c}\right], \tag{8.9}$$

where

$$r_n(t)=\sqrt{\left(r_n'+r_c\right)^2+\left(x_n'+x_c-vt\right)^2} \tag{8.10}$$

is the range history of the n-th scatterer within the observation time duration, μ is the chirp rate, $w_r\left[\tau-\frac{2r_n(t)}{c}\right]$ is the chirp envelope function centered at the round-trip time delay of the n-th scatterer, and N_S is the number of ideal scatterers. With reference to Figure 8.5, the CSA can be summarized by means of the following steps:

1. Azimuth Fourier transform and equalization of the range cell migration for each range cell. Because the range migration depends on the range, the first step of the algorithm

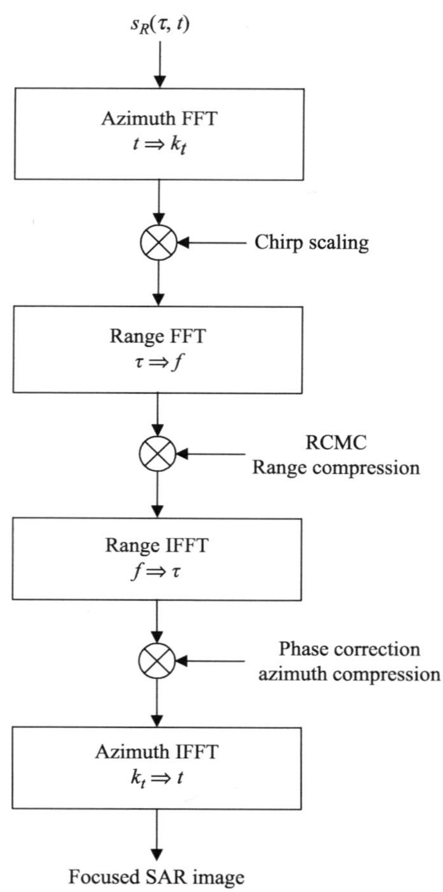

Figure 8.5 Chirp scaling algorithm flowchart.

consists of a range migration equalization. This operation aims at making the range migration look the same for all range cells, thereby eliminating the range migration dependence on range.

2. Range Fourier transform, range compression and range cell migration correction (RCMC).
3. Range inverse Fourier transform.
4. Removal of the residual phase term and azimuth compression.
5. Azimuth inverse Fourier transform.

8.2 Projection of SAR Image onto Wavenumber Domain—Inversion Mapping

ISAR processors accept raw data at their input and not SAR images. Since a SAR image is obtained by processing a very large amount of data, which may contain several targets, each one with its own motions, subimages containing separate targets have to be cut from the entire SAR image and treated separately. The solution discussed in this chapter consists of inverting each SAR subimage to obtain equivalent raw data, which contains only the target's echo with some residual background clutter/noise. Such raw-like data are then used as the input of an ISAR processor, which produces well-focused ISAR images. This concept is depicted in Figure 8.1. In this section the problem of obtaining raw-like data from a SAR subimage is addressed. Specifically, four inversion techniques are proposed and analyzed: the inverse Ω-k (IOK), the inverse PFA (IPFA), the inverse CS (ICS), and the inverse range-Doppler (IRD). To obtain equivalent raw data from SAR images, a wavenumber domain equivalent to the original one must be defined for each SAR subimage. As the resolution of the whole SAR image and that of the subimage are the same, the observation time and the bandwidth must be the same. On the other hand, the PRF and the frequency spacing in the equivalent raw data are different and can be calculated as follows:

$$\begin{cases} PRF = N/T_{obs} \\ \Delta f = B/M \end{cases}, \tag{8.11}$$

where N is the number of samples in the cross-range direction, and M is the number of samples in the range direction of each subimage.

8.2.1 Inverse Ω-k Algorithm

The Ω-k algorithm makes use of an interpolation to obtain very high-resolution SAR images. Unfortunately, such an operation on the data makes the Ω-k algorithm not perfectly invertible, which means that some errors (or artifacts) are introduced in the data after attempting any type of inversion. The IOK algorithm is obtained by inverting each single step of the direct algorithm, as shown in Figure 8.6 [26]. Since all steps in the direct algorithm are invertible except for the interpolator, we will expect that artifacts may be introduced by the interpolation inversion, which is implemented by means of an additional interpolator. Additional comments on the nonperfect inversion of the Ω-k algorithm can be found at the end of this chapter. An additional interpolator is necessary to remap the rectangular domain

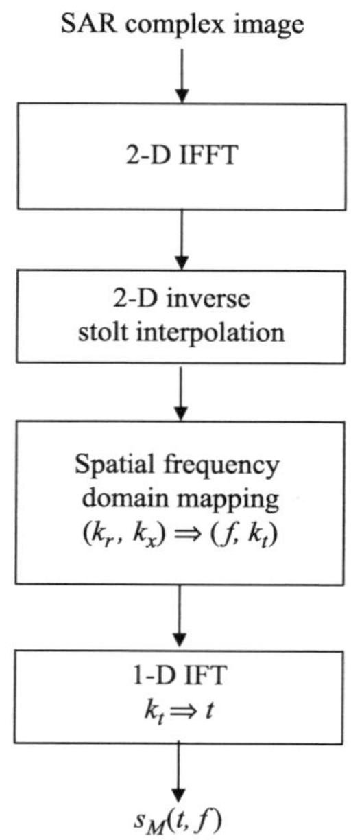

Figure 8.6 Inverse Ω-k algorithm.

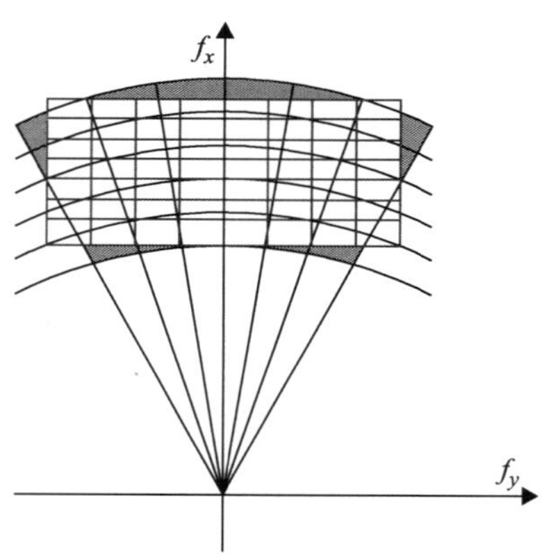

Figure 8.7 Illustration of the sample grids in the PFA.

onto the polar domain. This step involves a loss of information because some regions of the circular sector cannot be calculated using an interpolator (data in the gray region shown in Figure 8.7). Although an extrapolator could be used to estimate the data in those regions, some introduced artifacts may degrade the image quality. There is also no need to reinsert the motion compensation phase term because the subsequent ISAR processing can take care of the residual motion compensation needed to refocus moving targets.

8.2.2 Polar Formatting Inversion

The PFA uses the assumption that radar pulses spherical wavefronts can be assumed planar around a reference point in the scene [27]. This assumption is quite satisfied when the swath size is much smaller than the distance between the radar and the focusing point. The most relevant drawback of the PFA is that it requires an interpolation step that is typically time-consuming. The IPFA can be obtained by tracing back the steps used to form the image via the PF algorithm. Specifically, a 2-D IFFT is applied to the SAR subimage followed by a 2-D inverse Stolt interpolation and a spatial frequency domain mapping. In the IPFA flowchart (Figure 8.8), the steps involved in the polar format algorithm are run in the opposite order (from the bottom to the top). The same consideration regarding the reinsertion of the motion compensation phase term applies in this case as in the case of the Ω-k algorithm.

8.2.3 Chirp Scaling Inversion

Contrasted with the PFA, the CSA does not need an interpolation step, which results in a more computationally efficient algorithm. The CSA addresses the problem of equalizing the range migration of all the point scatterers that compose the target. Since all the scatterers follow the same trajectory, they can be compensated for by a known phase term (details can be found in [28]). A flowchart of the inverse chirp scaling algorithm (ICSA) is shown in Figure 8.9, where the steps involved in the chirp scaling algorithm are run in the opposite order. It should also be noted that the inversion of the RCMC step does not take place, as the subsequent ISAR processing will compensate only the residual motion added in by moving targets. This is perfectly equivalent to the case of the Ω-k and PF algorithms.

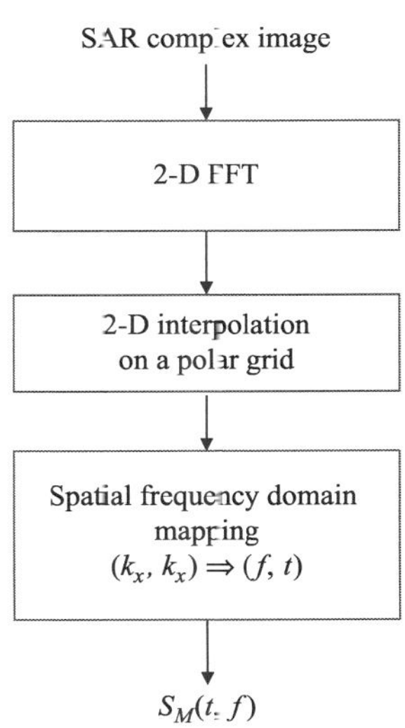

Figure 8.8 Inverse polar format algorithm.

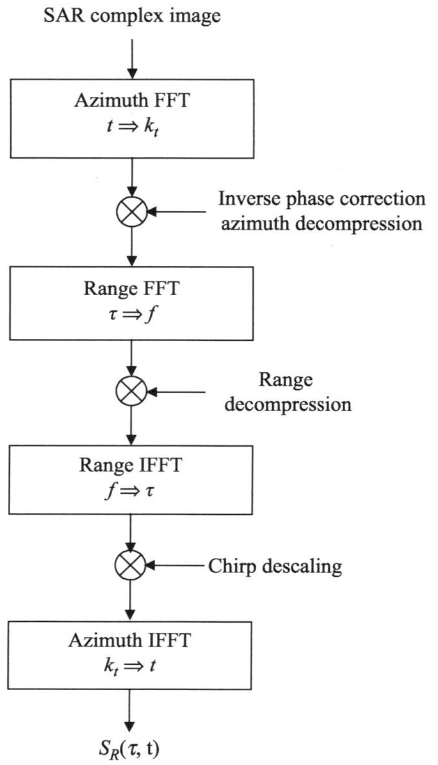

Figure 8.9 A block diagram of the inverse chirp scaling algorithm.

8.2.4 Fourier Inversion

The Range-Doppler technique represents an accurate and computationally effective tool for SAR/ISAR image reconstruction when the total aspect angle variation is not too large and when the effective rotation vector is sufficiently constant during the observation time. Under this constraint, the polar grid in the spatial frequency domain can be assumed to be an almost regularly sampled rectangular grid. Therefore, a 2-D FFT can be used to reconstruct the image. In this case, the inversion algorithm, namely, inverse range-Doppler (IRD), consists of a Fourier inversion, which is usually implemented via a 2-D IFFT.

8.3 Examples

The results provided in this section are obtained by applying the proposed technique to Cosmo Skymed (CSK) spotlight SAR data. These types of images used here have been formed using the Ω-k algorithm. Since the inverse Ω-k has a higher computational cost with respect to the Fourier inversion algorithm, it becomes important to evaluate the benefits of applying the former to justify its use. To this purpose, a subimage containing the target is back-transformed into the data domain by means of both the inverse Ω-k and the Fourier inversion algorithm. A comparison analysis between the two ISAR images is then carried out by considering the visual image quality and the image contrast (IC). As previously stated, for long recorded data the target rotation vector cannot be considered constant. It is therefore necessary to select one or more shorter integration time intervals to obtain one or more focused ISAR images of the target. To compare the results obtained using the inverse Fourier algorithm with the results obtained using the correct inverse SAR algorithm, it is necessary to compare the IC values of a large set of ISAR images.

8.3.1 ISAR Processing

The ISAR processor used to refocus SAR subimages of moving targets is based on a range-Doppler (RD) approach, as defined in Chapter 1.

To compensate for the residual target's motion, autofocus is needed. In fact, the platform motion can be considered already compensated for by the SAR processor to form the focused SAR image. From a geometrical point of view, the residual target motion is the relative motion of the target with respect to the SAR center scene, which is supposed to be stationary after the SAR motion compensation. The autofocus algorithm used to remove such residual phase term is the ICBA, as described in Chapter 5.

A time-moving window of fixed length has been used to form subaperture ISAR images in cases where time windowing was necessary because targets are undergoing complex motions during the CPI.

8.3.2 SAR Images of Moving Targets

The refocusing algorithm has been tested on maritime targets since in the available SAR images some ships were clearly visible and no ground truth was needed to validate the results. Specifically, a CSK SAR image acquired on April 23, 2008, covering the area of Messina (in the island of Sicily, Italy), has been processed. A quick-view image is shown in Figure 8.10, where two moving ships are highlighted with red boxes.

8.3.3 Refocusing Results

Results obtained by applying ISAR image processing to SAR data are presented in this subsection. Specifically, the ICBA autofocus technique is first applied to the entire subimage data to refocus moving targets with simple motions. The time windowing technique is then applied to a case of a moving target with complex motions, which cannot be simply refocused by applying the autofocus technique to the entire data. Finally, cross-range scaling is applied to the refocused ISAR images to obtain fully scaled ISAR images, which can lead directly to the target's size estimation.

ISAR Autofocus Results

As previously stated, in this section results will be shown in terms of image contrast values and image quality form a visual point of view. Specifically, original SAR images and

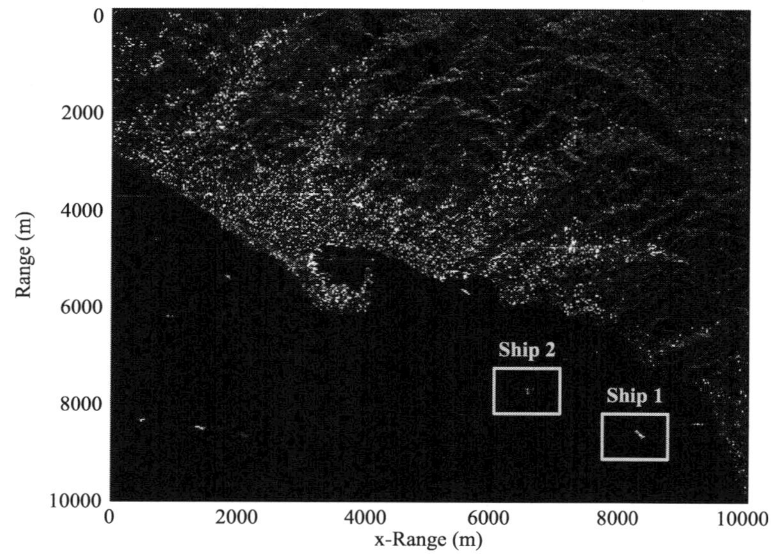

Figure 8.10 CSK SAR image of Messina.

refocused ISAR images will be shown for each selected target. As a first case study, we will consider the problem of refocusing targets with simple motions, which usually do not require any time windowing. These are typical cases where the target's motion is mainly due to its translational movement, and therefore the target's rotation vector can be assumed constant during the observation time. Figures 8.11 and 8.12 prove the effectiveness of the proposed refocusing algorithm. Specifically, both the refocused ISAR image obtained by means of the IRD and IOK are shown to demonstrate that both inversion algorithms are effective. Moreover, the obtained IC values are shown in Table 8.1 with the aim of quantifying the effect of the refocus algorithm.

By observing the previous results, the following remarks can be made:

1. The proposed ISAR processing is able to produce well-focused radar images of moving targets when it is applied to data obtained by using both the IRD and the IOK inversion algorithms. Therefore, there is no evident reason to prefer one inversion algorithm to another one in terms of image focus (measurable from the image magnitude).
2. Although the difference in terms of IC does not lead to very evident differences in the ISAR image magnitudes (from a visual inspection point of view), IRD may have introduced phase errors because it is an approximated way of inverting the SAR image, and IOK could have also introduced phase errors because of the application of an interpolation, which is not an invertible operation. It should be said that phase errors introduced by the inversion algorithm may affect useful information for phase-related applications/postprocessing, such as interferometry [29] and super-resolution [30,31].

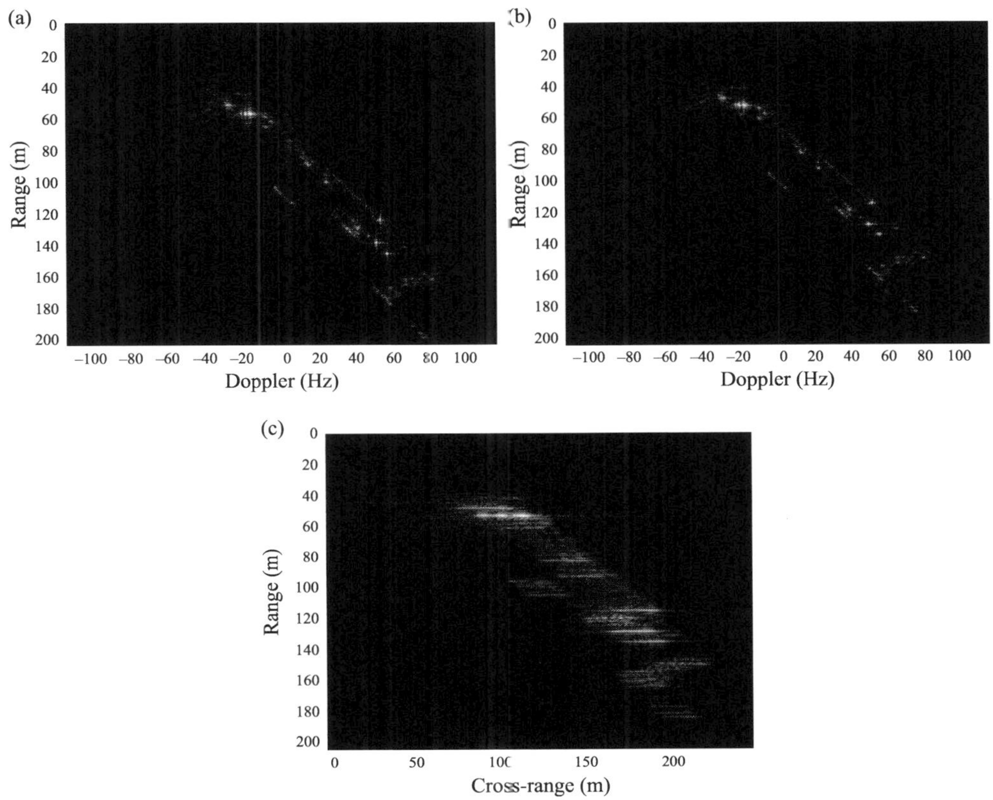

Figure 8.11 Ship no. 1 - (a) Refocused ISAR image with inverse Fourier transform; (b) refocused ISAR image with inverse Ω-k; (c) original (unfocused) SAR image.

Time Windowing Results

As stated already, for long recorded data the target rotation vector may not be constant during the observation time. Therefore, the range-Doppler technique used to form the ISAR images cannot produce well-focused images. This issue can be overcome using a time windowing technique (subaperture formation). This technique allows for the CPI to be controlled to keep the target rotation vector constant during the reduced CPI. Differing from the previous results, in the following example the necessity of using a time windowing algorithm is rather evident. The target is a small maneuvering target. As seen from Figure 8.13, the ISAR processing applied to the entire CPI is not able to produce well-focused target images. Furthermore, it is quite clear from Figure 8.13 that the target experiences an evident yaw motion during the observation time.

By applying the time windowing technique, an increase in the IC value can be measured, which typically indicates a better image reconstruction. Specifically, by setting the CPI equal to $\Delta t = 0.2$ sec and by separating the ISAR image frames with a time step

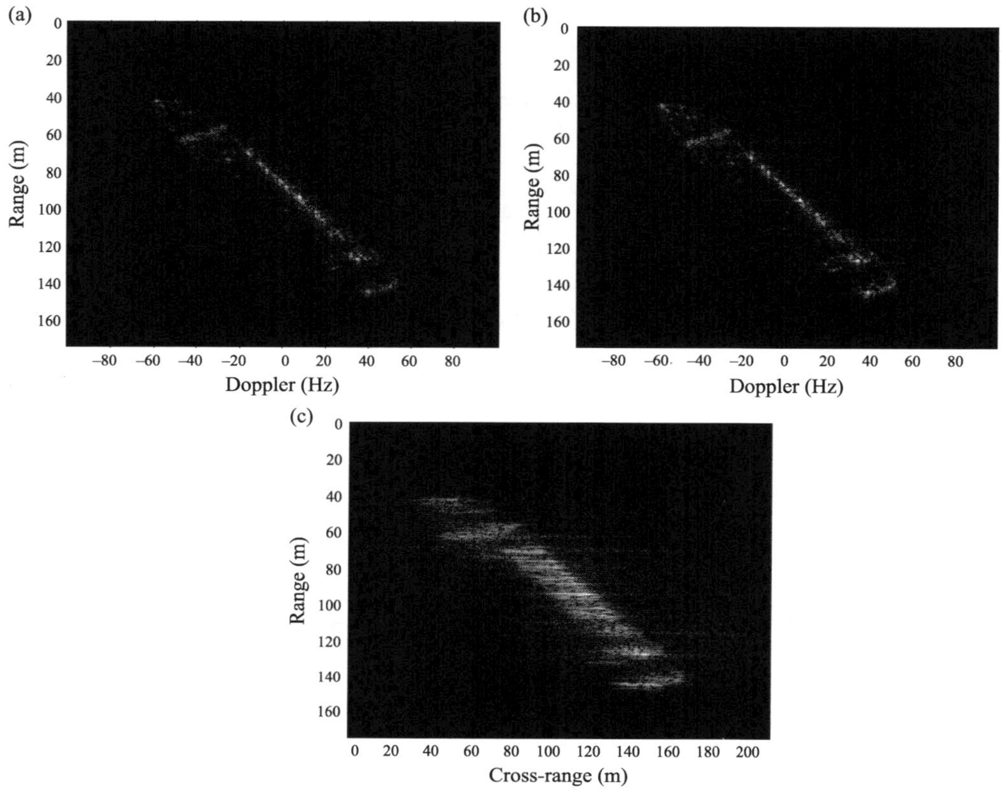

Figure 8.12 Ship no. 2 - (a) Refocused ISAR image with inverse Fourier transform; (b) refocused ISAR image with inverse Ω-*k*; (c) original (unfocused) SAR image.

Table 8.1 Image Contrast Values

IC	SAR	ISAR-IRD	ISAR-IOK
Ship no. 1	26.97	64.96	75.19
Ship no. 2	9.74	26.02	18.61

$T_{step} = 0.1$ sec, we can obtain many frames from a temporal sequence of ISAR images. Figure 8.14 illustrates four ISAR image frames obtained using a Fourier inversion. In the figure the vessel orientation in the range-Doppler domain changes from frame to frame, indicating that the target is either maneuvering or undergoing oscillating motions.

The ISAR images obtained by applying the automatic time windowing algorithm described in Chapter 6 are shown in Figure 8.15 for both the Fourier inversion and the IOK. The IC values of the two images are $IC_{\mathrm{IRD}} = 10.66$ and $IC_{\mathrm{IOK}} = 13.31$, respectively.

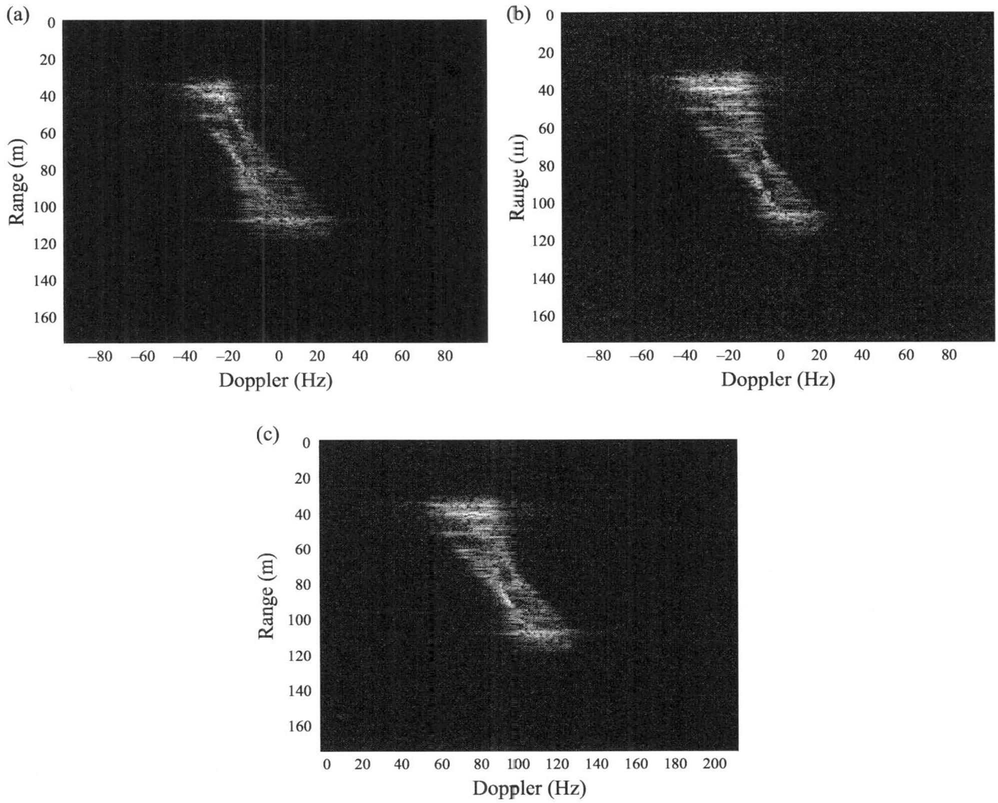

Figure 8.13 Ship with complex motions (a) attempted refocus with inverse Fourier transform; (b) attempted refocus with inverse Ω-k; (c) original SAR image.

Although good images are obtained by manually choosing the window length, better results and better Doppler resolutions are typically obtained using the automatic time windowing algorithm.

Cross-range Scaling Results

Figures 8.16 and 8.17 show the full-scale ISAR images of the refocused targets obtained using the IRD algorithm. Figures 8.18 and 8.19 show the estimated chirp rates in dots and the regression straight line, whose slope is proportional to the effective rotation vector modulus square value. As it can be noted, for each target the chirp rate estimates fit the least square error (LSE) straight line very tightly. It is worth noting that, although the actual length of the ships is unknown, the estimates obtained by applying the cross-range scaling algorithm are all likely values. Also, targets are projected onto the ISAR image plane. Therefore, any estimated distance between two points on the target is actually the result of a projection, thereby producing an underestimated value. The ISAR image plane is unknown, as it

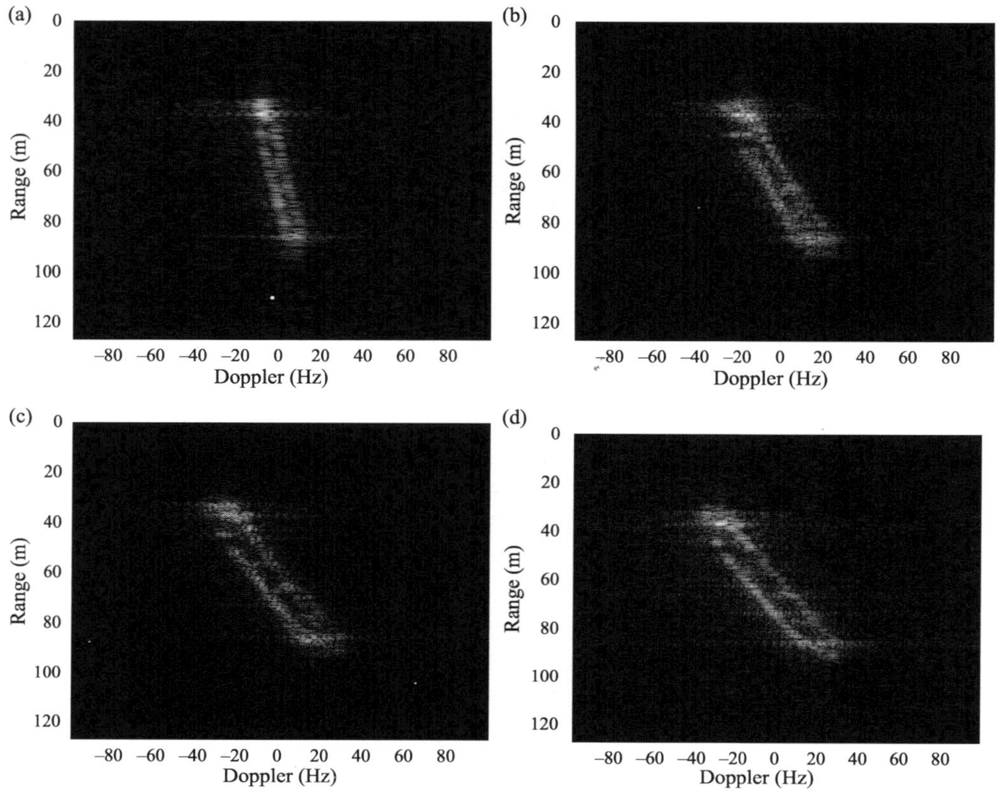

Figure 8.14 Four time windowed ISAR image frames (a), (b), (c), and (d) with each 0.2 sec of CPI and 0.1 sec time separation taken from a temporal sequence of ISAR images.

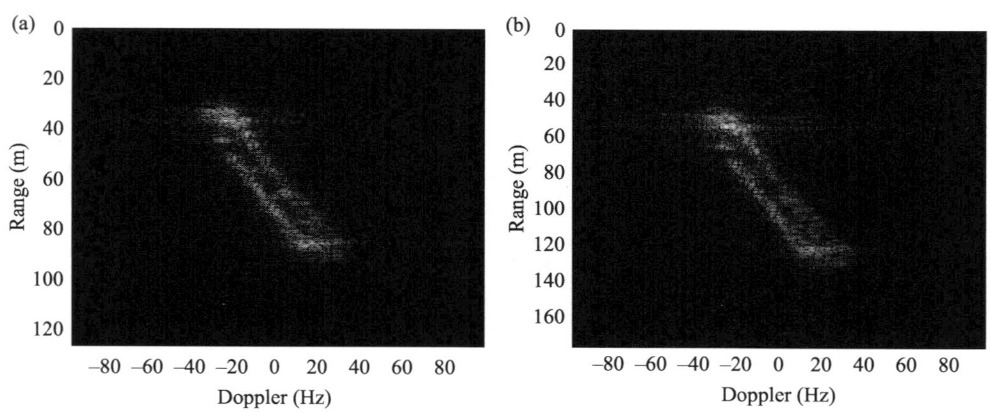

Figure 8.15 Best focused time windowed ISAR images (a) with inverse Fourier transform and (b) inverse Ω-k.

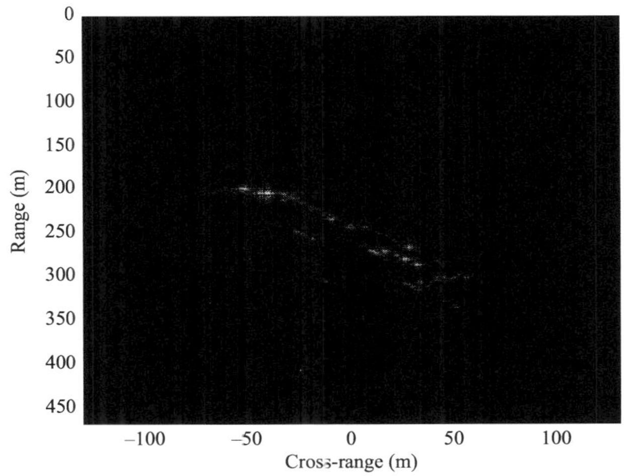

Figure 8.16 Fully scaled ISAR image of ship no. 1.

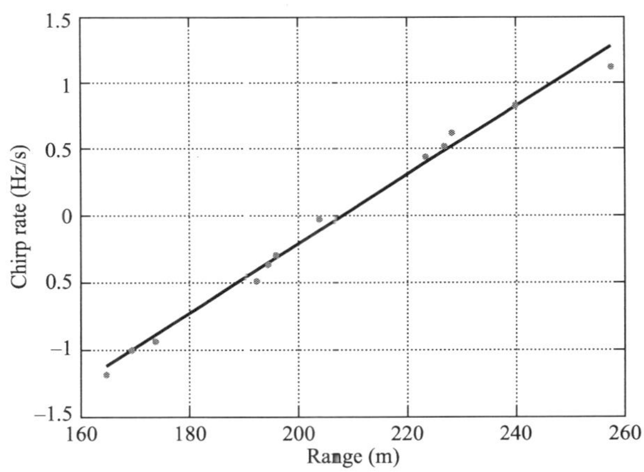

Figure 8.17 Range/chirp rate scatterplot and regression line of ship no. 1.

depends on the orientation of the effective rotation vector, and therefore such a projection cannot be determined a priori.

The adopted cross-range scaling algorithm is based on phase information. The results showed that the phase information is nicely preserved both when using the IRD and the IOK after inverting the SAR image back to the data domain. The phase quality after the applications of the image refocus algorithm may also be good enough for ISAR interferometry applications, although there is still no evidence to prove this.

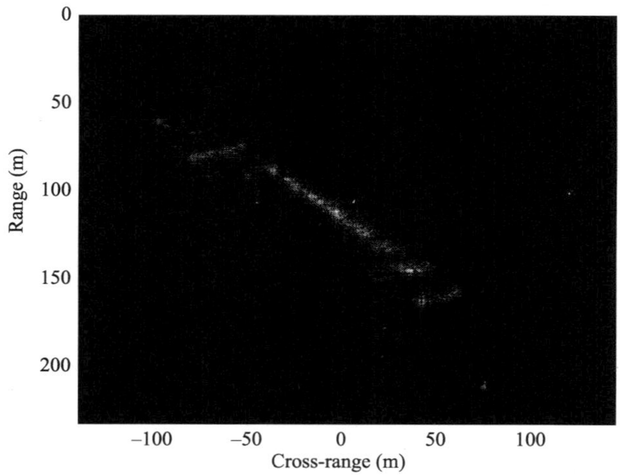

Figure 8.18 Full-scale ISAR image of ship no. 2.

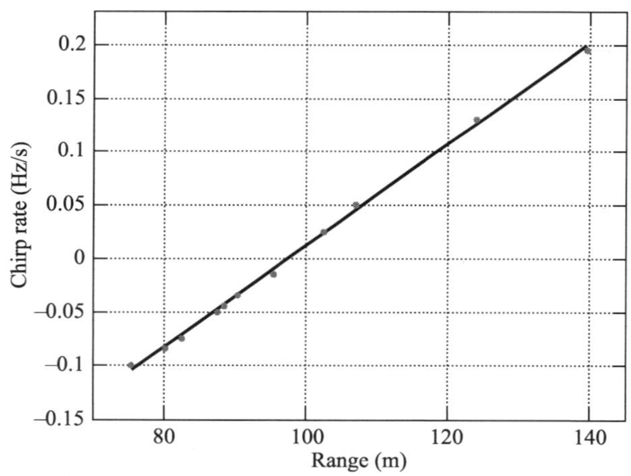

Figure 8.19 Range/chirp rate scatterplot and regression line of ship no. 2.

References

[1] Carrara, W.C., R.S. Goodman, and R.M. Majewsky, *Spotlight Synthetic Aperture Radar: Signal Processing Algorithms*, Artech House, 1995.

[2] Sjogren, T.K., V.T. Vu, M.I. Pettersson, A. Gustavsson, and L.M.H. Ulander, "Moving Target Relative Speed Estimation and Refocusing in Synthetic Aperture Radar Images," *IEEE Transactions on Aerospace and Electronic Systems*, vol. 48, no. 3, pp. 2426–2436, 2012.

[3] Jao, J.K., "Theory of Synthetic Aperture Radar Imaging of a Moving Target," *IEEE Transactions on Geoscience and Remote Sensing*, vol. 39, no. 9, pp. 1984–1992, 2001.

[4] Martorella, M., E. Giusti, F. Berizzi, A. Bacci, and E. Dalle Mese, "ISAR Based Techniques for Refocusing Non-cooperative Targets in SAR Images," *IET Radar Sonar and Navigation*, vol. 6, no. 5, pp. 332–340, 2012.

[5] Davis, M.E., *Foliage Penetration Radar*, SciTech Publishing, 2012.

[6] Tello, M., C. Lopez-Martinez, and J.J. Mallorqui, "A Novel Algorithm for Ship Detection in SAR Imagery Based on the Wavelet Transform," *IEEE Geoscience and Remote Sensing Letters*, vol. 2, no. 2, pp. 201–205, 2005.

[7] Jao, J.K., C.E. Lee, and S. Ayasli, "Coherent Spatial Filtering for SAR Detection of Stationary Targets," *IEEE Transactions on Aerospace and Electronic Systems*, vol. 35, no. 2, pp. 614–626, 1999.

[8] Pettersson, M.I., "Optimum Relative Speed Discretisation for Detection of Moving Objects in Wideband SAR," *IET Radar, Sonar & Navigation*, vol. 1, no. 3, pp. 213–220, 2007.

[9] Blacknell, D., "Contextual Information in SAR Target Detection," *IEE Proceedings—Radar, Sonar and Navigation*, vol. 148, no. 1, pp. 41–47, 2001.

[10] Rignot, E.J.M., and J.J. van Zyl, "Change Detection Techniques for ERS-1 SAR Data," *IEEE Transactions on Geoscience and Remote Sensing*, vol. 31, no. 4, pp. 896–906, 1993.

[11] Lombardo, P., and C.J. Oliver, "Maximum Likelihood Approach to the Detection of Changes between Multitemporal SAR Images," *IEE Proceedings—Radar, Sonar and Navigation*, vol. 148, no. 4, pp. 200–210, 2001.

[12] Ulander, L.M., H.M. Lundberg, W. Pierson, and A. Gustavsson, "Change Detection for Low-Frequency SAR Ground Surveillance," *IEE Proceedings—Radar, Sonar and Navigation*, vol. 152, no. 6, pp. 413–420, 2005.

[13] Preiss, M., D.A. Gray, and N.J.S. Stacy, "Detecting Scene Changes Using Synthetic Aperture Radar Interferometry," *IEEE Transactions on Geoscience and Remote Sensing*, vol. 44, no. 8, pp. 2041–2054, 2006.

[14] Ender, J.H.G., and A.R. Brenner, "PAMIR—A Wideband Phased Array SAR/MTI System," *IEE Proceedings—Radar, Sonar and Navigation*, vol. 150, no. 3, pp. 165–172, 2003.

[15] Damini, A., M. McDonald, and G.E. Haslam, "X-Band Wideband Experimental Airborne Radar for SAR, GMTI and Maritime Surveillance," *IEE Proceedings—Radar, Sonar and Navigation*, vol. 150, no. 4, pp. 305–312, 2003.

[16] Cerutti-Maori, D., and J.H.G. Ender, "Performance Analysis of Multistatic Configurations for Spaceborne GMTI Based on the Auxiliary Beam Approach," *IEE Proceedings—Radar, Sonar and Navigation*, vol. 153, no. 2, pp. 95–103, 2006.

[17] Carrara, W., R. Goodman, and R. Majewski, *Spotlight Synthetic Aperture Radar: Signal Processing Algorithms*, Artech House, 1995.

[18] Cafforio, L., C.C. Prati, and F. Rocca, "SAR Data Focusing Using Seismic Migration Techniques," *IEEE Transactions on Aerospace and Electronic Systems*, vol. 27, no. 2, pp. 194–207, 1991.

[19] Papoulis, A., *Signal Analysis*, McGraw-Hill, 1984.

[20] Silver, S., *Microwave Antenna Theory and Design*, Dover, 1965.

[21] Li, A., and O. Loffeld, "Two-Dimensional SAR Processing in the Frequency Domain," *Proceedings of the IEEE International Geoscience and Remote Sensing Symposium*, pp. 1065–1068, 1991.

[22] Chen, C.W., "Modified polar format algorithm for processing spaceborne SAR data," *Radar Conference, 2004. Proceedings of the IEEE*, pp. 44,49, 26–29, April 2004.

[23] Raney, R.K., H. Runge, R. Bamler, I.G. Cumming, and F.H. Wong, "Precision SAR Processing Using Chirp Scaling," *IEEE Transaction on Geoscience and Remote Sensing*, vol. 32, no. 4, pp. 786–799, 1994.

[24] Moreira, A., and Y. Huang, "Airborne SAR Processing of Highly Squinted Data Using a Chirp Scaling Approach with Integrated Motion Compensation," *IEEE Transactions on Geoscience and Remote Sensing*, vol. 32, no. 5, pp. 1029–1040, 1994.

[25] Moreira, A., J. Mittermayer, and R. Scheiber, "Extended Chirp Scaling Algorithm for Air and Spaceborne SAR Data Processing in Stripmap and Scansar Imaging Modes," *IEEE Transactions on Geoscience and Remote Sensing*, vol. 34, no. 5, pp. 1123–1136, 1996.

[26] Martorella, M., E. Giusti, F. Berizzi, A. Bacci, and M.E. Dalle Mese, "ISAR Based Techniques for Refocusing Non-cooperative Targets in SAR Images," *IETRadar, Sonar and Navigation*, vol. 6, no. 5, pp. 332–340, 2012.

[27] Curlander, J.C., *Synthetic Aperture Radar: Systems and Signal Processing*, Wiley, 1991.

[28] Khwaja, A.S., L. Ferro-Famil, and E. Pottier, "SAR Raw Data Generation Using Inverse SAR Image Formation Algorithms," *Geoscience and Remote Sensing Symposium, 2006. IGARSS 2006. IEEE International Conference on*, pp. 4191,4194, July 31 2006–Aug. 4 2006.

[29] Battisti, N., and M. Martorella, "Intereferometric phase and target motion estimation for accurate 3D reflectivity reconstruction in ISAR systems," *Radar Conference, 2010 IEEE*, pp. 108,112, 10–14 May 2010.

[30] Martorella, M., N. Acito, and F. Berizzi, "Statistical CLEAN Technique for ISAR Imaging," *IEEE Transactions on Geoscience and Remote Sensing*, vol. 45, no. 11, pp. 3552–3560, 2007.

[31] Tsao, J., and B. Steinberg, "Reduction of Sidelobe and Speckle Artifacts in Microwave Imaging: The CLEAN Technique," *IEEE Transactions on Antennas and Propagation*, vol. 36, no. 4, pp. 543–556, 1988.

CHAPTER 9

FMCW ISAR

As introduced in Chapter 2, continuous wave (CW) radar can only measure Doppler frequency shifts of a moving target without range resolution. To have range resolution, the transmitted signal must have enough bandwidth through modulation of the transmitted continuous waveform. Frequency modulated continuous wave (FMCW) is one of such modulation schemes.

FMCW radar systems have a simple architecture and thus are lightweight, low-power, and low-cost. Range information from an FMCW radar is obtained by the measurement of its beat frequency between the transmitted and the received signals, which can be simply performed using the fast Fourier transform (FFT). In principle, FMCW radar is capable of measuring targets at range extremely close to the radar transmit and receive antennas. Also, the range resolution of the FMCW radar is determined by the frequency bandwidth of the FMCW signal. Thus, using wideband FMCW signal, the range resolution can be very high. Another advantage of the FMCW system is its rectangular shaped power spectrum, which is desirable in low probability of intercept (LPI) radars.

FMCW radar was first used in radio altimeters and has been successfully used in short-range applications, such as automotive radars. However, the major disadvantage of FMCW radar is the coupling between the transmitter and the receiver, which limits the dynamic range of the FMCW radar for the use as a general-purpose synthetic aperture imaging system. After improving the dynamic range, the capability of the FMCW radar for synthetic aperture radar (SAR) imaging has been demonstrated in [1–5].

In this chapter, we will focus on the use of FMCW radar to form inverse synthetic aperture radar (ISAR) images of noncooperative targets.

9.1 FMCW Radar for SAR and ISAR Imaging

As demonstrated in [1–5], FMCW waveforms have been used in SAR imaging. The image processing required for forming SAR images differs from the conventional SAR processing [3,6]. Such a difference is dictated by the stop-and-go assumption. As we know, the traditional radar signal model is based on the stop-and-go model, which neglects the relative movement between radar and target during pulse propagation. However, the stop-and-go assumption collapses in the FMCW systems.

The usual SAR signal processing can be restored once the frequency offset introduced in the data has been corrected. In a SAR configuration, where the radar-target geometry is known a priori, such correction becomes reasonably straightforward. An example of the effects brought in by the stop-and-go assumption can be appreciated by looking at Figure 9.1, where the phase history of a target's scatterer is plotted for the case when the stop-and-go assumption is (erroneously) applied (stepped line) and is compared with the true phase history (smooth line). Details regarding how such a phase history is obtained are provided in Section 9.2.4.

In ISAR systems [7,8], the radar-target geometry is usually not known a priori. Therefore, the required data correction cannot be performed in a simple manner, as in the SAR case. In this chapter, first the received signal model for FMCW ISAR systems is defined based on physical principles. Then the image formation issue is addressed, and the imaging system point spread function (PSF) is analytically derived. The PSF will highlight the differences between pulse-Doppler radar and FMCW radar in ISAR imaging. Specifically, the image distortions introduced by FMCW radars are analyzed while forming an ISAR range-Doppler (RD) image.

ISAR autofocus is also detailed in this chapter to account for the continuous nature of the received signal [9–13]. Such a problem, which is typically addressed in ISAR scenarios as the problem of target radial motion compensation, becomes more complex in

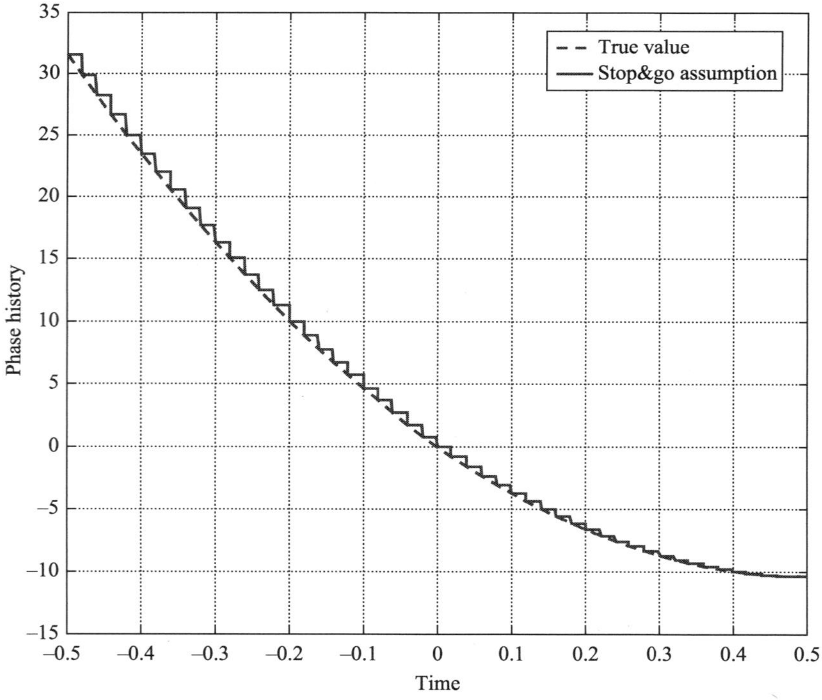

Figure 9.1 Phase history: Comparison between a true phase history and that obtained when the stop-and-go assumption is made.

FMCW ISAR as the number of phase terms to be estimated is considerably higher than that in pulsed radars. This is because the phase is not stationary within a single sweep time. Therefore, at each sample of the received signal, the phase must be estimated. The parametric algorithm defined in this chapter proposes a solution that allows for the phase history to be estimated without increasing the computational complexity. It must be mentioned that the problem of estimating the phase of all samples of the received signal also arises in pulsed stepped-frequency ISAR systems, where the stop-and-go assumption is also not valid [14,15,16].

9.2 FMCW ISAR Received Signal Model

The received signal model is defined by considering the radar-target geometry defined in Figure 9.2, and the transmitted signal waveform is assumed to be a sawtooth FMCW signal, as described in Chapter 2. Therefore, the instantaneous transmitted frequency can be expressed for each frequency sweep, as follows:

$$f(t) = f_c + \mu t, \tag{9.1}$$

where f_c is the carrier frequency, $\mu = B/T_s$ is the frequency sweep rate, B is the bandwidth of frequency sweep, T_s is the sweep repetition interval, and $t \in \left[-\frac{T_s}{2}, \frac{T_s}{2}\right]$ is the time interval covered by a single frequency sweep.

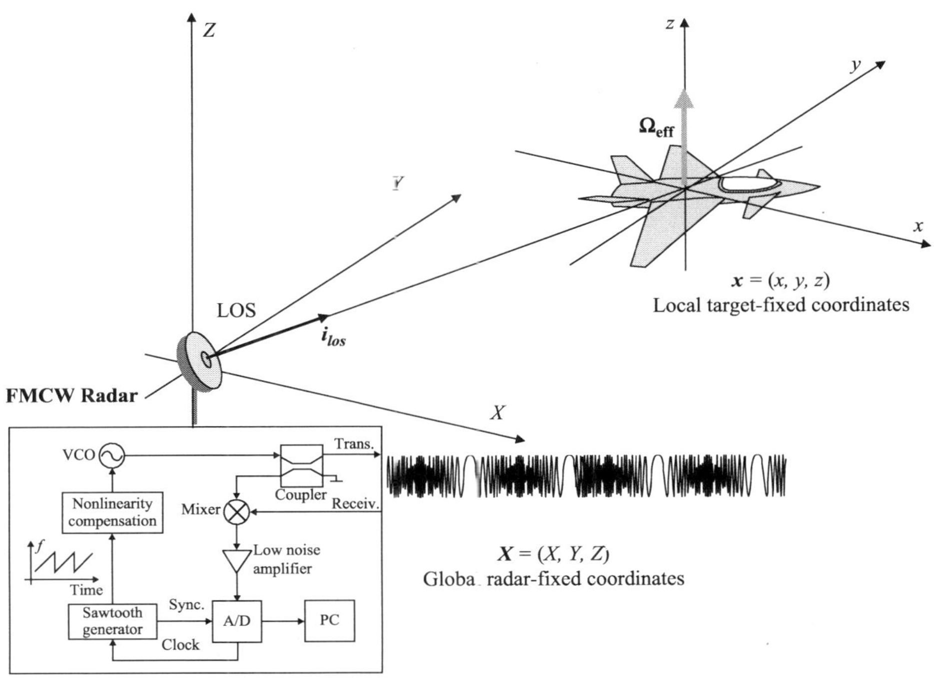

Figure 9.2 Radar-target geometry of an FMCW ISAR.

Let $s_T(t)$ be the analytical transmitted signal

$$s_T(t) = \sum_{n=-N}^{N} s_T(t_s, n), \tag{9.2}$$

where $t \in \left[-\frac{T_{obs}}{2}, \frac{T_{obs}}{2}\right]$ is obtained by adjoining all the time segments of duration, T_s. It should be noted that the observation time is a multiple of the FMCW period, specifically $T_{obs} \in (2N+1)T_s$, where n represents the n-th sweep with the constraint $n \in [-N, N]$, the number of transmitted sweeps within T_{obs} is equal to $2N+1$ and

$$s(t_s, n) = A_T \cdot \exp\left\{j2\pi\left(f + \frac{\mu}{2}t_s\right)t_s\right\} \cdot \text{rect}\left(\frac{t_s - nT_s}{T_s}\right), \tag{9.3}$$

where A_T is a complex value that takes into account the signal energy and the initial phase, and $\text{rect}(t/T_s)$ is equal to 1 when $|t| \le T_s/2$ and equal to 0 otherwise. Since the FMCW transmitted signal a periodic signal with period equal to T_s, the received signal in free space conditions can be expressed as

$$s_R(t) = \sum_{n=-N}^{N} s_R(t_s, n), \tag{9.4}$$

where

$$s_R(t_s, n) = K \int \rho(\boldsymbol{X}) s_T(t_s - \tau(\boldsymbol{X}, t_s, n), n) d\boldsymbol{X}, \tag{9.5}$$

where K is a complex amplitude depending on the radar equation, $\tau(\boldsymbol{X}, t_s, n) = 2R(\boldsymbol{X}, t_s, n)/c$, c is the speed of wave propagation, and $\rho(\boldsymbol{X})$ is the target's reflectivity function, where vector $\boldsymbol{X} = (X, Y, Z)$ represents the global radar coordinate system. $R(\boldsymbol{X}, t_s, n)$ is the modulus of $\boldsymbol{R}(\boldsymbol{X}, t_s, n)$, which is the vector that locates the position of an arbitrary point on the target. The time-varying coordinates $\boldsymbol{R}(\boldsymbol{X}, t_s, n)$ are defined with respect to the global radar coordinate system. It is worth pointing out that the received signal is expressed as a function of the two variables t_s and n to be able to refer to the fast time, t_s, and the frequency sweep, n, separately. It should also be noted that $|t_s| \le T_s/2$. In a typical ISAR scenario, we can assume that the target size is much smaller than the radar-to-target distance. When this assumption is made, the distance between an arbitrary scatterer and the radar can be approximated as follows:

$$R(\boldsymbol{x}, t_s, n) \simeq R_0(t_s, n) + \boldsymbol{x}^T \cdot \boldsymbol{i}_{los}(t_s, n), \tag{9.6}$$

where $\boldsymbol{x} = (x, y, z)$ is defined with respect to a target body-fixed coordinate system, $R_0(t_s, n)$ is the distance between an arbitrary point, O, on the target and the radar at the n-th sweep and at time, t_s, $\boldsymbol{i}_{\text{los}}(t_s, n)$ is the column unit vector of $\boldsymbol{R}_0(t_s, n)$. By defining

$$\tau(\boldsymbol{x}, t_s, n) \triangleq \tau_0(t_s, n) + \tau'(\boldsymbol{x}, t_s, n), \tag{9.7}$$

where $\tau_0(t_s, n) = 2R_0(t_s, n)/c$ and $\tau'(x, t_s, n) = 2\boldsymbol{x}^T \boldsymbol{i}_{\text{los}}(t_s, n)/c$, (9.5) can be rewritten as follows:

$$s_R(t_s, n) = K \int \rho(\boldsymbol{x}) s_T(t_s - \tau_0(t_s, n) + \tau'(\boldsymbol{x}, t_s, n), n) d\boldsymbol{x}, \tag{9.8}$$

where $\rho(\boldsymbol{x})$ is the reflectivity function expressed with respect to the target body-fixed coordinate system.

9.3 FMCW ISAR Processing

Conventional ISAR systems use coherent pulsed radars [7,8,12,17]. As already stated, when a pulsed radar is used, the stop-and-go approximation can be made to simplify the image formation processing. In the case of FMCW ISAR, the stop-and-go assumption is no longer valid because of the continuous nature of the transmitted waveform. In this scenario, the target motion must be considered also with respect to the time, t_s, which is also called the fast time. The effect of the stop-and-go assumption becomes more evident in the case of target's fast motion. To account for the main differences between pulsed and CW ISAR, the radial motion compensation problem is first analyzed followed by the analysis of the PSF when using RD image formation.

9.3.1 Radial Motion Compensation

For a relatively short observation time, T_{obs}, and relatively smooth target motions, the radar-target distance can be expressed by means of a quadratic form as follows:

$$R_0(t) = R_0 + v_R t + \frac{1}{2} a_R t^2, \tag{9.9}$$

where v_R and a_R are, respectively, the radial velocity and acceleration of the target. A parametric autofocus algorithm has to estimate both the target motion parameters, v_R and a_R, to compensate for the radial motion. Because the last step of the classical RD image reconstruction technique is a two-dimensional (2-D) Fourier transform, R_0, produces a shift only along the range coordinate and does not affect the image focusing [12], it can therefore be ignored. During the n-th sweep interval, the radar-target distance becomes a function of time, t_s, as defined by

$$\begin{aligned} R_0(t_s, n) &= R_0 + v_R(t_s + nT_s) + \frac{1}{2} a_R (t_s + nT_s)^2 \\ &= R_0 + v_R nT_s + \frac{1}{2} a_R (nT_s)^2 + (v_R + a_R nT_s) t_s. \end{aligned} \tag{9.10}$$

The approximation in (9.10) is effective when the following inequality applies:

$$\frac{a_R}{2} T_s \lll v_R. \tag{9.11}$$

By defining the target's position and velocity at time instants, nT_s, as follows:

$$R_0(n) = R_0 + v_R nT_s + \frac{1}{2} a_R (nT_s)^2, \tag{9.12}$$

and

$$v_R(n) = v_R + a_R nT_s, \tag{9.13}$$

then (9.10) can be rewritten as

$$R_0(t_s, n) = R_0(n) + v_R(n) t_s. \tag{9.14}$$

After demodulation, the baseband signal can be written as follows

$$s_B(t_s, n) = \gamma s_T^*(t_s, n) s_R(t_s, n), \tag{9.15}$$

where $\gamma \in \mathbb{R}^+$ is an amplitude factor introduced by the demodulator (typically $\gamma \ll 1$). By substituting (9.3) and (9.8) into (9.15) the beat signal can be expressed as follows [18]:

$$\begin{aligned} s_B(t_s, n) = K' \int \rho(\boldsymbol{x}) \exp\left\{j2\pi\left[f_c + \frac{\mu}{2}t_s - \frac{\mu}{2}\tau(\boldsymbol{x}, t_s, n)\right][t_s - \tau(\boldsymbol{x}, t_s, n)]\right\} \\ \cdot \exp\left\{j2\pi\left(f_c + \frac{\mu}{2}t_s\right)t_s\right\} \cdot w[t_s, \tau(\boldsymbol{x}, t_s, n), n] d\boldsymbol{x}, \end{aligned} \tag{9.16}$$

where $K' = \gamma K$ is a complex amplitude, A_T^2 is included in K, and

$$w[t_s, \tau(\boldsymbol{x}, t_s, n), n] = w_T(t_s, n) w_R[t_s, \tau(\boldsymbol{x}, t_s, n), n], \tag{9.17}$$

where $w_T(t_s, n) = \mathrm{rect}\left(\frac{t_s - nT_s}{T_s}\right)$ and $w_R[t_s, \tau(\boldsymbol{x}, t_s, n), n] = \mathrm{rect}\left(\frac{t_s - nT_s - \tau(\boldsymbol{x}, t_s, n)}{T_s}\right)$.

Since the radar parameters are usually designed such that $\tau(\boldsymbol{x}, t_s, n) \ll T_s$, the product of the time windows in (9.17) can be simplified as

$$w_T(t_s, n) w_R[t_s, \tau(\boldsymbol{x}, t_s, n), n] = w_T(t_s, n). \tag{9.18}$$

Therefore, the baseband signal can be rewritten as

$$\begin{aligned} s_B(t_s, n) = K' w_T(t_s, n) \int \rho(\boldsymbol{x}) \exp\left\{j2\pi\left[f_c + \frac{\mu}{2}t_s - \frac{\mu}{2}\tau(\boldsymbol{x}, t_s, n)\right][t_s - \tau(\boldsymbol{x}, t_s, n)]\right\} \\ \cdot \exp\left\{j2\pi\left(f_c + \frac{\mu}{2}t_s\right)t_s\right\} d\boldsymbol{x}. \end{aligned} \tag{9.19}$$

By substituting (9.7), into (9.19) the baseband signal can be rewritten as follows:

$$s_B(t_s, n) = K' w_T(t_s, n) \exp\{j\varphi_0(t_s, n)\} \int \rho(\boldsymbol{x}) \exp\{j[\varphi_1(t_s, n) + \varphi_2(t_s, n)]\} d\boldsymbol{x}, \tag{9.20}$$

where

$$\varphi_0(t_s, n) = -2\pi(f_c + \mu t_s) \cdot \tau_0(t_s, n) + \pi\mu\tau_0^2(t_s, n), \tag{9.21}$$

$$\varphi_1(\boldsymbol{x}, t_s, n) = -2\pi(f_c + \mu t_s) \cdot \tau'(\boldsymbol{x}, t_s, n), \tag{9.22}$$

and

$$\varphi_2(\boldsymbol{x}, t_s, n) = \pi\mu[\tau'(\boldsymbol{x}, t_s, n)]^2 + 2\pi\mu\tau_0(t_s, n)\tau'(\boldsymbol{x}, t_s, n). \tag{9.23}$$

By using (9.11) and (9.18), (9.21) can be written as follows:

$$\varphi_0(t_s, n) = -2\pi(f_c + \mu t_s) \cdot \left[\tau_0(n) + \frac{2}{c}v_R(n)t_s\right] - \pi\mu\left[\tau_0(n) + \frac{2}{c}v_R(n)t_s\right]^2. \tag{9.24}$$

By defining $f_D(n) = \frac{2f_c}{c} v_R(n)$ and $f_B(n) = \mu\tau_0(n)$, $\varphi_0(t_s,n)$ can be rewritten as

$$\varphi_0(t_s,n) \simeq -2\pi f_c\tau_0(n) - 2\pi f_D(n)t_s - 2\pi f_B(n)t_s\left[1 - \frac{f_D(n)}{f_c}\right]$$
$$-\pi\mu\frac{f_D(n)}{f_c}t_s^2\left[2 - \frac{f_D(n)}{f_c}\right] + \pi\mu\tau_0^2(n). \tag{9.25}$$

When the inequality $f_D(n) \ll f_c$ is satisfied for each value of n, $\varphi_0(t_s, n)$ can be finally rewritten as follows:

$$\varphi_0(t_s,n) \simeq -2\pi f_c\tau_0(n) - 2\pi f_D(n)t_s - 2\pi f_B(n)t_s - 2\pi\mu\frac{f_D(n)}{f_c}t_s^2 + \pi\mu\tau_0^2(n), \tag{9.26}$$

where

$$\tau_0(n) = 2R_0(n)/c. \tag{9.27}$$

The term $\pi\mu\tau_0^2(n)$ is known as the residual video phase (RVP). The compensation of the RVP is possible, although in most cases the effect of this term is negligible [5,19]. Since the target's radial speed can be assumed constant during the sweep time, the Doppler frequency shift can also be assumed constant in the same interval. Therefore, the frequency response is shifted only in range without being affected by significant distortions. However, the variations of the Doppler frequency shift and that of the phase term $\varphi_0(t_s, n)$ cause a range migration that must be accounted for. To compensate for the target radial motion, the phase term outside the integral in (9.20), namely, $\varphi_0(t_s, n)$, must be estimated and removed. In ISAR scenarios, where the target is usually noncooperative, this operation is done by means of autofocus algorithms. This aspect will be detailed in the next section, where we will introduce a revisited version of the ICBA technique (detailed in Chapter 5 for pulsed radar). This revisited version will be named FMCW-ICBA.

9.3.2 Autofocus Algorithm—FMCW-ICBA

The estimation of $\varphi_0(t_s, n)$ resorts to the estimation of the target radial motion parameters, namely, v_R and a_R. As demonstrated in [12], the parameter R_0 does not affect the image focus. Assuming a perfect radial motion compensation, the baseband signal can be expressed as

$$s_{BC}(t_s,n) = K'\int \rho(\mathbf{x})\exp\{j[\varphi_1(t_s,n) + \varphi_2(t_s,n)]\}d\mathbf{x}, \tag{9.28}$$

where $w_T(t_s, n)$ has been omitted for notation simplicity.

Let $\mathbf{\Theta} = [\mu_1, \mu_2]$ be the vector containing the unknowns and $\varphi_0(t_s, n, \mathbf{\Theta})$ be the relative phase term obtained by substituting the unknowns with those in $\mathbf{\Theta}$. The radial motion compensation problem can be recast as an optimization problem where the image contrast (IC) is maximized with respect to the unknown vector $\mathbf{\Theta}$, as defined by

$$\hat{\mathbf{\Theta}} = \arg\max_{\mathbf{\Theta}}\{IC(\mathbf{\Theta})\}, \tag{9.29}$$

where

$$IC(\mathbf{\Theta}) \triangleq \frac{\sqrt{Avg\left\{[I(\mathbf{\Theta}) - Avg\{I(\mathbf{\Theta})\}]^2\right\}}}{Avg\{I(\mathbf{\Theta})\}}, \tag{9.30}$$

where $Avg\{\cdot\}$ indicates the average over the image pixels, and $I(\mathbf{\Theta})$ is the ISAR image magnitude that is formed after compensating the beat signal with the phase term $\varphi_0(t_s, n, \mathbf{\Theta})$. This is formalized by

$$I(\mathbf{\Theta}) = RD\{s_B(t_s, n)\exp[-j\varphi_0(t_s, n, \mathbf{\Theta})]\}, \tag{9.31}$$

where $RD\{\cdot\}$ indicates the operation of image formation using an RD approach, as clarified in the Section 9.5. Such radial motion compensation will be addressed as FMCW-ICBA to distinguished it from the conventional ICBA, which was defined in Chapter 5.

The problem of radial motion compensation becomes more complex in FMCW radar since the number of phase terms to be estimated becomes equal to the product $N'M'$, where $N' = 2N + 1$ is the number of slow-time samples (or sweeps) and $M' = T_s F_{samp}$ is the number of fast-time samples (or number of samples within a sweep). It must be specified that F_{samp} is the sampling frequency used to digitize the received signal. The sampling frequency, F_{samp}, is defined according to the desired range ambiguity, ΔR, where the latter depends on the sampling frequency by means of $\Delta R \frac{c}{2F_{samp}}$. An example of phase history plot is shown in Figure 9.2, where the true phase history $\varphi_0(t_s, n, \mathbf{\Theta})$ is compared with that obtained when a stop-and-go assumption is made. Although the stop-and-go assumption reduces the number of points where the phase history is estimated, this introduces errors that produce slightly defocussed images, as will be shown in Section 9.5.

9.3.3 Range-Doppler Image Formation and Point Spread Function Analysis

As described in Chapter 3, the RD method can be considered the most conventional method used for reconstructing ISAR images. Such a technique is based on the assumption that the Doppler frequency generated by the motion of each scatterer, relative to a reference point taken on the target, is constant within the observation time. The assumption of stationary Doppler components is usually satisfied when the effective rotation vector, $\mathbf{\Omega}_{\mathbf{eff}}(t)$, is constant within the observation time and the CPI is kept short. Under these assumptions, the expression of the received signal model given in (9.12) is considered to derive the imaging system PSF.

For a rigid body, we can assume that the target rotation occurs around an axis passing through point O. Let $\theta(t_s, n)$ be the target aspect angle at any given time, t_s, and in the sweep, n, and let T_x be defined such that the z-axis is aligned with $\mathbf{\Omega}_{\mathbf{eff}}$. The latter is not constraining the problem since the choice of the orientation of the reference system embedded in the target is fully arbitrary.

Considering these assumptions, the scalar product between the vector defining the position of a point scatterer and the unit vector of the line of sight (LO)S at the n-th sweep and at a generic instant time, t_s, can be written as follows:

$$\mathbf{x}^T \cdot \mathbf{i}_{los}(n) = x\cos[\theta(t_s, n)] + y\sin[\theta(t_s, n)]. \tag{9.32}$$

By substituting (9.22), (9.23), and (9.7) into (9.28) and by assuming that $f_D(n) \ll f_0$ for any n, the baseband signal after motion compensation can be approximated as follows, where $w_T(t_s, n)$ has been omitted for notation simplicity:

$$s_{BC}(t_s, n) = K' \int_S \rho'(\boldsymbol{x}) \exp\{j[x \cdot q_1(t_s, n) + y \cdot q_2(t_s, n)]\} d\boldsymbol{x}. \tag{9.33}$$

It is worth noting that the vector $\boldsymbol{x}$ is now reduced to a two-dimensional vector containing only the x and y components. Also, the reflectivity function $\rho'(\boldsymbol{x})$ is the projected version of the reflectivity function $\rho(\boldsymbol{x})$ onto the image plane. The functions $q_1(t_s, n)$ and $q_2(t_s, n)$ are derived in the following mathematical passages.

The phase term inside the integral on (9.28) can be rewritten as

$$\varphi_1(t_s, n) + \varphi_2(t_s, n) = -\frac{4\pi}{c}\left\{f_c + \frac{\mu}{2}[t_s - \tau'(\boldsymbol{x}, t_s, n)]\right\}\boldsymbol{x}^T \cdot \boldsymbol{i}_{los}(t_s, n). \tag{9.34}$$

Moreover, by using (9.32) in (9.34) the following expression can be obtained:

$$\varphi_1(t_s, n) + \varphi_2(t_s, n) = -\frac{4\pi}{c}\left\{f_c + \frac{\mu}{2}[t_s - \tau'(\boldsymbol{x}, t_s, n)]\right\}\{x \cos[\theta(t_s, n)] + y \sin[\theta(t_s, n)]\}. \tag{9.35}$$

Therefore, the functions $X_1(t_s, n)$ and $X_2(t_s, n)$ can be defined as follows:

$$q_1(t_s, n) = \frac{2}{c}\{f_c + \mu[t_s - \tau_0(n) - \tau'(\boldsymbol{x}, t_s, n)/2]\}\cos[\theta(t_s, n)], \tag{9.36}$$

and

$$q_2(t_s, n) = \frac{2}{c}\{f_c + \mu[t_s - \tau_0(n) - \tau'(\boldsymbol{x}, t_s, n)/2]\}\sin[\theta(t_s, n)]. \tag{9.37}$$

It must be clarified that the integral in (9.33) is a 2-D integral over a domain S defined in the image plane formed by the x and y-axes. Also, $\tau'(\boldsymbol{x}', t_s, n) = \tau'(\boldsymbol{x}, t_s, n)$ since the LOS is always orthogonal to the z-axis (as a result of the choice of an arbitrary reference system embedded in the target). When the effective rotation vector, $\boldsymbol{\Omega}_{\text{eff}}(t)$, can be considered constant, the analytical expression of the aspect angle $\theta(t_s, n)$ is linear with respect to time:

$$\theta(t_s, n) \simeq \Omega_{\text{eff}}(t_s + nT_s). \tag{9.38}$$

The RD method can be successfully implemented by means of a Fourier transform when the polar grid in the Fourier domain can be approximated by a rectangular grid. Such an approximation can be assumed valid when [7,8,17]

$$\left|\theta(t_s, n) - \frac{\pi}{2}\right| \ll 1 \text{ with } n \in [-N, N] \text{ and } |t_s| \leq T_s/2. \tag{9.39}$$

As detailed in Chapter 3, Section 3.1.5, when a large rotation occurs (when the condition in (9.35) is not satisfied), a rotational motion compensation is typically needed. Under the assumptions expressed in (9.38) and (9.39), (9.36) and (9.37) can be rewritten as follows:

$$q_1(t_s, n) = \frac{2}{c}\Omega_{\text{eff}}(t_s + nT_s)\{f_c + \mu[t_s - \tau_0(n) - \tau'(\boldsymbol{x}, t_s, n)/2]\}, \tag{9.40}$$

and

$$q_2(t_s, n) = \frac{2}{c}\{f_c + \mu[t_s - \tau_0(n) - \tau'(\mathbf{x}, t_s, n)/2]\}. \tag{9.41}$$

A closed-form derivation of the system PSF can be obtained in the case of narrowband FMCW ISAR systems. An FMCW waveform can be considered a narrowband one when the following assumption can be made:

$$\gamma[t_s - \tau_0(n) - \tau'(\mathbf{x}, t_s, n)/2] \ll f_c. \tag{9.42}$$

After applying (9.42), (9.40) and (9.41) can be rewritten in a simpler form as follows:

$$q_1(t_s, n) = \frac{2f_c}{c}\Omega_{\text{eff}}(t_s + nT_s), \tag{9.43}$$

and

$$q_2(t_s, n) = \frac{2}{c}(f_c + \mu t_s). \tag{9.44}$$

By considering the relationship between time and frequency for an FMCW waveform (as detailed in Chapter 2, Section 2.2.2), $q_1(t_s, n)$ and $q_2(t_s, n)$ can be expressed as

$$Q_1(f, n) \simeq \frac{2\Omega_{\text{eff}} f_0}{\mu c}(f - f_c) + \frac{2\Omega_{\text{eff}} f_c n T_s}{c}, \tag{9.45}$$

and

$$Q_2(f, n) \simeq 2f/c. \tag{9.46}$$

where the use of the capital letters Q_1 and Q_2 indicates the direct dependance on the variable "f" instead of the variable "t".

A relationship between Q_1 and Q_2 is obtained by substituting (9.46) into (9.45). The result is

$$Q_1(Q_2, n) = \frac{\Omega_{\text{eff}} f_c}{\mu} Q_2 - \frac{2\Omega_{\text{eff}} f_c^2}{\mu c} + \frac{2\Omega_{\text{eff}} f_c n T_s}{c}. \tag{9.47}$$

The window $W(Q_1, Q_2)$ defines the domain where the Fourier transform of the reflectivity function exists. The analytical expression of $W(Q_1, Q_2)$ is

$$W(Q_1, Q_2) = \begin{cases} 1 & \text{when } -\dfrac{C_2}{2} + C_3 \le Q_2 \le \dfrac{C_2}{2} + C_3; \\ & C_1 Q_2 + C_4 - C_5 \le Q_1 \le C_1 Q_2 + C_4 + C_5, \\ 0 & \text{otherwise} \end{cases} \tag{9.48}$$

where

$$C_1 \triangleq \tan\varphi = \frac{\Omega_{\text{eff}} f_c}{\mu}, \tag{9.49a}$$

$$C_2 \triangleq \frac{2B}{c}, \tag{9.49b}$$

$$C_3 \triangleq \frac{2f_c}{c}, \tag{9.49c}$$

$$C_4 \triangleq \frac{2\Omega_{\text{eff}} f_c^2}{\mu c}, \tag{9.49d}$$

$$C_5 \triangleq \frac{2\Omega_{\text{eff}} f_c n T_s}{c}. \tag{9.49e}$$

A pictorial view of the Fourier domain window $W(Q_1, Q_2)$ is shown in Figure 9.3, and in the gray area the function is equal to 1.

To apply the RD operator and calculate the PSF, the motion compensated beat signal in (9.33) is rewritten as a function of the spatial frequency coordinates:

$$s_{BC}(t_s, n) = W(Q_1, Q_2)\int_S \rho'(x, y)\exp\{j2\pi[xQ_1 + yQ_2]\}dxdy = W(Q_1, Q_2)G'(Q_1, Q_2). \tag{9.50}$$

where $G'(Q_1, Q_2)$ is the 2-D FT of $\rho'(x, y)$. The complex valued ISAR image can be obtained by applying the RD operator, which consists of a 2-D IFT:

$$I_c(x, y) = \rho'(x, y) \otimes \otimes w(x, y), \tag{9.51}$$

where the operator $\otimes\otimes$ represents a 2-D convolution. It is worth noting that, as already demonstrated for the case of pulsed radar, also in the case of FMCW ISAR, the ISAR image can be interpreted as the linear convolution of the projection of the reflectivity function onto the image plane with the system PSF, namely, $w(x, y)$. Such a characteristic is a consequence of the geometrical properties of ISAR imaging systems and should never depend on the waveform used. The window $W(Q_1, Q_2)$ defines a parallelepiped in the spatial frequency plane. The characteristic angle, φ, depends on the modulus of the effective rotation vector, as shown in (9.45a). The window $W(Q_1, Q_2)$ approximates a rectangle when the effective rotation vector modulus is small. The narrowband FMCW ISAR system PSF is the 2-D FT of (9.48):

$$w(x, y) = C_2 C_5 \cdot \operatorname{sinc}(C_5 x) \cdot \operatorname{sinc}(C_2[C_1 x + y]) \cdot \exp\{-j2\pi x C_4\} \cdot \exp\{-j2\pi C_4[C_1 x + y]\}. \tag{9.52}$$

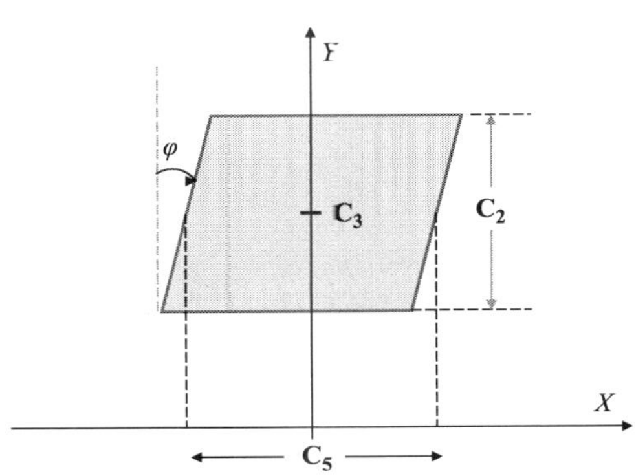

Figure 9.3 FMCW ISAR Fourier domain acquisition window.

In typical ISAR scenarios, the distortion caused by the nonrectangular Fourier domain is not strong enough to seriously alter the sinc-like shape of an ideal ISAR PSF. A numerical simulation is produced and shown in the next section to provide an example of FMCW ISAR imaging.

9.4 Example of FMCW ISAR Autofocusing

In this section we will show some FMCW ISAR simulated results, with the main goal of showing the effect of the implementation of an autofocus algorithm designed for FMCW ISAR when compared with a standard pulse-Doppler ISAR autofocus algorithm.

Figures 9.4 through 9.8 show the ISAR images of the target moving along a straight line with an angle formed with the radar LOS (i.e., the aspect angle) equal to 50° and at a velocity $v = [50, 100, 150, 200, 250]$ m/s. The ISAR images are obtained by using both the FMCW ISAR autofocus algorithm and the standard (pulsed radar) autofocus algorithm. The number

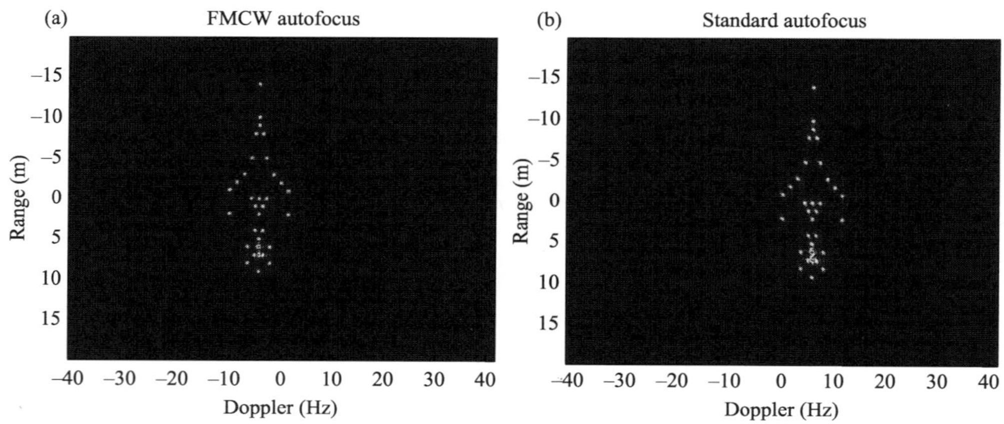

Figure 9.4 Comparison between (a) FMCW and (b) standard autofocus at target velocity 50 m/s.

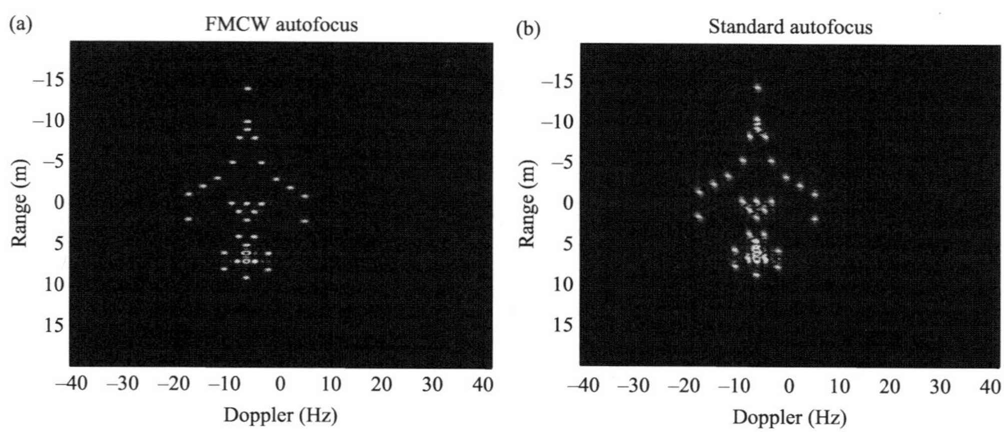

Figure 9.5 Comparison between (a) FMCW and (b) standard autofocus at target velocity 100 m/s.

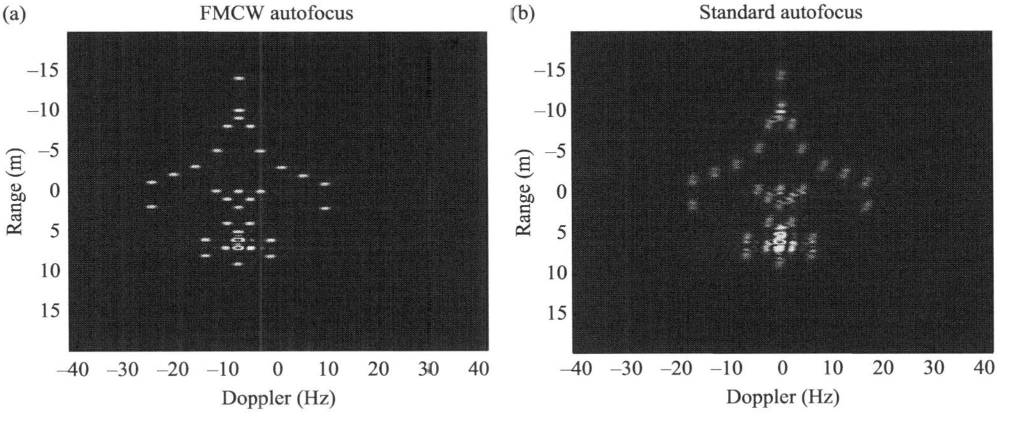

Figure 9.6 Comparison between (a) FMCW and (b) standard autofocus at target velocity 150 m/s.

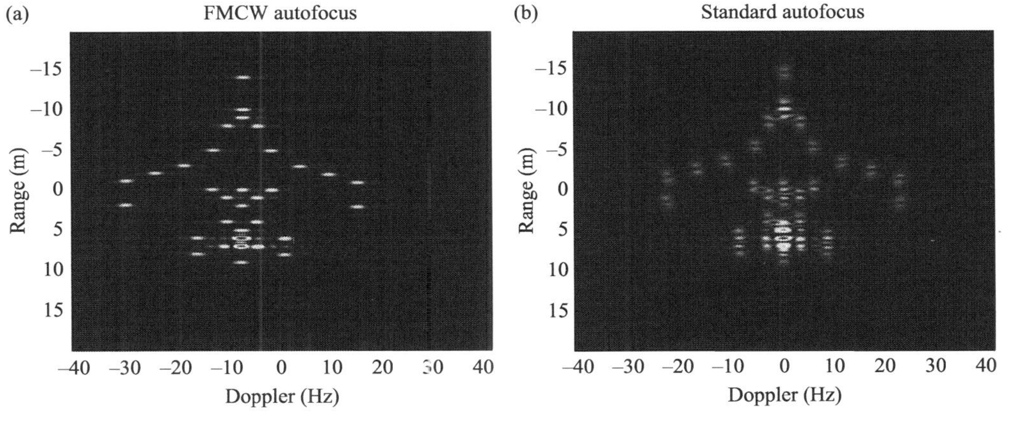

Figure 9.7 Comparison between (a) FMCW and (b) standard autofocus at target velocity 200 m/s.

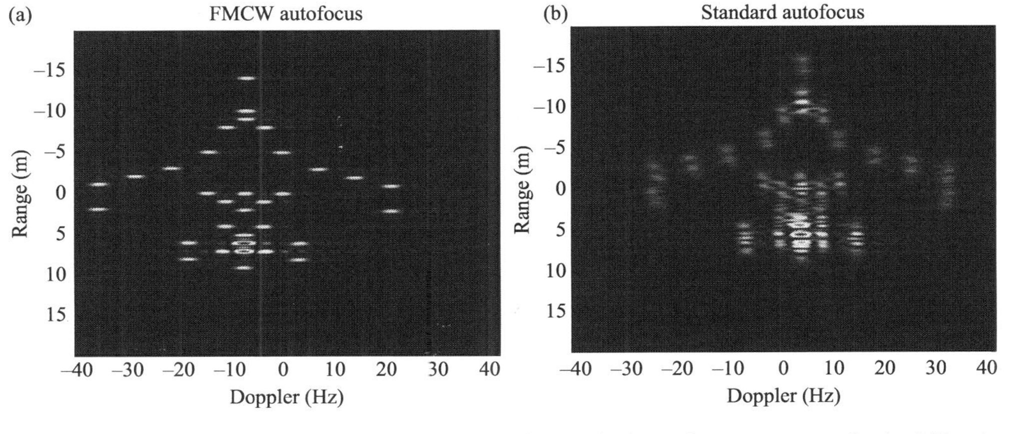

Figure 9.8 Comparison between (a) FMCW and (b) standard autofocus at target velocity 250 m/s.

of transmitted sweeps depends on the target velocity, v, and it is set to have a cross-range resolution equal to $\Delta r_{cr} = 1.45$ m. It must be pointed out that the stop-and-go approximation is no longer valid when FMCW radars are used, especially in the case of high target radial velocity. In particular, the radial target motion within each sweep causes a spreading of the PSF mainly along the range direction. The PSF spread along the range direction is more evident in Figure 9.9, where two range sections relatively to the ISAR images presented in Figures 9.5a and 9.5b are shown. The image entropy (IE) value has also been used to assess the image quality. Therefore, the IE values of the ISAR images obtained by using both the FMCW ISAR autofocusing algorithm ($IE^{(FM)}$) and the conventional autofocusing algorithm ($IE^{(C)}$) are displayed in Table 9.1. Moreover, the differences between the estimated and the

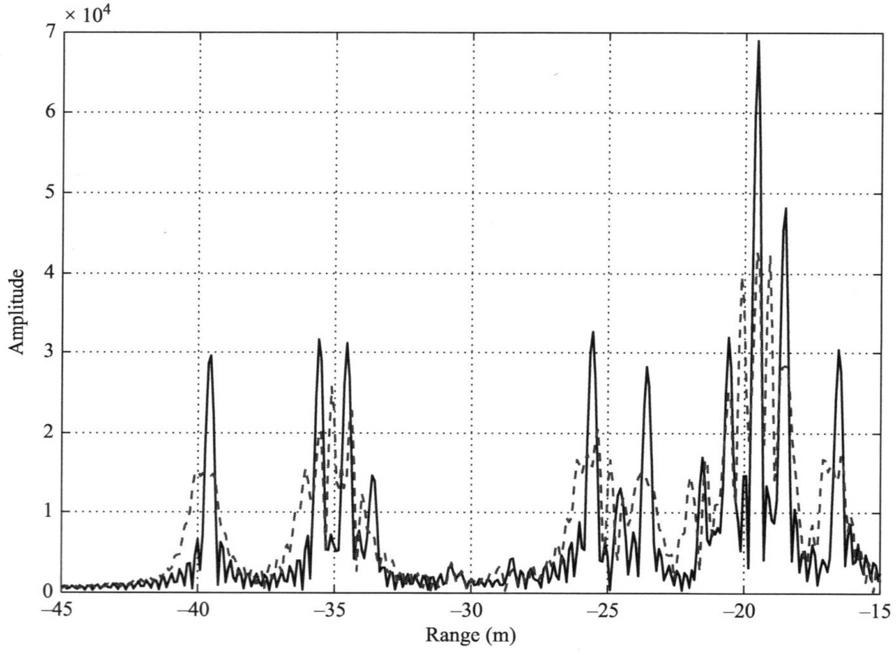

Figure 9.9 Range sections at cross-range = 0 (m) of ISAR images in Figure 9.5a and Figure 9.5b—solid line and dish line, respectively.

Table 9.1 Results in Terms of Velocity, Acceleration Differences, and Image Entropy

v [m/s]	$\left(\left\|\delta_{v_R}^{(FM)}\right\|, \left\|\delta_{a_R}^{(FM)}\right\|\right)$	$\left(\left\|\delta_{v_R}^{(C)}\right\|, \left\|\delta_{a_R}^{(C)}\right\|\right)$	$IE^{(FM)}, IE^{(C)}$
50	(0.005, 0.021)	(0.179, 0.019)	(4.948, 7.465)
100	(0.073, 0.02)	(0.745, 0.04)	(4.638, 9.283)
150	(0.065, 0.043)	(0.169, 0.057)	(4.626, 8.833)
200	(0.027, 0.083)	(0.106, 0.095)	(4.614, 9.217)
250	(0.11, 0.116)	(1.118, 0.015)	(4.592, 9.497)

actual values of target radial motion parameters are also shown in Table 9.1. Specifically, $\left(\left|\delta_{v_R}^{(FM)}\right|, \left|\delta_{a_R}^{(FM)}\right|\right)$ are the differences between the actual radial velocity and acceleration and the estimated ones obtained by using the FMCW ISAR autofocusing algorithm. Instead, the couple of values $\left(\left|\delta_{v_R}^{(C)}\right|, \left|\delta_{a_R}^{(C)}\right|\right)$ are the differences between the actual radial velocity and acceleration and the estimated ones obtained by using the conventional autofocusing algorithm. As it can be noted, the FMCW ISAR autofocusing algorithm estimates both v_R and a_R with higher accuracy. The corresponding ISAR images therefore have higher contrast values and lower entropy values.

References

[1] Charvat, G.L., and L.C. Kempel, "Synthetic Aperture Radar Imaging Using a Unique Approach to Frequency-Modulated Continuous-Wave Radar Design," *IEEE Antennas and Propagation Magazine*, vol. 48, no. 1, pp. 171–177, 2006.

[2] Nouvel, J.F., S. Roques, and O.R. du Plessis, "A Low-Cost Imaging Radar: DRIVE on Board ONERA Motorglider," *IEEE 2007 IGARSS* pp. 5306–5309, 2007.

[3] Meta, A., P. Hoogeboom, and L.P. Ligthart, "Signal Processing for FMCW SAR," *IEEE Transactions on GRS*, vol. 45, no. 11, pp. 3519–3532, 2007.

[4] Edrich, M., and G. Weiss, "Second-Generation Ka-Band UAV SAR System," *EuRAD 2008*, pp. 479–482, 2008.

[5] De Wit, J.J.M., A. Meta, and P. Hoogeboom, "Modified Range-Doppler Processing for FM-CW Synthetic Aperture Radar," *IEEE Geoscience and Remote Sensing Letters*, vol. 3, no. 1, pp. 83–87, 2006.

[6] Burgos-Garcia, M., F. Perez-Martinez, and J. Gismero Menoyo, "Radar Signature of a Helicopter Illuminated by a Long LFM Signal," *IEEE Transactions on AES*, vol. 45, no. 3, pp.1104–1110, 2009.

[7] Ausherman, D.A., A. Kozma, J.L. Walker, H.M. Jones, and E.C. Poggio, "Developments in Radar Imaging," *IEEE Transactions on AES*, vol. 20, no. 4, pp. 363–400, 1984.

[8] Walker, J.L., "Range-Doppler Imaging of Rotating Objects," *IEEE Transactions on AES*, vol. 16, no. 1, pp. 23–52, 1980.

[9] Haywood, B., and R.J. Evans, "Motion Compensation for ISAR Imaging," *Proceedings of ASSPA*, Adelaide, Australia, pp. 113–117, 1989.

[10] Berizzi, F., M. Martorella, B. Haywood, E. Dalle Mese, and S. Bruscoli, "A survey on ISAR autofocusing techniques," *Image Processing, 2004. ICIP '04. 2004 International Conference on*, vol.1, pp. 9,12, 24–27 Oct. 2004.

[11] Li, X., G. Liu, and J. Ni, "Autofocusing of ISAR Images Based on Entropy Minimization," *IEEE Trans. on AES*, vol. 35, no. 4, pp. 1240–1252, 1999.

[12] Martorella, M., F. Berizzi, and B. Haywood, "A Contrast Maximization Based Technique for 2D ISAR Autofocusing," *IEE Proceedings on Radar, Sonar and Navigation*, vol. 152, no. 4, pp. 253–262, 2005.

[13] Wang, Y., H. Ling, and V.C. Chen, "SAR Motion Compensation via Adaptive Joint Time-Frequency Technique," *IEEE Trans. on AES*, vol. 34, no. 2, pp. 670–677, 1998.

[14] Lei, W., T. Long, and Y.Q. Han. "Text radar," *Signal Processing Proceedings, 2000. WCCC-ICSP 2000. 5th International Conference on*, vol. 3, pp. 1851,1855, 2000.

[15] Berizzi, F., M. Martorella, A. Cacciamano, and A. Capria, "A Contrast-Based Algorithm for Synthetic Range-Profile Motion Compensation," *IEEE Trans. on GRS*, vol. 46, no. 10, pp. 3053–3062, 2008.

[16] Karakasiliotis, A., A. Lazarov, P. Frangos, G. Boultadakis, and G. Kalognomos, "Two-Dimensional ISAR Model and Image Reconstruction with Stepped Frequency Modulated Signal," *IET Signal Processing*, Special Issue on ISAR, vol. 2, no. 3, pp. 277–290, 2008.

[17] Skolnik, M.I., *Radar Handbook*, 3d ed., McGraw Hill, 2008.

[18] Carrara, W.C., R.S. Goodman, and R.M. Majewsky, *Spotlight Synthetic Aperture Radar: Signal Processing Algorithms*, Artech House, 1995.

[19] Meta, A., and P. Hoogeboom, "Signal Processing Algorithms for FMCW Moving Target Indicator System Aperture Radar," *IEEE 2005 IGARSS*, July, Seoul, Korea, pp. 316–319, 2005.

CHAPTER 10

Bistatic ISAR

Bistatic inverse synthetic aperture radar (B-ISAR) imaging can theoretically be enabled in all bistatic radar configurations [1]. In this chapter, we will analyze the effects of bistatic geometry on the ISAR image formation. We will also study and understand B-ISAR imaging and how to implement it to overcome some limitations of monostatic ISAR, such as geometrical limitations, imaging of stealthy targets and to enable applications such as exploitation of bistatic synthetic aperture radar (SAR) systems, multistatic ISAR imaging, and passive ISAR imaging.

10.1 Basics of ISAR Imaging

10.1.1 Geometrical Limitations of Monostatic ISAR

To obtain ISAR images with a significant Doppler spread, the target is required to change its aspect angle with respect to the radar during the coherent processing interval (CPI). This requirement produces a set of geometrical cases where even if the target is moving with respect to the radar an ISAR image cannot be obtained. A simple case is given by a target moving along the radar's line of sight (LOS). In this case, the target aspect angle does not change in time, and hence an ISAR image cannot be produced. Such a problem is avoided when using a bistatic configuration, as also in this case a target's rotation with respect to the radar is produced. It may be argued that the same situation occurs when a target moves along the bistatic bisector (as will be clearer later). Nevertheless, the case that a target is flying directly at a monostatic radar is to be considered more likely as this could be intentional. Since the position of the receiver may be covert, in the case of a bistatic configuration a target would not know the direction of the bistatic bisector.

10.1.2 ISAR Imaging of Stealthy Targets

Stealthy targets are constructed to minimize the energy backscattered toward the radar. This makes them almost invisible to the eyes of a radar. Part of this is achieved by reflecting the electromagnetic energy toward directions other than that of the radar. Therefore, stealthiness usually refers only to monostatic radars. The use of a bistatic radar may enable the detection and, therefore, the imaging of stealthy targets, as the signal-to-noise ratio (SNR) and signal-to-clutter ratio (SCR) would likely increase with respect to the monostatic case.

10.1.3 Exploitation of Existing Bistatic SAR Systems

A number of bistatic SAR experiments have been conducted in the recent years to prove the effectiveness of bistatic radar imaging. The data collected by bistatic SAR systems could be processed as B-ISAR data; therefore, moving targets could be imaged using a B-ISAR processor.

10.1.4 Multistatic ISAR Imaging

Multistatic ISAR imaging may be achieved using one or more transmitters and a number of receivers. To maximize the gain out of such configurations, each receiver would benefit from acquiring the signal transmitted by other transmitters. This enables several bistatic configurations where the transmitters and the receivers are not colocated. To fully exploit multistatic configurations, the bistatic configuration must be first studied.

10.1.5 Passive ISAR Imaging

There is an increasing interest in the passive radar field, due to the possibility of exploiting illuminators of opportunity to detect and track targets. Limited bistatic range resolution has so far inhibited the interest in exploiting illuminators of opportunity to obtain B-ISAR configurations. Nevertheless, the possibility of exploiting modern digital broadcast communications may provide enough signal bandwidth to enable bistatic passive ISAR imaging.

Several techniques for image reconstruction have been proposed for bistatic synthetic aperture radar [2–5] that provide effective tools for radar imaging of stationary targets, such as in typical SAR applications. Unfortunately, such techniques do not apply in the case of ISAR because the target is moving in a noncooperative manner. On the other hand, ISAR imaging usually aims at providing images of relatively small targets when compared with SAR images, where the imaged area can reach the size of hundreds of square kilometers. This allows for the use of monostatic ISAR processors to be considered also when the geometry is bistatic. In this chapter, we will introduce bistatically equivalent monostatic (BEM) geometry, which will allow for a monostatic ISAR geometry to replace the bistatic one. This in turn enables any monostatic ISAR processing to be used in the B-ISAR case. Also, the limits of applicability of monostatic ISAR processing to a B-ISAR case are analyzed, both in the absence and presence of synchronization errors. The latter case should be considered because synchronization errors are usually an issue in bistatic radar systems.

It is worth mentioning the bistatic equivalence theorem as a tool for predicting bistatic from monostatic radar cross section (RCS) measurements [6,7]. Nevertheless, the equivalence stated in this chapter is related to geometrical properties of B-ISAR imaging and does not involve RCS properties.

A well-established method for analyzing radar imaging systems is the calculation of the point spread function (PSF; see Chapter 2 for its definition), which in our case will be addressed as bistatic PSF. It will be shown that the B-ISAR image PSF depends on the bistatic angle, which automatically introduces distortions in the B-ISAR image. The PSF of an imaging system also depends on the image formation processing adopted. Radial motion compensation followed by a range-Doppler technique will be considered as the ISAR image formation method, as it represents the standard procedure for obtaining ISAR images. Finally, multistatic ISAR is introduced to consider configurations where multiple transmitters and multiple receivers are combined to form a multistatic radar.

10.2 Geometry and Modeling

In this section, we introduce the terminology and theoretical aspects of B-ISAR imaging. The B-ISAR geometry is illustrated in Figure 10.1, where *Tx* represents the transmitter at *A*, and *Rx* represents the receiver at *B*.

As introduced in Chapters 2 and 3, the motion compensated received signal, for a generic radar configuration, can be written in a mathematical format. In this section, we will use a time-frequency format, which could be associated with a stepped-frequency type of radar (see Chapter 2). In formula, this is represented as follows:

$$S_R(f,t) = W(f,t)\int_V \rho(\boldsymbol{x})\exp\{-j\varphi(\boldsymbol{x},f,t)\}d\boldsymbol{x}, \tag{10.1}$$

where

$$W(f,t) = \text{rect}\left(\frac{t}{T_{obs}}\right)\text{rect}\left(\frac{f-f_0}{B}\right), \tag{10.2}$$

and where f_0 represents the carrier frequency, B is the transmitted signal bandwidth, T_{obs} is the observation time or the CPI, V is the spatial region where the target's reflectivity function, $\rho(\boldsymbol{x})$, is defined, and function rect(•) is the rectangular function. In a bistatic geometry, the function $\rho(\boldsymbol{x})$ represents the target's bistatic reflectivity function, and the phase term $\varphi(\boldsymbol{x}, f, t)$ in (10.1) can be written as follows:

$$\varphi(\boldsymbol{x},f,t) = \frac{2\pi f}{c}[R_t(t) + R_r(t) + \boldsymbol{x}\cdot\boldsymbol{i}_t(t) + \boldsymbol{x}\cdot\boldsymbol{i}_r(t)] = \frac{4\pi f}{c}[R_0(t) + K(t)\boldsymbol{x}\cdot\boldsymbol{i}_{BEM}(t)], \tag{10.3}$$

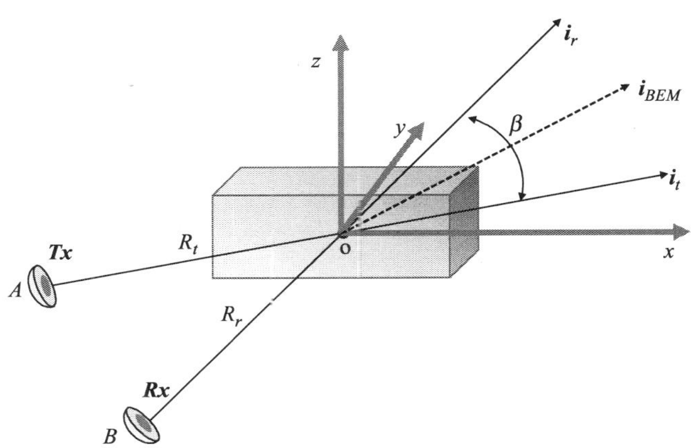

Figure 10.1 Bistatic ISAR geometry.

where

$$R_0(t) = [R_t(t) + R_r(t)]/2, \tag{10.4}$$

$$K(t) = |[\boldsymbol{i}_t(t) + \boldsymbol{i}_r(t)]/2|, \tag{10.5}$$

$$\boldsymbol{i}_{BEM}(t) = [\boldsymbol{i}_t(t) + \boldsymbol{i}_r(t)]/|\boldsymbol{i}_t(t) + \boldsymbol{i}_r(t)|. \tag{10.6}$$

In (10.3) to (10.6), c is the speed of light in free space, $R_t(t)$ is the distance from point A to O, $R_r(t)$ is the distance from point B to O, $\boldsymbol{i}_t(t)$ and $\boldsymbol{i}_r(t)$ are the unit vectors that indicate the target's LOS with respect to the transmitter and the receiver, respectively, the angle β between the $\boldsymbol{i}_t$ and $\boldsymbol{i}_r$ is called the bistatic angle as shown in Figure 10.1, and $\boldsymbol{x} = (x, y, z)$ is the vector that locates the position of a generic point on the target. By substituting (10.3) into (10.1) we can obtain the received signal model for the bistatic case:

$$S_R(f,t) = W(f,t)\int_V \rho(\boldsymbol{x})\exp\left\{-j\frac{4\pi f}{c}[R_0(t) + K(t)\boldsymbol{x}\cdot\boldsymbol{i}_{BEM}(t)]\right\}d\boldsymbol{x}. \tag{10.7}$$

The received signal model in (10.7) looks almost identical to the signal model in the monostatic case. Before introducing the BEM geometry, we will point out the differences between the bistatic and monostatic received signal models. These are found in:

- The function $\rho(\boldsymbol{x})$, which represents a bistatic reflectivity function rather than a monostatic one
- $R_0(t)$, which represents a BEM distance between point O (on the target) and the radar
- The BEM LOS $\boldsymbol{i}_{BEM}(t)$
- The presence of a distortion term $K(t)$

The presence of a bistatic reflectivity function instead of a monostatic one does not affect the image formation process but only the final image interpretation. The term $R_0(t)$ must be removed to successfully use conventional image formation, such as range-Doppler image formation. The fact that it is an equivalent distance does not change the problem as in the monostatic case. It represents an unknown function that must be estimated and removed. We can conclude that the first two differences do not involve any substantial changes in terms of ISAR image formation.

Particular attention must be paid to the last two differentiating factors, $\boldsymbol{i}_{BEM}(t)$ and $K(t)$. The former is important because it provides an interesting geometrical interpretation of B-ISAR imaging and the latter because it introduces image distortions due to the bistatic geometry.

10.3 Bistatically Equivalent Monostatic ISAR Geometry

The B-ISAR geometry can be reinterpreted in a convenient way by defining a BEM element. With the help of Figure 10.2, we can define the BEM element as a monostatic antenna element, which acts as a virtual transmitter and receiver (as in a monostatic radar). As a result of the definition of such a virtual element, the B-ISAR geometry can be substituted with a monostatic ISAR geometry by replacing the transmitting element, *Tx*, and the receiving element, *Rx*, with the BEM element. The BEM geometry allows for the interpretation of (10.3) as the received signal in a monostatic ISAR configuration, where $R_0(t)$ is

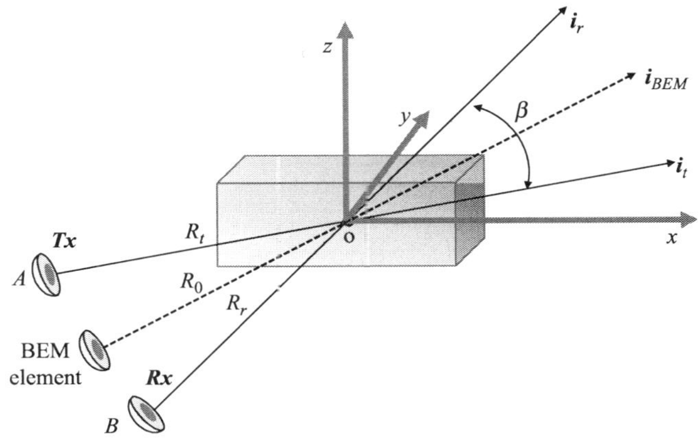

Figure 10.2 BEM ISAR geometry.

the distance from the BEM element to point O on the target, and $K(t)$ is a distortion term that takes into account the effects of the bistatic geometry.

10.4 Bistatic ISAR Image Formation

In the following subsections, we will detail all the steps required to form a B-ISAR image by implementing a bistatic range-Doppler image formation.

10.4.1 Radial Motion Compensation

Any of the techniques discussed for the case of monostatic ISAR systems may be used to perform the radial motion compensation in the bistatic case once the BEM signal model is adopted. The phase term $\varphi_0(t) = 4\pi f R_0(t)/c$ is equivalent to the phase term in the monostatic case since the virtual distance $R_0(t)$ is, as in the monostatic case, an unknown to be estimated and removed from the received signal. $K(t)$ does not affect the radial motion compensation. More details about radial motion compensation and image autofocus will be given later in Section 10.5 where synchronization errors are also introduced.

10.4.2 Bistatic ISAR Image Point Spread Function

$K(t)$ carries information about the bistatic geometry changes with respect to time. However, what significantly affects the PSF of the ISAR images is the change in the bistatic angle, θ, during the coherent processing interval (CPI). In this section, the ISAR image PSF will be derived for the bistatic case, and the distortion introduced by the bistatic geometry will be related to the bistatic angle variation. In deriving the PSF, two assumptions are made that will allow for the range-Doppler technique to be applied when reconstructing the ISAR image.

The two assumptions are (1) far-field condition and (2) short integration time. More specifically, the second assumption allows for the target rotation vector to be considered constant, whereas the combination of the two assumptions allows for the Fourier domain of the received signal to be rectangular. These assumptions are generally satisfied in ISAR scenarios where the resolutions required are not exceptionally high [8].

When the rotation velocity vector of the target is constant, the received signal reflected by a single ideal scatterer located at a generic point (x_0, y_0, z_0) can be rewritten, after translational motion compensation, as follows:

$$S_R(f,t) = A \cdot W(f,t) \cdot \exp\{-j\varphi(x_0, y_0, z_0, f, t)\}, \tag{10.8}$$

where $A = \rho(x_0, y_0, z_0)$ represents the reflectivity function of the ideal scatterer and

$$\varphi(x_0, y_0, z_0, f, t) = \varphi(x_0, y_0, f, t) = \frac{4\pi f}{c}[K(t)(x_0 \sin \Omega t + y_0 \cos \Omega t)]. \tag{10.9}$$

In (10.8) and (10.9), Ω is the norm (or modulus) of the effective rotation velocity vector of the target, and (x_0, y_0, z_0) are the coordinates of a generic scatterer on the target with respect to a body-fixed system centered on the target itself (Figures 10.1 and 10.2). The target's effective rotation velocity vector can be obtained from the target's rotation velocity vector by simply projecting the latter onto the plane orthogonal to $\boldsymbol{i_{BEM}}$. The equivalence in (10.9) is valid when the z-axis coincides with the effective rotation vector $\boldsymbol{\Omega_{\text{eff}}}$ (see Chapter 2). As the coordinate system embedded on the target can be chosen arbitrarily, we can assume that the equivalence in (10.9) is always satisfied. Under the aforementioned assumptions (1) and (2), bistatic angle changes are relatively small, even when a target covers relatively large distances during the CPI. In particular, the bistatic angle can be well approximated with a first-order Taylor-Maclaurin polynomial:

$$\beta(t) \cong \beta(0) + \dot{\beta}(0)t, \tag{10.10}$$

where

$$-T_{obs}/2 \leq t \leq T_{obs}/2, \tag{10.11}$$

and

$$\dot{\beta} = d\beta/dt. \tag{10.12}$$

As a result, $K(t)$ can also be well approximated with its first-order Taylor-Maclaurin polynomial. By using (10.5), the following equation is obtained:

$$K(t) \cong K_0(0) + K_1 t = \cos[\beta(0)/2] - \left[\dot{\beta}(0)/2\right] \cdot \sin[\beta(0)/2] \cdot t, \tag{10.13}$$

where the definitions of K_0 and K_1 are intrinsically defined in (10.13). Therefore, (10.10) becomes

$$\varphi(x_0, y_0, f, t) = \frac{4\pi f}{c}[K_0 + K_1 t](x_0 \sin \Omega t + y_0 \cos \Omega t). \tag{10.14}$$

For small integration angles or short CPI, the sinusoids can be approximated by means of linear functions as follows:

$$\varphi(x_0, y_0, f, t) = \frac{4\pi f}{c}[K_0 + K_1 t](x_0 \Omega t + y_0). \tag{10.15}$$

10.4.3 Image Formation

The image formation adopted is the range-Doppler technique, which uses two-dimensional (2-D) Fourier transforms: one along the frequency coordinate, f, for range compression; and one along the time coordinate, t, for cross-range compression. To obtain the PSF of the B-ISAR system, we calculate the two Fourier transforms analytically.

Range compression. The range compression is obtained by inverse Fourier transforming the signal in (10.8) as shown:

$$
\begin{aligned}
s_R(\tau,t) &= A\int_{-\infty}^{\infty} W(f,t)\exp\{-j\varphi(x_0,y_0,f\ t)\}\cdot\exp\{j2\pi f\tau\}df \\
&= A\cdot\delta\left(\tau-\frac{2}{c}[K_0+K_1t][x_0\Omega t+y_0]\right)\otimes_\tau w(\tau,t) \\
&= A\cdot B\cdot\text{sinc}\left\{B\left(\tau-\frac{2}{c}[K_0y_0-\Delta r_r(t)]\right)\right\}\exp\{j2\pi f_0\tau\}\text{rect}\left(\frac{t}{T_{obs}}\right) \\
&\quad\cdot\exp\left\{j\frac{4\pi f_0}{c}[K_0+K_1t][x_0\Omega t+y_0]\right\},
\end{aligned}
\tag{10.16}
$$

where

$$
w(\tau,t)=FT\{W(f,t)\}=B\cdot\exp\{j2\pi f_0\tau\}\text{rect}\left(\frac{t}{T_{obs}}\right)\text{sinc}(B\tau),\tag{10.17}
$$

and $\otimes_\tau$ is the convolution operator over the variable τ. As in (10.2), B is the transmitted signal bandwidth. Two effects are induced by the bistatic geometry on the generation of compressed range profiles:

1. The range position y_0 scaled by a factor K_0, so,the ideal scatterer is imaged in a new position $y_0' = K_0y_0$
2. A range migration induced by the bistatic angle variation during the CPI, which can be quantified as

$$
\Delta r_r(t) = K_0x_0\Omega t + K_1y_0t + K_1x_0\Omega t^2.\tag{10.18}
$$

While the first effect can be corrected a posteriori by rescaling the range coordinate, the second effect could be significantly detrimental. If range migration occurs, the position of one scatterer can be moved from one range cell to another during the integration time, thereby resulting in a distortion of the PSF. To avoid range migration, the following inequality must be satisfied:

$$
|\delta r(t)| = |K_0x_0\Omega t + K_1y_0t + K_1x_0\Omega t^2| < \frac{\Delta r_r}{2} = \frac{c}{4B},\quad \left(K_0, K_1, x, y, \Omega\ \&\ |t| < \frac{T_{obs}}{2}\right),\tag{10.19}
$$

where Δr_r is the radar range resolution.

It is worth pointing out that, in the monostatic case ($K_0 = 1$, $K_1 = 0$), the range migration term is equal to $|\delta r(t)| = |x_0\Omega t|$. Therefore, as $K_0 < 1$ in bistatic configurations, we can conclude that the bistatic case may produce a higher range migration effect only when there is a significant bistatic angle variation component (i.e., large values of K_1).

When the constraint in (10.19) is satisfied, (10.16) can be rewritten as follows:

$$s_R(\tau,t) = A\cdot\exp\left\{j\frac{4\pi f_0}{c}[K_0+K_1t][x_0\Omega t+y_0]\right\}\delta\left(\tau-\frac{2}{c}K_0y_0\right)\otimes_\tau w(\tau,t). \tag{10.20}$$

Cross-range compression. Cross-range compression is achieved by taking the Fourier transform of the range-compressed signal in (10.20) along the time variable, *t*. The result is a complex image in the time-delay (range) and Doppler domains. Therefore, the PSF can be calculated as follows:

$$\begin{aligned} PSF(\tau,\nu,x_0,y_0) &= A\cdot B\cdot\operatorname{sinc}\left(B\left[\tau-\frac{2}{c}K_0y_0\right]\right)\exp\{j2\pi f_0\tau\} \\ &\quad\cdot\int_{-\infty}^{\infty}\operatorname{rect}\left(\frac{t}{T_{obs}}\right)\exp\left\{j\frac{4\pi f_0}{c}[K_0+K_1t][x_0\Omega t+y_0]\right\}\exp\{-j2\pi t\nu\}dt \\ &= \delta\left(\tau-\frac{2}{c}K_0y_0\right)\delta\left(\nu-\frac{2f_0\Omega}{c}K_0x_0-\delta_{cr}\right)\otimes_\tau\otimes_\nu w(\tau,\nu)\otimes_\nu D(\nu), \end{aligned} \tag{10.21}$$

where $\otimes_\nu$ is the convolution operator over ν, and

$$D(\nu) = FT\left\{\exp\left(-j\frac{4\pi f_0K_1\Omega x_0}{c}t^2\right)\right\} \tag{10.22}$$

is a chirp-like distortion term. It should also be noted that there is a shift along the Doppler coordinate introduced by the bistatic angle change rate, K_1:

$$\delta_{cr} = 2f_0K_1y_0/c. \tag{10.23}$$

The PSF of the focusing point (i.e., the point O) is represented by

$$w(\tau,\nu) = A\cdot B\cdot T_{obs}\cdot\exp(j2\pi f_0\tau)\cdot\exp\left(j\frac{4\pi f_0}{c}K_0y_0\right)\operatorname{sinc}(T_{obs}\nu)\cdot\operatorname{sinc}(B\tau). \tag{10.24}$$

Under the assumption of a constant target's rotation vector and bistatic angle, β, the PSF of the B-ISAR image in (10.24) is space invariant. Desirable consequences derive from this property that will be discussed next.

10.5 Bistatic ISAR Image Interpretation

In Section 10.5, we addressed the problem of forming B-ISAR images using a range-Doppler method. We have also demonstrated that, under the assumption of constant target's rotation vector and bistatic angle, the PSF of the B-ISAR image is space invariant. Before commenting on the consequences of this property, we will demonstrate that the B-ISAR image,

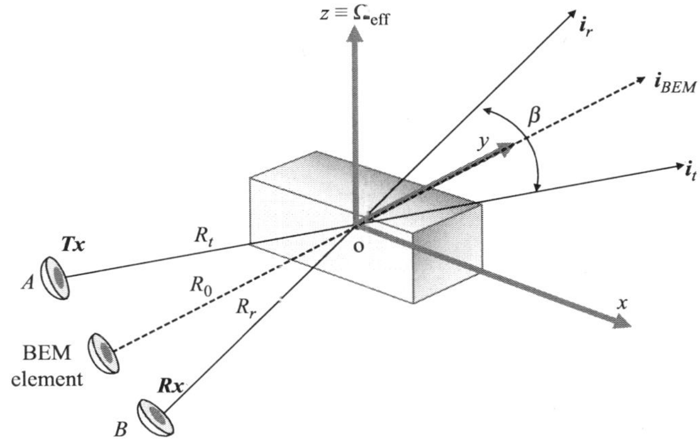

Figure 10.3 B-ISAR image projection plane.

under the same assumptions, can be interpreted as a filtered projection of the target's reflectivity function onto a plane, namely, the image projection plane (IPP).

To demonstrate this, we will refer to Figure 10.3, where the coordinate system embedded in the target is chosen such that the y-axis is aligned with the BEM LOS and the z-axis is aligned with the target's effective rotation vector velocity, and to equation (10.7), where the received signal is rewritten by using the BEM notation. Any target's rotation velocity vector can be written as the sum of two components, one aligned with the BEM LOS, which in this case coincides with the y-axis, and one orthogonal to it, which in this case is chosen to be aligned with the z-axis. The motion induced by a rotation vector of any point, $\boldsymbol{x}$, on the target can be calculated by solving the differential equation system $\dot{\boldsymbol{x}}(t) = \boldsymbol{\Omega} \times \boldsymbol{x}(t)$, with the initial condition $\boldsymbol{x}(0) = \boldsymbol{x}_0$. It can be readily proven that a target's rotation velocity vector aligned with the y-axis does not produce any Doppler component as $\dot{y}$ would be equal to zero. Any Doppler effect would be produced only by the component of Ω that is aligned with the z-axis, namely, the target's effective rotation velocity vector. Thus, the consequent effect is that the radar echo would carry information only relative to the x- and y-coordinates, as the third coordinate, z, does not produce any Doppler component and therefore does not affect the range coordinate within the limits of the far-field condition. In other words, the received signal produced by a point scatterer that rotates with respect to a vector aligned with the z-axis does not depend on the position of the scatterer along the same axis. This is equivalent to saying that, in the eyes of the radar, a three-dimesional (3-D) target can be assumed to be virtually the same as the projection of the same target onto the plane identified by the x-axis and y-axis. Such a plane is typically referred to as the image plane, as the consequent mapping of the projected 2-D reflectivity function onto the ISAR image becomes a $R^2 \rightarrow R^2$ mapping, that is, identifiable with a homomorphism. This also justifies equation (10.9).

The space-invariance of the PSF guarantees that each target's scattering center generates an image of itself that can be seen as a PSF located in the position of the scattering center (projected onto the image plane). Therefore, in the case of B-ISAR imaging, we can conclude that a B-ISAR image is a scaled version (by a factor K_0) of the projected and filtered reflectivity function.

To compare the bistatic with the monostatic case, we can finally state that the mechanism with which the B-ISAR image is formed is the same as that of the monostatic case except for a scaling operation and a different (bistatic rather than monostatic) physical scattering mechanism.

Moreover, the space-invariant characteristic of the PSF in B-ISAR image makes sure that the B-ISAR image is not distorted and that the resolution properties remain constant in any region of the image. This characteristic is desirable as the target's image shape; therefore, the relative position of scatterers is not altered, which is often a requirement for classification and recognition purposes.

10.6 The Effect of Synchronization Errors on B-ISAR Imaging

A bistatic radar system must be suitably synchronized to effectively enable B-ISAR imaging. In the most general case, three levels of synchronization must be achieved: space, time, and phase synchronization.

Space synchronization is needed because antennas with high gain have a narrow beam pattern, which results in a relatively small illuminated area. In a bistatic system, therefore, the transmitting and the receiving antenna beams must be directed toward the same area to maximize the received signal signal-to-noise ratio (SNR) and consequently the SNR in the formed ISAR image. The transmitting and receiving subsystems must be cooperative and controlled by a central system to make sure that this condition is met. In dynamic scenarios, this level of synchronization may be a difficult task.

Time synchronization allows for correct target ranging and time gating for the selected data subset to be processed in order to form an ISAR image. In software-defined radar systems, a finer time synchronization is also needed because the same time reference is used to synchronize analog-to-digital converters (ADC) and field programmable gate arrays (FPGAs) for signal preprocessing, including signal down-conversion, demodulation, and filtering.

Phase synchronization is needed to align radio frequency (RF) oscillators for signal down-conversion in standard radar RF front ends. An oscillator's frequency may drift due to temperature changes and other physical parameters. As two separate oscillators are used to up-convert (transmitting station) and down-convert (receiving station) the RF signal, coherence must be guaranteed by synchronizing the two oscillators. This latter problem is usually the dominant issue in radar systems that require a high level of coherency, such as radar imaging systems [9,10]. In the next subsection, a detailed analysis is shown that quantifies the effects of phase synchronization errors in the formation of ISAR images.

10.6.1 Phase Synchronization Errors and Image Distortion Analysis

The phase term, $\varphi(\mathbf{x}, f, t)$, defined in (10.3) was considered to be dependent only on the radar-target configuration. In this section, such a result is extended to a more realistic scenario where phase synchronization errors are present. Firstly, synchronization errors are modeled as the sum of two components: a frequency drift and a frequency jitter:

$$\Delta f(t) = \eta_0 + \eta_1 t + \delta F(t), \tag{10.25}$$

where η_0 is a frequency offset, η_1 is the coefficient of a linear frequency drift, and $\delta F(t)$ is a random frequency jitter that accounts for noise-like phase jitters in both the transmitter and receiver local oscillator.

Oscillator synchronization errors introduce an effect on the phase term in the received signal. Such an effect can be analyzed separately by considering the phase term associated with the target's radial motion and the phase term associated with the target's rotational motion. Specifically, the former term can be written as follows:

$$\varphi_0(f,t) = -\frac{4\pi[f+\Delta f(t)]}{c}R_0(t) = -\frac{4\pi[f+\eta_0+\eta_1 t+\delta F(t)]}{c}\cdot\frac{R_t(t)+R_r(t)}{2}, \quad (10.26)$$

whereas the latter term can be expressed as follows:

$$\varphi(x_0,y_0,f,t) = \frac{4\pi}{c}\{[f+\eta_0+\eta_1 t+\delta F(t)]K(t)[x_0\sin\Omega t + y_0\cos\Omega t]\}, \quad (10.27)$$

where the same consideration as in (10.9) can be made regarding the dependence of (10.27) on two of the three scatterer coordinates.

Autofocus in the presence of synchronization errors. Parametric autofocus algorithms are often employed in ISAR imaging because they provide more accurate solutions to the radial motion compensation problem [11]. The effects of bistatic synchronization errors are analyzed here for a specific algorithm, namely, the image contrast based autofocus (ICBA) as introduced in Chapter 5, although the conclusions may be extended to any parametric and nonparametric autofocus technique. To use such parametric algorithms, a phase term model of bistatically equivalent radial motion must be derived. A quadratic polynomial phase term is generally used as an approximation of the radial motion in the monostatic case:

$$[R_t(t)+R_r(t)]/2 = \alpha_0+\alpha_1 t+\alpha_2 t^2. \quad (10.28)$$

The validity of the monostatic approximation in the bistatic case can be simply demonstrated. In fact, the bistatic phase history associated with an arbitrary scatterer on the target can be seen as the semi-sum of two monostatic phase histories because they would be seen separately from points *Tx* and *Rx*. Therefore, if the monostatic phase histories can be approximated with *n*-th order polynomials, then the semi-sum of these phase histories can be approximated with a polynomial of the same order.

Therefore, by substituting (10.28) into (10.26) we obtain

$$\begin{aligned}\varphi_0(f,t) &= -\frac{4\pi[f+\eta_0+\eta_1 t+\delta F(t)]}{c}\cdot\left(\alpha_0+\alpha_1 t+\alpha_2 t^2\right)\\ &\cong -\frac{4\pi}{c}\cdot\{(f+\eta_0)\alpha_0+[(f+\eta_0)\alpha_1+\eta_1\alpha_0]t+[(f+\eta_0)\alpha_2+\eta_1\alpha_1]t^2+\eta_1\alpha_2 t^3\}.\end{aligned} \quad (10.29)$$

It is worth noting that the term $\delta F(t)$ is usually much smaller than the frequency offset and linear drift and therefore it will be neglected from now on. Moreover, phase noise is also present in monostatic radar systems and thus is not seen as a problem introduced by the bistatic geometry. In addition, the resultant phase noise variance in a B-ISAR system would be the sum of the *Tx* and *Rx* oscillator phase noise variances (independency of the oscillators), so it should be considered slightly larger than in the monostatic case.

Furthermore, $\varphi_0(f, t)$ can be seen as a consequence of a virtual radial motion, which differs from the actual radial motion because of the presence of synchronization errors. The presence of a phase offset and a phase drift will not significantly affect the estimation of the bistatically equivalent radial motion parameters. A three-parameter estimation problem can be stated by rewriting (10.29) as follows:

$$\varphi_0(f,t) = -\frac{4\pi f}{c}\left(\xi_0 + \xi_1 t + \xi_2 t^2 + \xi_3 t^3\right), \tag{10.30}$$

where

$$\begin{aligned} \xi_0 &= (f+\eta_0)\alpha_0 \\ \xi_1 &= (f+\eta_0)\alpha_1 + \eta_1\alpha_0 \\ \xi_2 &= (f+\eta_0)\alpha_2 + \eta_1\alpha_1 \\ \xi_3 &= \eta_1\alpha_2 \end{aligned}. \tag{10.31}$$

As also mentioned in Chapter 5, for the monostatic case ξ_0 can be ignored since it induces a range shift only in an ISAR image. Also, the way of correcting the phase via parametric autofocus techniques requires multiplying the received signal in the frequency/slow-time domain with a term that is linearly dependent on the frequency. For the sake of clarity and completeness, the general form for the compensation phase term up to the order N is shown as

$$\varphi_c(f,t) = \frac{4\pi f}{c}\sum_{k=1}^{N}\gamma_k t^k. \tag{10.32}$$

Because the dependence on the frequency in (10.30) is not linear, the solution obtained by using (10.30) can be considered as only an approximated solution. Nevertheless, after rewriting (10.30) into a form that reflects that of (10.32) and by neglecting ξ_0, we obtain

$$\varphi_0(f,t) = -\frac{4\pi f}{c}\left(\xi_1' t + \xi_2' t^2 + \xi_3' t^3\right), \tag{10.33}$$

where

$$\begin{aligned} \xi_1' &= \left(1+\frac{\eta_0}{f}\right)\alpha_1 + \frac{\eta_1\alpha_0}{f} \\ \xi_2' &= \left(1+\frac{\eta_0}{f}\right)\alpha_2 + \frac{\eta_1\alpha_0}{f} \\ \xi_3' &= \frac{\eta_1\alpha_2}{f} \end{aligned}. \tag{10.34}$$

The following considerations can be made:

1. Since $\eta_0 \ll f$, the first and second offset parameters can be reasonably approximated with

$$\xi_1' \simeq \alpha_1 + \frac{\eta_1\alpha_0}{f} \tag{10.35}$$

and

$$\xi_2' \simeq \alpha_2 + \frac{\eta_1\alpha_0}{f}. \tag{10.36}$$

2. As shown in Chapter 5, relative to the monostatic ISAR case, the second term in (10.35) produces a shift only along the Doppler coordinate. Therefore, the first term in (10.35) can be approximated with $\xi_1' \approx \alpha_1$.
3. When the following inequality is satisfied:

$$\left|\frac{4\pi}{c}\left(\eta_1\alpha_1 T_{obs}^2 + \eta_1\alpha_2 T_{obs}^3\right)\right| < \frac{\pi}{4}, \tag{10.37}$$

the focusing parameters ξ_2' in (10.36) and ξ_3' in (10.34) can be approximated with the following:

$$\begin{aligned} \xi_2' &\simeq \alpha_2 \\ \xi_3' &= 0 \end{aligned}. \tag{10.38}$$

To ensure that the inequality in (10.38) is satisfied, the following more restrictive constraints are applied:

$$\begin{aligned} |\eta_1| &< \frac{c}{32|\alpha_1|T_{obs}^2} \\ |\eta_2| &< \frac{c}{32|\alpha_2|T_{obs}^2} \end{aligned}. \tag{10.39}$$

Depending on the type of oscillators used, the frequency drift may vary from 10^{-6} to 10^{-12} or better within a temperature range of 50° to 100°. Low-cost quartz oscillators with no temperature control have frequency drifts of 10^{-6} over a range of 50° to 100°, whereas temperature-controlled quartz oscillators can reduce the drift to 10^{-7} over the same temperature range. Oven-controlled quartz oscillators may improve to 10^{-8}, and expensive rubidium or cesium oscillators may reach frequency drifts of 10^{-12} or better. Moreover, in typical ISAR applications, the CPI (or T_{abs}) is in the order of one second. The inequalities in (10.39) are usually satisfied unless targets are moving very fast, integration time is long, and oscillators have large drifts. The straightforward result is that, in practice, the parametric autofocus techniques proposed in the literature for monostatic ISAR can be applied to B-ISAR to remove the bistatically equivalent radial motion, provided that the constraints in (10.39) are satisfied. Moreover, nonparametric autofocus techniques, such as the prominent point processing (PPP) and phase gradient autofocus (PGA), as described in Chapter 5, would not be affected at all by oscillator frequency offsets and drifts since they would treat the phase to be compensated as a random process. The consequent result is that also nonparametric autofocus techniques proposed for the monostatic case are applicable to the bistatic case.

B-ISAR image formation in the presence of synchronization errors. After performing radial motion compensation via the single-step ICBA technique introduced in Chapter 5, in this section we analyze the effects of synchronization errors on the B-ISAR image formation.

The term $K(t)$ carries information about the change in time of the bistatic geometry. Changes in the bistatic angle during the CPI significantly affect the PSF of the B-ISAR image as demonstrated at the beginning of this chapter. In this section, the PSF will be derived when synchronization errors are also present. In deriving the PSF, two assumptions are made that will allow the range-Doppler technique to be used when reconstructing the

ISAR image after the radial motion compensation: (1) far-field condition and (2) short integration time. Such assumptions are generally satisfied in typical ISAR scenarios where the resolutions required are not very high. When the target rotation vector is constant, the received signal reflected by a single ideal scatterer located at a generic point can be written as in (10.9), where the phase term, after motion compensation, is represented in the next equation (10.40).

It is worth pointing out that (10.27) contains both the effects of the bistatic angle changes and synchronization errors. By considering the approximation in (10.13), (10.27) may be rewritten as follows:

$$\varphi(x_0, y_0, f, t) = \frac{4\pi}{c}\{[f + \eta_0 + \eta_1 t + \delta F(t)] \cdot [K_0 + K_0 t] \cdot [x_0 \sin \Omega t + y_0 \cos \Omega t]\}. \quad (10.40)$$

When the target undergoes small rotations (typically occurring when short integration time is used) and by neglecting the random phase term $\delta F(t)$, (10.40) can be approximated by

$$\varphi(x_0, y_0, f, t) \cong \frac{4\pi}{c}\{[f + \eta_0 + \eta_1 t] \cdot [K_0 + K_0 t] \cdot [x_0 \Omega t + y_0]\}. \quad (10.41)$$

Range compression. The range-compressed signal is obtained by calculating the Fourier transform (FT) with respect to f as follows:

$$\begin{aligned} s'_R(\tau, t) &= A \int_{-\infty}^{\infty} W(f, t) \exp\{-j\varphi(x_0, y_0, f, t)\} \cdot \exp\{j2\pi f \tau\} df \\ &= A \cdot \exp\left\{-j\frac{4\pi(\eta_0 + \eta_1 t)}{c}[K_0 + K_1 t][x_0 \Omega t + y_0]\right\} \\ &\quad \delta\left(\tau - \frac{2}{c}[K_0 + K_1 t][x_0 \Omega t + y_0]\right) \otimes_\tau w(\tau, t) \\ &= A \cdot B \cdot \text{sinc}\left\{B\left(\tau - \frac{2}{c}[K_0 y_0 - \Delta r_r(t)]\right)\right\} \exp\{j2\pi f_0 \tau\} \text{rect}\left(\frac{t}{T_{obs}}\right) \\ &\quad \cdot \exp\left\{j\frac{4\pi}{c}(f_0 + \eta_0 + \eta_1 t)(K_0 + K_1 t)(x_0 \Omega t + y_0)\right\}, \end{aligned} \quad (10.42)$$

where $w(\tau, t)$ is defined in (10.17), and $\Delta r(t)$ in (10.18). It must be pointed out that both the range compression and migration components are (1) dependent on the bistatic angle and (2) independent of synchronization errors. Therefore, the same constraint as in (10.19) should be applied to avoid range migration.

When the constraint in (10.19) is satisfied, the range-compressed signal in (10.42) can be rewritten as follows:

$$\begin{aligned} s'_R(\tau, t) &\cong A \cdot B \cdot \text{sinc}\left[B\left(\tau - \frac{2}{c}K_0 y_0\right)\right] \exp\{j2\pi f_0 \tau\} \text{rect}\left(\frac{t}{T_{obs}}\right) \\ &\quad \cdot \exp\left\{j\frac{4\pi}{c}(f_0 + \eta_0 + \eta_1 t)(K_0 + K_1 t)(x_0 \Omega t + y_0)\right\} \end{aligned}. \quad (10.43)$$

Cross-range compression. The cross-range compression is obtained by calculating the FT with respect to the slow-time variable, t. Therefore the ISAR image PSF can be calculated as follows:

$$PSF(\tau, v, x_0, y_0) = A \cdot B \cdot \operatorname{sinc}\left(B\left[\tau - \frac{2}{c}K_0 y_0\right]\right)\exp\{j2\pi f_0 \tau\}$$

$$\cdot \int_{-\infty}^{\infty} \operatorname{rect}\left(\frac{t}{T_{obs}}\right)\exp\left\{j\frac{4\pi}{c}(f_0 + \eta_c + \eta_1 t)(K_0 + K_1 t)(x_0\Omega t + y_0)\right\}\exp\{-j2\pi t v\}dt,$$

$$= \delta\left(\tau - \frac{2}{c}K_0 y_0\right)\delta\left(v - \frac{2f_0\Omega}{c}K_0 x_0\right)\otimes_\tau \otimes_v w(\tau, v)\otimes_v D(v) \tag{10.44}$$

where $w(\tau, t)$ is defined in (10.24) and

$$D(v) = \int_{-\infty}^{\infty} \exp\left\{\frac{4\pi}{c}(\gamma_0 + \gamma_1 t + \gamma_2 t^2 + \gamma_3 t^3)\right\}\exp(-j2\pi v t)dt, \tag{10.45}$$

with

$$\gamma_0 = (f_0 - \eta_0)K_0 y_0, \tag{10.46}$$

$$\gamma_1 = -\eta_0 K_0 \Omega x_0 - [\eta_1 K_0 - (f_0 - \eta_0)K_1 y_0], \tag{10.47}$$

$$\gamma_2 = (f_0 - \eta_0)K_1 \Omega x_0 - \eta_1(K_0 \Omega x_0 + K_1 y_0), \tag{10.48}$$

and

$$\gamma_3 = -\eta_1 K_1 \Omega x_0. \tag{10.49}$$

The distortion term in (10.45) can be rewritten in terms of a linear phase term and a quadratic/cubic phase term as follows:

$$D(v) = \exp\{j\theta_0\}D_1(v)\otimes_v D_2(v), \tag{10.50}$$

where the linear term is

$$D_1(v) = \int_{-\infty}^{\infty} \exp\left\{\frac{4\pi}{c}\gamma_1 t\right\}\exp(-j2\pi v t)dt = \delta\left(v - \frac{2}{c}\gamma_1\right), \tag{10.51}$$

the quadratic/cubic term is

$$D_2(v) = \int_{-\infty}^{\infty} \exp\left\{\frac{4\pi}{c}(\gamma_2 t^2 + \gamma_3 t^3)\right\}\exp(-j2\pi v t)dt, \tag{10.52}$$

and

$$\theta_0 = \frac{4\pi}{c}(f_0 - \eta_0)K_0 y_0. \tag{10.53}$$

The term γ_0 does not produce any distortions or shift since the phase term, $4\pi\gamma_0/c$, is constant.

10.6.2 Comments on the Bistatic PSF in the Presence of Synchronization Errors

It should be noted that the position of the scattering center in the Doppler-delay time domain, represented by the two delta functions, is scaled with respect to the monostatic case by the bistatic factor, K_0. Some more remarks follow:

1. The PSF of the range-Doppler B-ISAR image, in the presence of synchronization errors, can be interpreted as a distorted version of the monostatic PSF.
2. A linear space-variant distortion term, $D_1(v)$, causes scattering center shifts along the Doppler coordinate that depend on their spatial position (both range and cross-range coordinates). Although this term looks like a shift term, it provokes an image distortion because it introduces a spatial-dependent scatterer's misplacement.
3. A nonlinear space-variant distortion term, $D_2(v)$, causes chirp-like and third-order distortions that result in PSF smearing (typical defocusing effect).

A detailed distortion analysis follows that aims at quantifying the effects of $D_1(v)$ and $D_2(v)$.

10.6.3 B-ISAR Image Distortion Analysis

Linear, quadratic (i.e., chirp-like) and cubic distortions are considered separately to provide a quantitative analysis of the distortions introduced by the joint action of bistatic angle changes and phase synchronization errors.

Linear distortions. The linear distortion term can be quantified in terms of relative Doppler shift and can be compared with the Doppler resolution for assessing its impact on the B-ISAR image. Specifically, the Doppler induced shift caused by the linear distortion term can be expressed by

$$\Delta_1 \cong -\frac{2}{c}\{\eta_0 K_0 \Omega x_0 - [\eta_1 K_0 - (f_0 - \eta_0)K_1 y_0]\} = -\frac{2}{c}\{\eta_0 K_0 \Omega x_0 - (\eta_1 K_0 - f_0 K_1)y_0\}, \tag{10.54}$$

where the approximation is accurate since $\eta_0 \ll f_0$.

It should be noted that the Doppler shift (or the cross-range shift) depends on both scatterer spatial coordinates (in the projected image plane). Nevertheless, it can be argued that the linear distortion in (10.54) can be neglected when its absolute value is smaller than that of the Doppler resolution cell:

$$|\Delta_1| < 1/T_{obs}. \tag{10.55}$$

By applying (10.55) and after simple manipulations, we can rewrite (10.54) as follows:

$$|\eta_0 K_0 \Omega x_0 - (\eta_1 K_0 - f_0 K_1)y_0| < |\eta_0 K_0 \Omega x_0| + |\eta_1 K_0 y_0| + |f_0 K_1 y_0| < \frac{c}{2T_{obs}}. \tag{10.56}$$

It can be noted that, in the worst bistatic scenario ($K_0 = 1$, which actually represents the monostatic case) and even considering poor performance oscillators, the following two components can be disregarded:

$$|\eta_0 x_0| < \frac{c}{2|\Omega| T_{obs}}, \tag{10.57}$$

and

$$|\eta_1 y_0| < \frac{c}{2T_{obs}}. \tag{10.58}$$

Therefore, the remaining constraint can be expressed as follows:

$$|K_1 y_0| < \frac{c}{2 f_0 T_{obs}}, \tag{10.59}$$

which is purely a geometrical constraint. In other words, synchronization errors, even when coupled with bistatic angle changes, do not cause significant linear distortions. It is also important to note that the Doppler displacement caused by the dominant component (10.59) is linearly dependent on the range coordinate, y_0, and on the bistatic angle change rate, K_1:

$$\Delta_1 \cong f_0 K_1 y_0. \tag{10.60}$$

Therefore, given a bistatic angle change rate, the Doppler displacement becomes proportional to the range coordinate. This is an important result as the linear distortion given by the term in (10.60) only causes a linear range-dependent Doppler shift. The visible effect for small values of the Doppler shift in (10.60) is an apparent target rotation in the B-ISAR range-Doppler domain, although it is not a physical rotation.

Quadratic distortions. The quadratic term causes chirp-like distortions in the B-ISAR image. To limit the distortions to an acceptable level, the following constraint must be taken into account:

$$|\Delta_2| < 1/T_{obs}^2, \tag{10.61}$$

where

$$\Delta_2 \cong -\frac{2}{c}\eta_2 = -\frac{2}{c}\{(f_0 - \eta_0)K_0\Omega x_0 - \eta_1(K_0\Omega x_0 + K_1 y_0)\}. \tag{10.62}$$

By approximating $f_0 - \eta_0 \approx f_0$ and by considering the worst bistatic scenario ($K_0 = 1$), the following more restrictive constraint can be formulated

$$\left|\frac{2}{c} f_0 K_1 \Omega x_0\right| + \left|\frac{2}{c}\eta_1 K_0 \Omega x_0\right| + \left|\frac{2}{c}\eta_1 K_1 y_0\right| < \frac{1}{T_{obs}^2}, \tag{10.63}$$

which can be broken into three separate constraints by constraining each term to be less than a third of $\frac{1}{T_{obs}^2}$, as follows:

$$|K_1 x_0| < \frac{c}{6 f_0 |\Omega| T_{obs}^2}, \tag{10.64}$$

$$|\eta_1 K_0 x_0| < \frac{c}{6|\Omega| T_{obs}^2}, \tag{10.65}$$

and

$$|\eta_1 K_1 y_0| < \frac{c}{6T_{obs}^2}. \tag{10.66}$$

It must be pointed out that:

(a) The constraint in (10.64) is a geometrical-only constraint, since it only depends on the bistatic angle;
(b) The constraint in (10.65) only depends on synchronization errors; and
(c) The constraint in (10.66) depends on both the bistatic angle and synchronization errors.

It is important to note that the inequalities in (10.64) and (10.65) can be neglected unless very rapid bistatic angle changes and very large synchronization errors occur.

Cubic distortions. To limit cubic distortions, the following constraint applies:

$$|\Delta_3| < 1/T_{obs}^3, \tag{10.67}$$

where

$$\Delta_3 \cong \frac{2}{c}\gamma_3 = -\frac{2}{c}\eta_1 K_1 \Omega x_0. \tag{10.68}$$

Also, the quantity Δ_3 can be neglected unless very rapid bistatic angle changes occur in conjunction with very large synchronization errors. The inequality in (10.67) is rewritten in (10.69) to highlight the dependence on both synchronization errors and bistatic angle changes

$$|\eta_1 K_1 x_0| < \frac{c}{2|\Omega| T_{obs}^3}. \tag{10.69}$$

Although the joint action of bistatic geometry changes and phase synchronization errors cause distortions, for these to be noticeable in the B-ISAR image, particular scenarios with strong bistatic angle changes and large synchronization errors must be considered.

In the rare event of significant quadratic distortions, the quadratic distortion term can be nullified by using a time-frequency–based image formation (see Chapter 3) instead of the Fourier transform to form the image along the Doppler coordinate, as formally demonstrated in a number of papers [12–14].

10.7 Examples

In this section we present a few examples that aim at (1) verifying the effectiveness of monostatic ISAR autofocus techniques when used for B-ISAR, and (2) analyzing the distortion effects caused by both bistatic angle changes and synchronization errors when using a monostatic ISAR processor.

The simulator used for generating ISAR data makes use of the point-scatterer model of target and generates radar returns by assuming that the transmitted signal is a stepped-frequency waveform. The point-scatterer model resembles an airplane, and it is shown in Figure 10.4. The waveform parameters and the bistatic geometry used change according to the specific experiment and will be detailed for each single experiment.

10.7.1 B-ISAR Autofocus

First, the simulation setup is defined and then the results are shown in three cases, which include a case with no synchronization errors and two cases with strong and severe

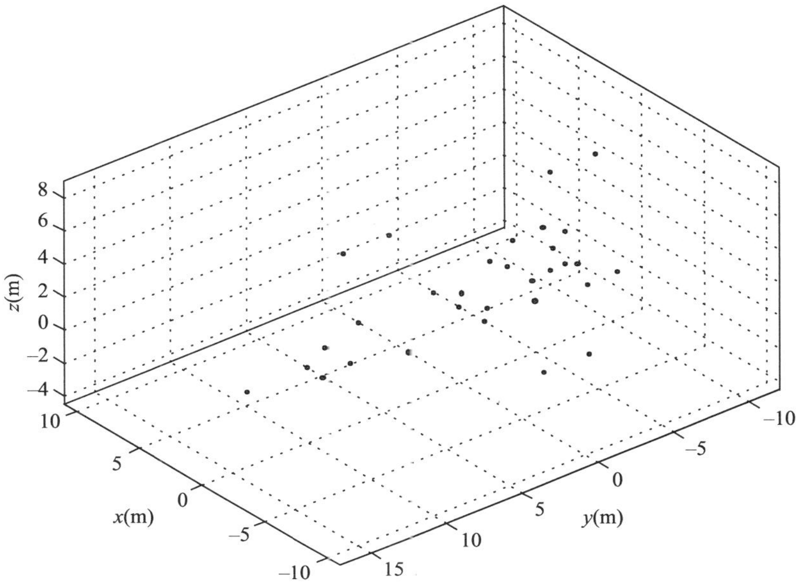

Figure 10.4 Point-scatterer model of an airplane.

Table 10.1 Simulation Parameters

Parameter	Value
Center frequency: f_0	10 GHz
Bandwidth: B	500 MHz
Number of frequencies	128
Number of bursts	128
Frequency offset: η_0	20 KHz
Coefficient of linear frequency drift: η_1	0, 0.1 MHz (Case 1), 100 MHz (Case 2)

synchronization errors. It is worth noting that the bistatic angle changes do not affect the ISAR image autofocus process. Therefore, although defined for the sake of completeness, the bistatic geometry will not play a fundamental role in this example.

Simulation setup. The waveform and phase synchronization error parameters are summarized in Table 10.1, and the bistatic geometry is defined in Figure 10.5. The target moves along a rectilinear trajectory, forming an angle of 45° with respect to the LOS of the *Tx* with a speed of 500 km/hr. Both *Tx* and *Rx* stations are stationary. A frequency drift such as that of Case 2 has been selected for the sole purpose of demonstrating the effects of synchronization errors on the ISAR image autofocus, since any low-cost oscillator is able to keep the frequency drift well below that value. The CPI is set equal to $T_{obs} = 0.8$ sec.

Simulation results. The simulation results prove what was predicted by the theory. In fact, a defocusing effect appears (as shown in Figure 10.8 when compared with Figures 10.6 and 10.7) only in the case of a severe frequency drift (Case 2, $\eta_1 = 100$ MHz).

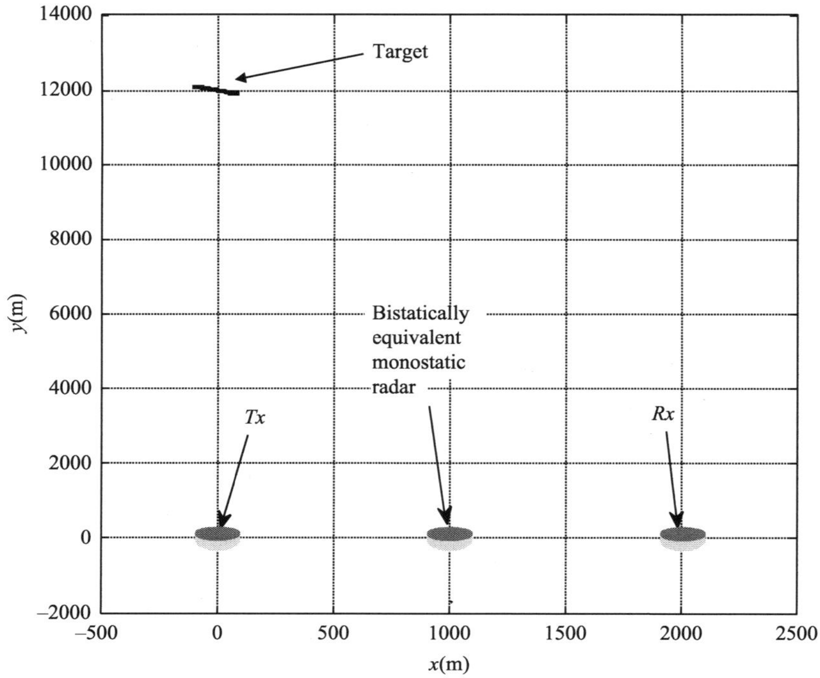

Figure 10.5 Simulation geometry—autofocus.

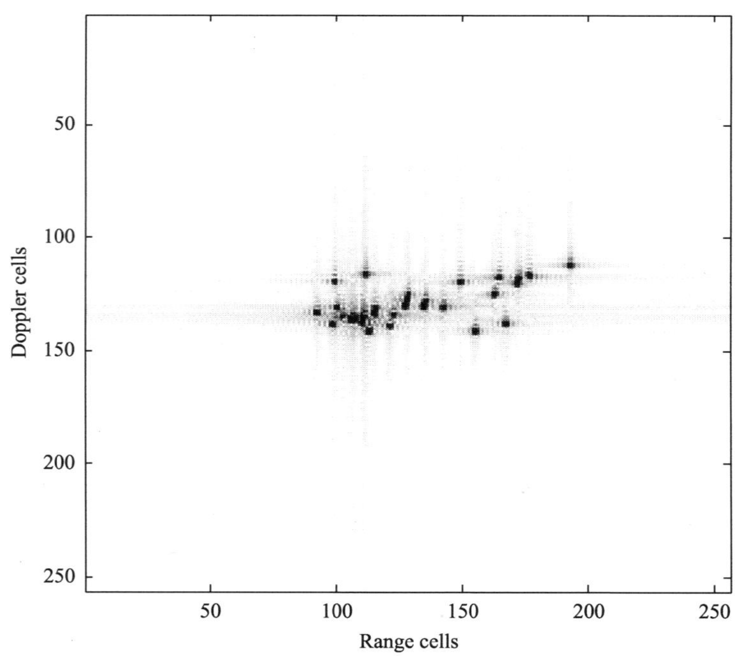

Figure 10.6 B-ISAR image—no synchronization errors.

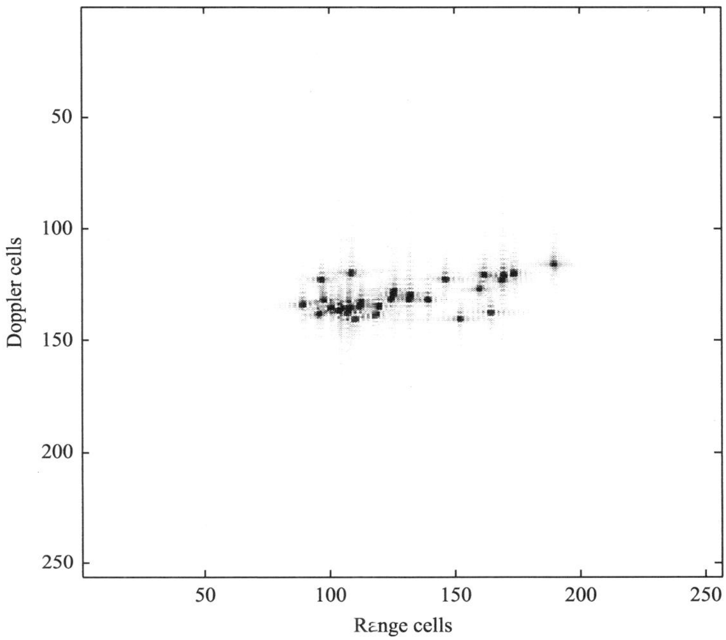

Figure 10.7 B-ISAR image—Case 1.

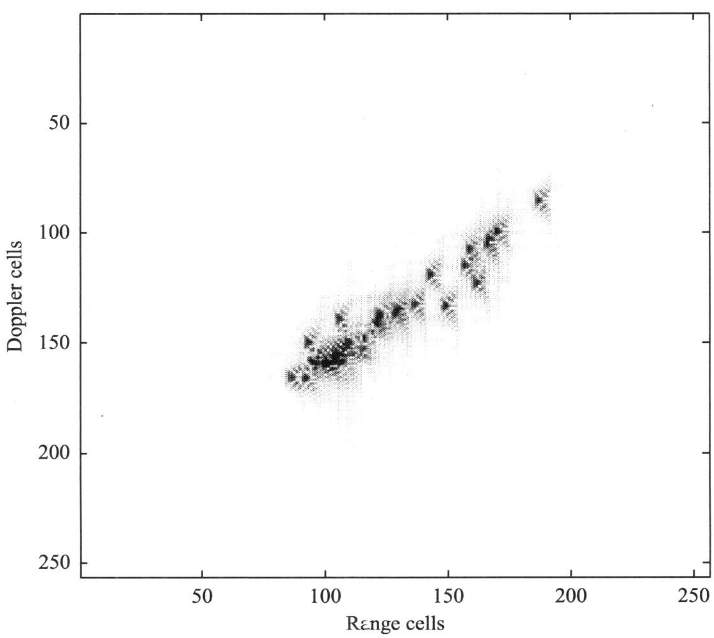

Figure 10.8 B-ISAR image—Case 2.

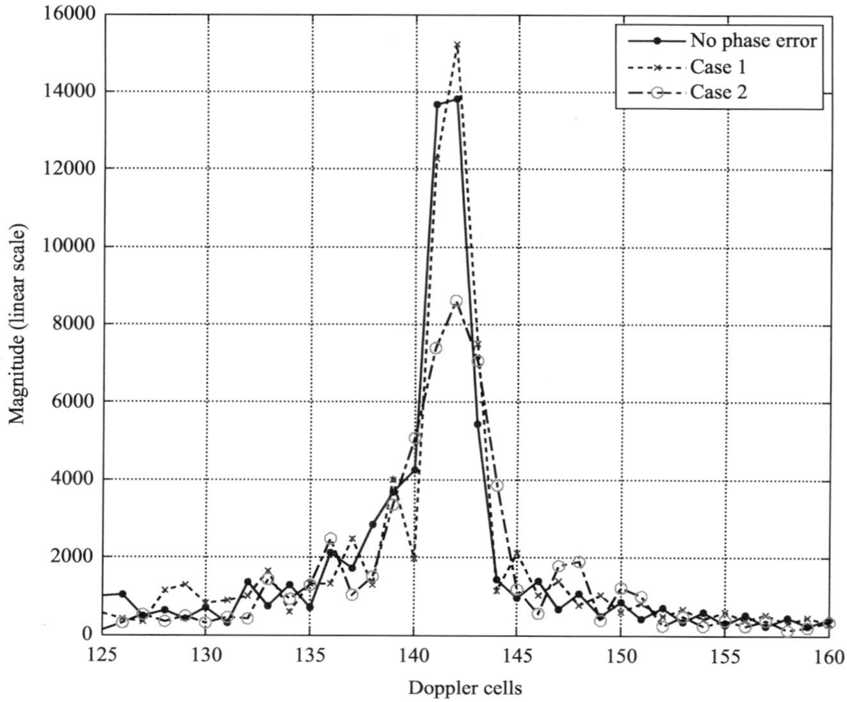

Figure 10.9 Doppler section superposition.

Table 10.2 IC and IE

Case	IC	IE
No sync error	6.1	4.4
Case 1	6.1	4.4
Case 2	3.9	5.7

To highlight the effects of phase synchronization errors, a superposition of Doppler sections relative to a single scatterer is shown in Figure 10.9. The section relative to Case 1 practically coincides with that of the case with no synchronization errors. The section relative to Case 2 shows a significant main lobe attenuation and sidelobe increase. To quantify this result, the image contrast (IC) and the image entropy (IE) have been calculated for the two cases and are shown in Table 10.2. The numerical results obtained in Case 1 are the same as with no synchronization errors. Also, the severe frequency drift in Case 2 causes a linear distortion since the constraint in (10.56) is not satisfied. The clear effect is that of an apparent target rotation, such as that displayed in Figure 10.8 compared with Figures 10.6 and 10.7.

10.7.2 Bistatic ISAR Point Spread Function Distortion Analysis

Two cases are analyzed in this subsection to support the theoretical results relative to linear and quadratic distortions introduced by the joint action of bistatic angle changes and phase synchronization errors. Specifically, the first experiment is set up such that only linear distortions affect the B-ISAR image, whereas in the second experiment both linear and quadratic distortions are shown.

Simulation setup—linear distortions. Error parameters of the waveform and phase synchronization are the same as shown in Table 10.1, relative to Case 1, whereas the bistatic geometry is described in Figure 10.10. The target moves along a linear trajectory forming an angle of 45° with the *Tx* LOS at a speed of 1,000 km/hr. Both *Tx* and *Rx* are static. The short distance between target and both *TX* and *RX* has been chosen such that the bistatic angle changes are strong enough to cause linear distortions. The CPI is set equal to $T_{obs} = 0.3$ sec.

Simulation results—linear distortions. The strong bistatic angle variation does not satisfy the constraint in (10.56). The linear distortion effect, as highlighted in Figure 10.11, is that of shifting scattering centers along the Doppler coordinate without causing any energy spread. To prove that a Doppler shift has occurred, the ISAR image that would be produced by a monostatic radar colocated with the BEM element (BEM ISAR image) is shown in Figure 10.12. In the case of no synchronization errors and no bistatic angle changes, the two ISAR images should be identical, apart from a scaling factor, K_0. As theoretically predicted,

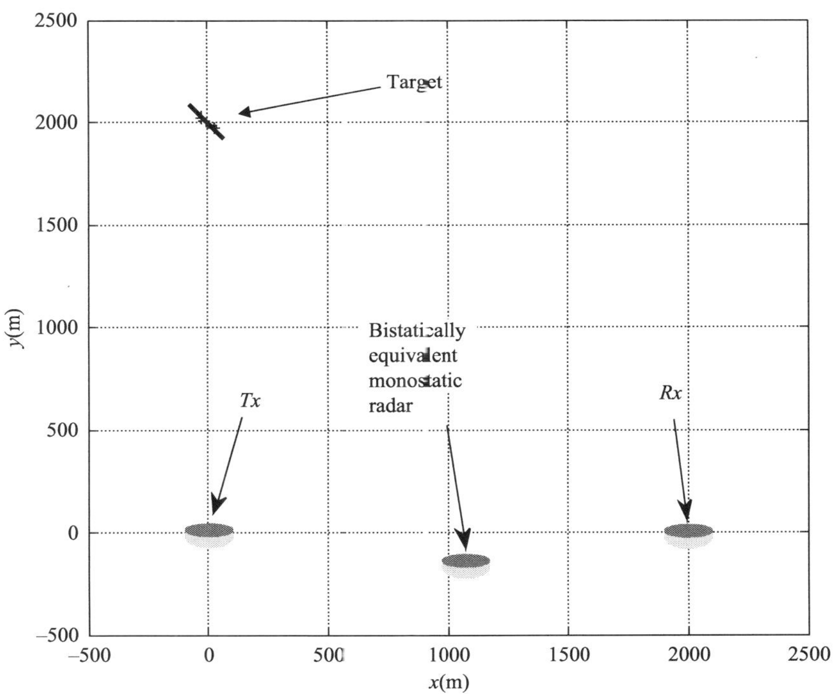

Figure 10.10 Simulation geometry—linear distortions.

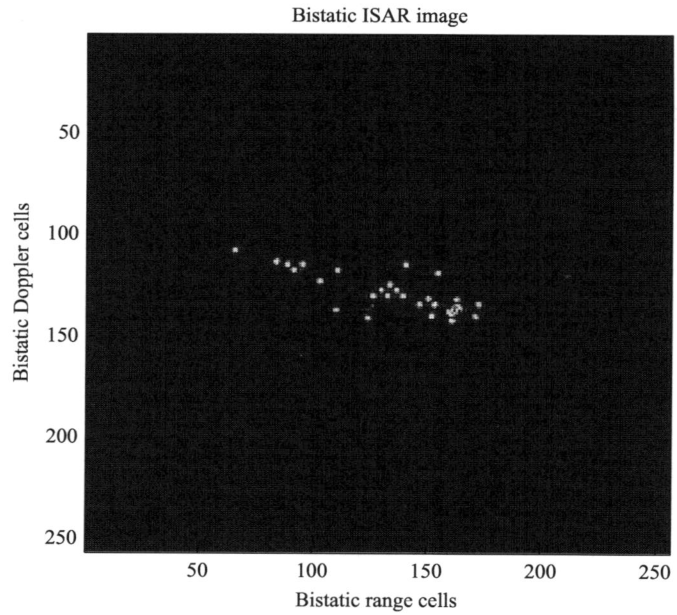

Figure 10.11 Linear distortion analysis—B-ISAR image.

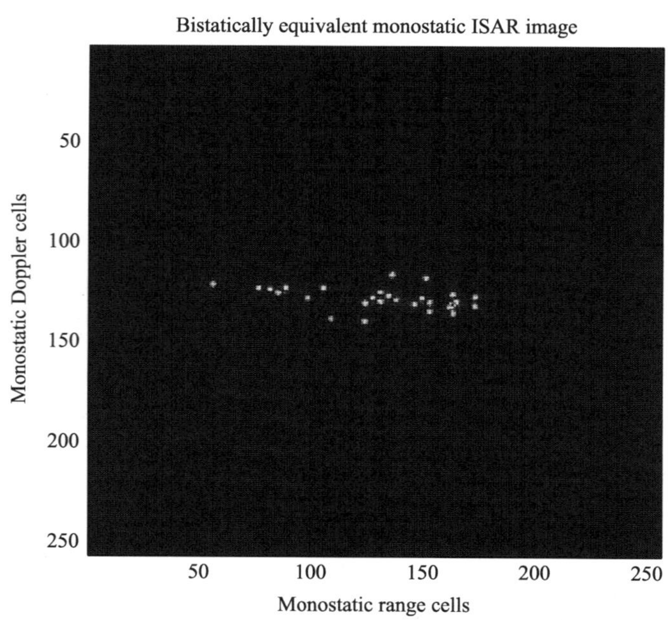

Figure 10.12 Linear distortion analysis—monostatic ISAR image obtained with an antenna located at the BEM position.

the Doppler shift is proportional to the range of the scattering center. The visual effect of this is an apparent rotation in the range-Doppler image plane. The image compression introduced by the bistatic angle is also visible when comparing the two images. The compression factor can be roughly estimated by measuring the range distance between the two farthest scatterers in both images and calculating the ratio. The estimated value is equal to 0.91, against the predicted value of $K_0 = 0.92$. To show that linear distortions do not cause any energy spread across the Doppler coordinate, the Doppler sections obtained from the two images relative to the same scatterer have been superposed after removing the Doppler shift, as shown in Figure 10.13.

Simulation setup—quadratic distortions. The waveform and synchronization error parameters are the same as shown in Table 10.1 relative to Case 1, whereas the bistatic geometry is described in Figure 10.14. In this case the target is moving along the same trajectory as in the previous case with a velocity of 2,000 km/hr. In addition, the *Rx* moves along a rectilinear trajectory of –45° with a velocity of 420 km/hr. The CPI is set equal to $T_{obs} = 0.25$ sec. It is worth noting that the bistatically equivalent monostatic element also moves along a rectilinear trajectory. These scenarios are not very common and are shown here for the purpose of creating a case where quadratic distortions are significant.

Simulation results—quadratic distortions. When setting the radar parameters as described in the previous subsection and the *Tx* target *Rx* geometry as described in Figure 10.14, the constraint in (10.62) is not satisfied. This causes an energy spread across the Doppler

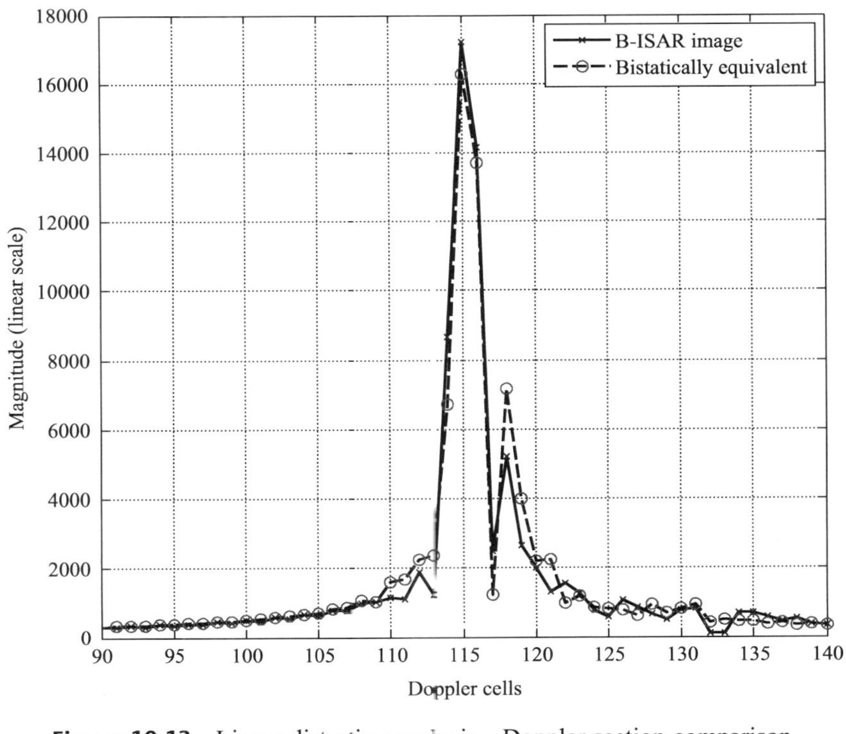

Figure 10.13 Linear distortion analysis—Doppler section comparison.

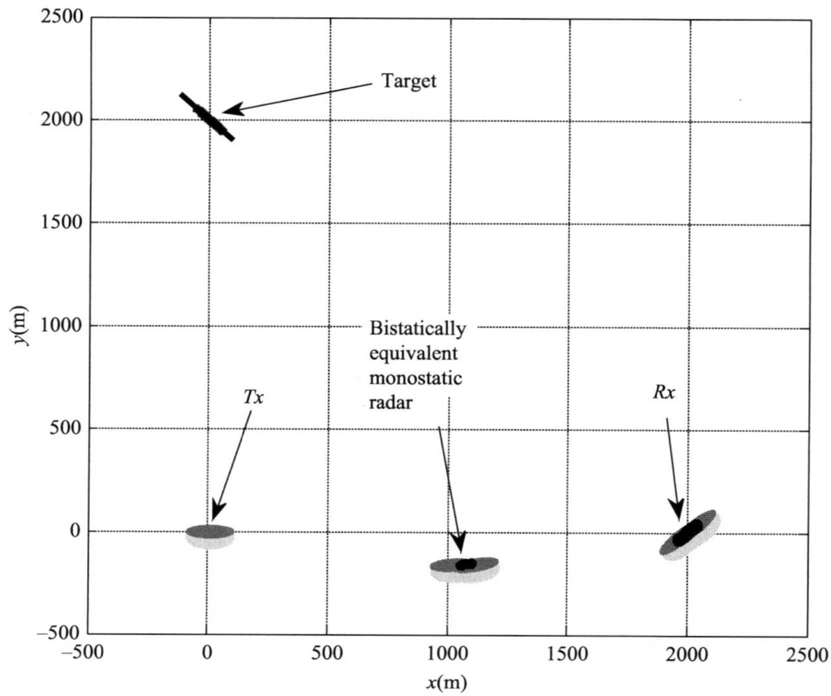

Figure 10.14 Simulation geometry—quadratic distortions.

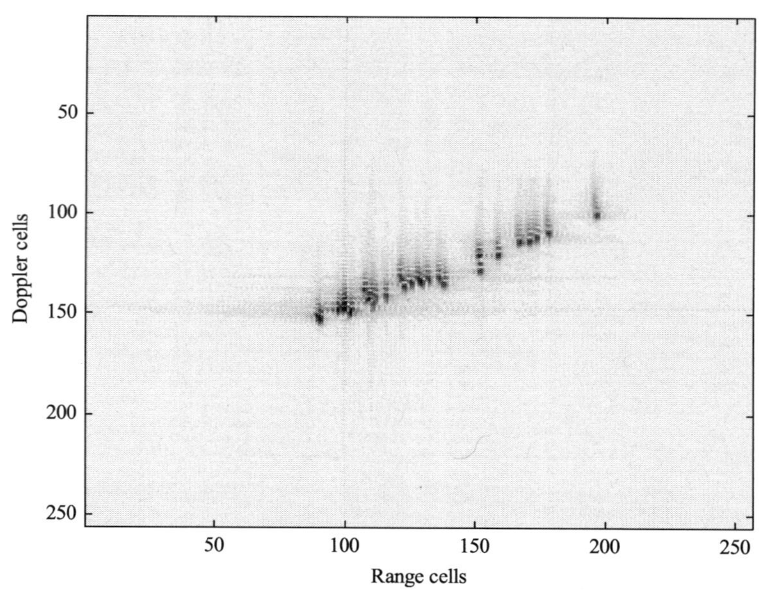

Figure 10.15 Quadratic distortion analysis—B-ISAR image.

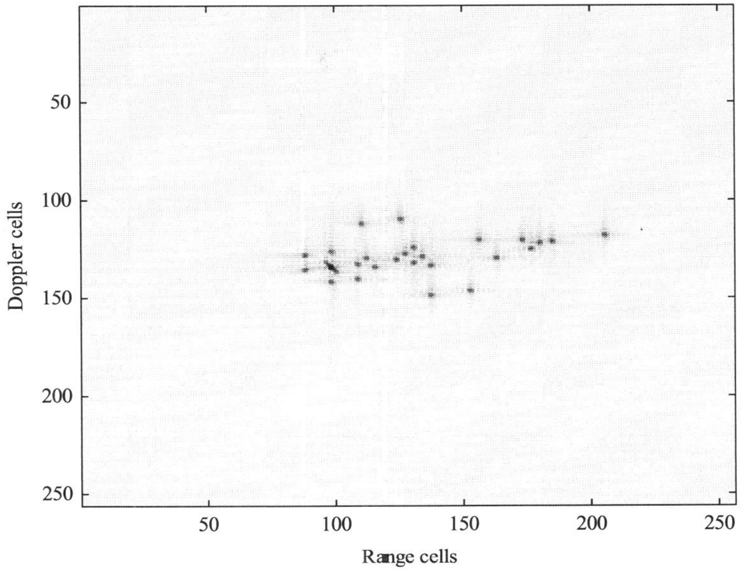

Figure 10.16 Quadratic distortion analysis: BEM ISAR image.

coordinate, as shown in the B-ISAR image in Figure 10.15. For comparison, the BEM ISAR image is shown in Figure 10.16. Since the linear distortion constraint in (10.57) is not met, a Doppler shift also causes a linear distortion. To highlight the distortion effects, a superposition of the Doppler sections relative to a single scatterer is shown in Figure 10.17. Both a main lobe reduction and a sidelobe increase are clearly noticeable, which is the main effect of the quadratic distortion.

10.8 Multistatic ISAR

B-ISAR can be extended to multistatic configurations, where generally more than one transmitter and more than one receiver are simultaneously used to generate multiple monostatic and B-ISAR configurations. In a multiple-input, multiple-output (MIMO)—like multistatic geometry with N sensors, N^2 configurations are generated: N monostatic and $N(N-1)$ bistatic. In Figure 10.18, three transmitting/receiving sensors generate three monostatic configurations and six bistatic configurations. Therefore, potentially nine ISAR images could be produced, one per configuration.

A MIMO-like configuration can be enabled provided that a full MIMO radar system is available beforehand. Therefore, waveform orthogonality and multichannel receivers must be already implemented in the system.

A simpler configuration can be obtained when a single transmitter is used together with a number of receivers. Such a configuration goes under the name of single-input, multiple-output (SIMO) configuration. In this case, when a transmitter/receiver and N receiver are used, a monostatic and N bistatic configurations are enabled. This particular configuration

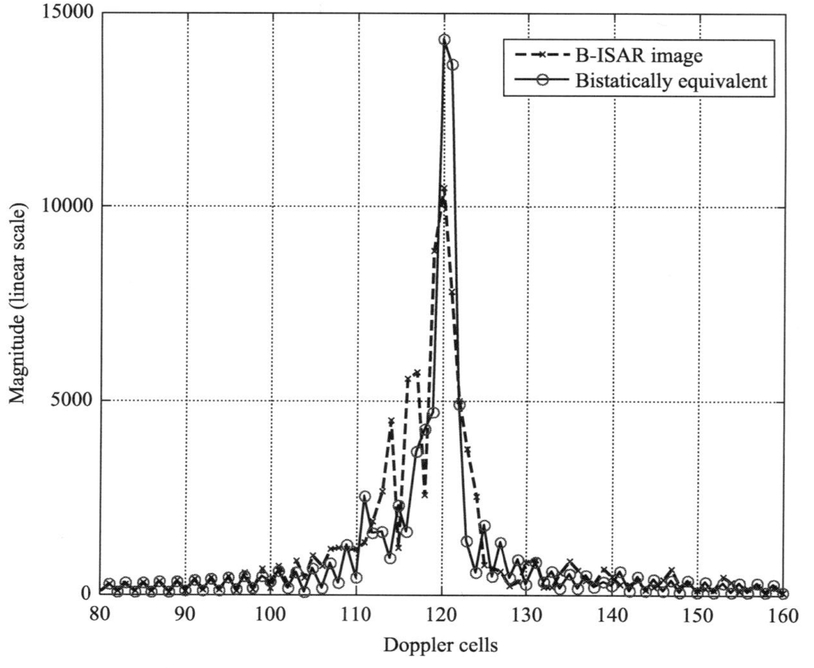

Figure 10.17 Quadratic distortion analysis—Doppler section.

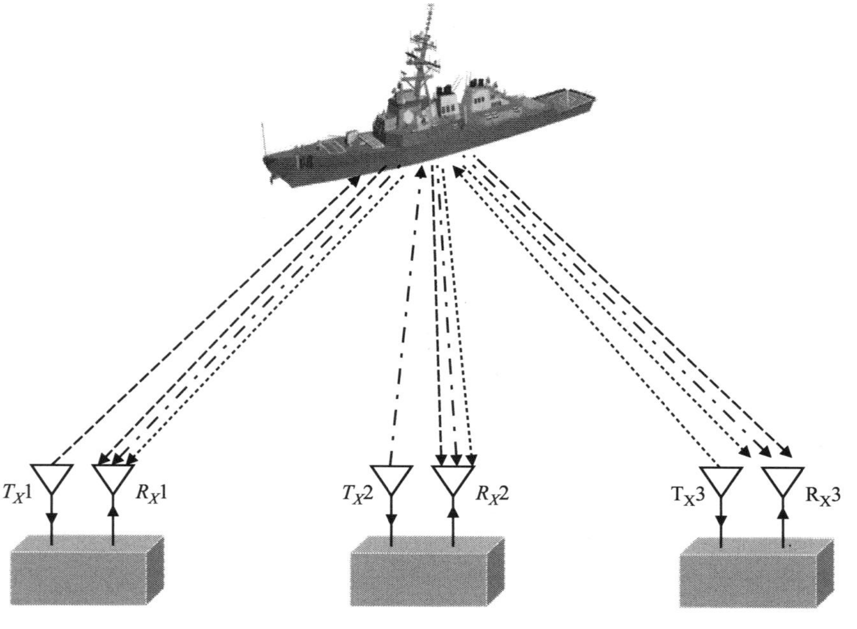

Figure 10.18 Multistatic geometry.

becomes interesting because costs can be reduced using only one transmitter and several lower-cost receivers. The receive-only stations have a very low probability of intercept (LPI) because they do not transmit electromagnetic (EM) waves.

One advantage of using a multistatic ISAR (M-ISAR) imaging systems relies on the rapidity and effectiveness of obtaining ISAR images that are suitable for classification and recognition purposes. A well-focused ISAR image that shows some desirable IPP, such as front, side, or top view, is hard to obtain with a time-limited data set because this depends on the target's own motions. Usually, the radar is forced to illuminate the target for a long time to make sure that a time window exists for the formation of such a desired ISAR image. The probability of obtaining a desired IPP can be maximized by placing the illuminator in a suitable position (optimal position), as discussed in [15]. Therefore, having a number of illuminators (which could also be some virtual BEM illuminators) provides the basis for increasing such probability or, equivalently, maintaining the same probability when reducing the illumination time of a target.

Another advantage of using M-ISAR is related to the fact that only some of the target's scatterers are visible from a given viewpoint, whereas others may be masked or scatter a small quantity of the EM energy toward the radar. Therefore, having several radars to look at the same target provides the means to collect more information and therefore extract more reliable and complete sets of features for classification and recognition purposes. Also, bistatic configurations, which derive from the use of a M-ISAR system, allow bistatic RCS of scatterers to be measured, which could also be used as an effective method for defeating stealth technologies.

An example of the use of M-ISAR is provided in Figures 10.19a and 10.19b, where a monostatic and a B-ISAR image of the same target are shown. An incoherent summation of

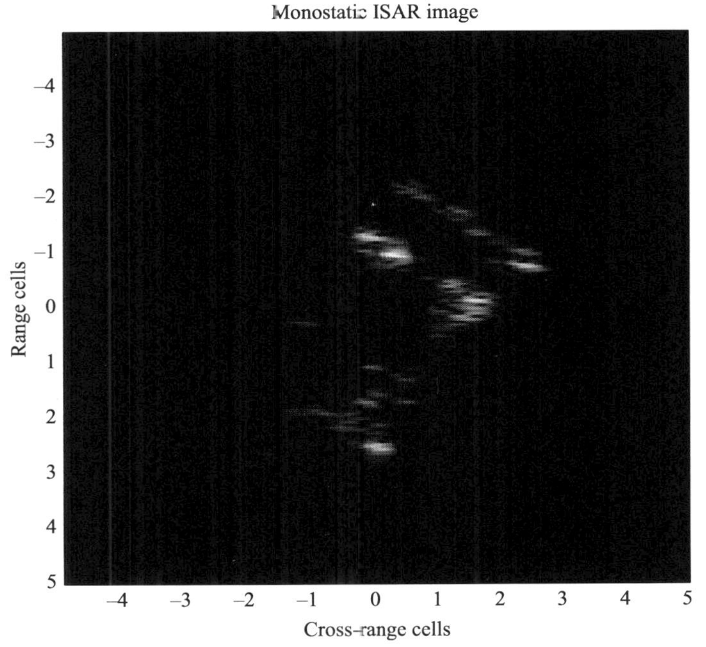

Figure 10.19a Monostatic ISAR image.

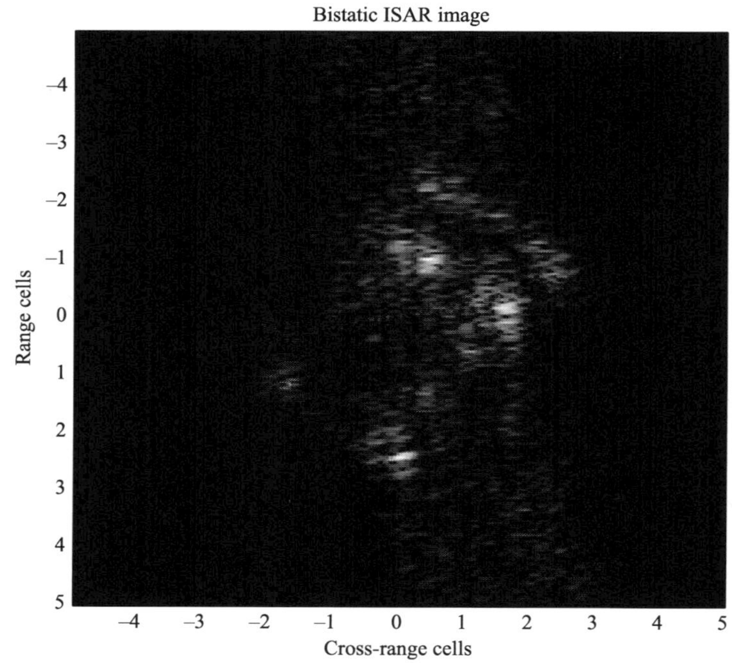

Figure 10.19b Bistatic ISAR image.

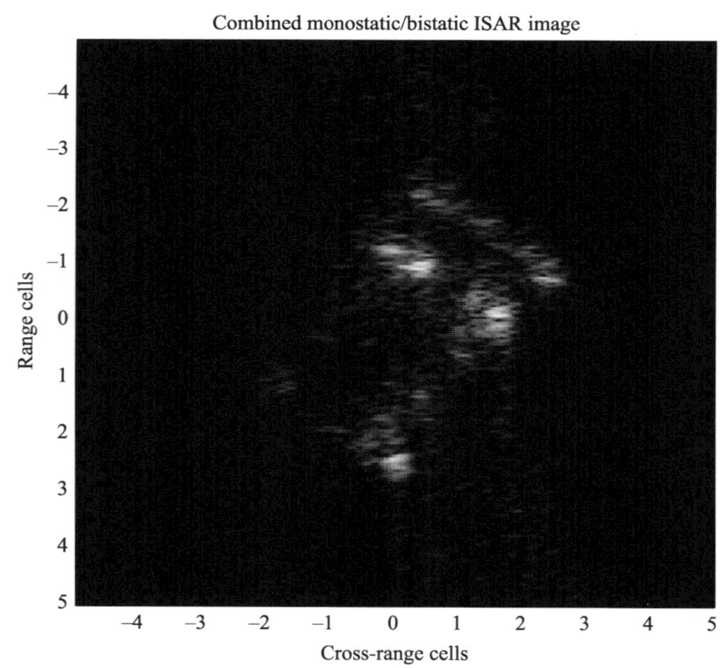

Figure 10.19c Combined monostatic + B-ISAR image by incoherent summation of monostatic and B-ISAR images.

the two images is also shown in Figure 10.19c, to highlight the effects of combining monostatic and bistatic information in a single composed image. It should be remarked that no gain in resolution is achieved in this case since the superposition is purely incoherent. The resulting resolution in the combined image is a value in between the monostatic and bistatic resolutions [16].

Another example of multistatic ISAR imaging can be found in experiments conducted to verify the possibility of exploiting the sea surface multipath to emulate bistatic and multistatic radar configurations, as detailed in [17]. Examples of applications of ISAR imaging that relate to the exploitation of emulated multistatic ISAR imaging are also provided in Chapter 12.

References

[1] Martorella, M., J. Palmer, J. Homer, B. Littleton, and I.D. Longstaff, "On Bistatic Inverse Synthetic Aperture Radar," *IEEE Trans on AES*, vol. 43, no. 3, pp. 1125–1134, 2007.

[2] Soumekh, M., "Bistatic Synthetic Aperture Radar Inversion with Application in Dynamic Object Imaging," *IEEE Trans on Signal Processing*, vol. 39, no. 9, pp. 2044–2055, 1991.

[3] Loffeld, O., H. Nies, V. Peters, and S. Knedlik, "Models and Useful Relations for Bistatic SAR Processing," *IEEE 2003 IGARSS*, Toulouse, France, Jul. 21–25, vol. 3, pp. 1442–1445, 2003.

[4] Ender, J.H.G., I. Walterscheid, and A.R. Brenner, "New Aspects of Bistatic SAR: Processing and Experiments," *IEEE 2004 IGARSS*, Anchorage, AK, Sep. 20–24, vol. 3, pp. 1758–1762, 2004.

[5] D'Aria, D., A.M. Guarnieri, and F. Rocca, "Focusing Bistatic Synthetic Aperture Radar Using Dip Move Out," *IEEE Trans on Geoscience and Remote Sensing*, vol. 42, no. 7, pp. 1362–1376, 2004.

[6] Crispin, J.W. Jr., and K.M. Siegel, *Methods of Radar Cross-Section Analysis*, Chap. 5, Academic Press, 1968.

[7] Willies, N.J., *Bistatic Radar*, 2d ed., SciTech Publishing, 2005.

[8] Martorella, M., and F. Berizzi, "Time Windowing for Highly Focussed ISAR Image Reconstruction," *IEEE Trans on AES*, vol. 41, no. 3, pp. 992–1007, 2005.

[9] Krieger, G., and M. Younis, "Impact of Oscillator Noise in Bistatic and Multistatic SAR," *IEEE Geoscience and Remote Sensing Letters*, vol. 3, no. 3, pp. 424–428, 2006.

[10] Lopez-Dekker, P., J.J. Mallorqui, P. Serra-Morales, and J. Sanz-Marcos, "Phase Synchronization and Doppler Centroid Estimation in Fixed Receiver Bistatic SAR Systems," *IEEE Trans on Geoscience and Remote Sensing*, vol. 46, no. 11, pp. 3459–3471, 2008.

[11] Berizzi, F., M. Martorella, B. Haywood, E. Dalle Mese, and S. Bruscoli, "A Survey on ISAR Autofocusing Techniques," *Proceedings of the IEEE 2004 ICIP*, Singapore, 2004.

[12] Thayaparan, T., L. Stankovic, C. Wernik, and M. Dakovic, "Real-Time Motion Compensation, Image Formation and Image Enhancement of Moving Targets in ISAR and SAR Using S-Method Based Approach," *IET Signal Processing*, vol 2, no. 3, pp. 247–264, 2008.

[13] Chen, V.C., and H. Ling, *Time-Frequency Transforms for Radar Imaging and Signal Analysis*, Artech House, 2002.

[14] Berizzi, F., E.D. Mese, M. Diani, and M. Martorella, "High-Resolution ISAR Imaging of Maneuvering Targets by Means of the Range Instantaneous Doppler Technique: Modeling and Performance Analysis," *IEEE Trans on Image Processing*, vol. 10, no. 12, pp. 1880–1890, 2001.

[15] Martorella, M., "Optimal Sensor Positioning for Inverse Synthetic Aperture Radar," *IEEE Trans on AES*, vol. 49, no. 1, pp. 648–658, 2013.

[16] Brisken, S., and J.G. Worms, "Incoherent Addition of ISAR Images from Spatially Distributed Receivers for Classification Purposes," 11th International Radar Symposium (IRS), 16–18 June, 2010.

[17] Palmer, J., M. Martorella, J. Homer, B. Littleton, and I.D. Longstaff, "ISAR Imaging Using an Emulated Multistatic Radar System," *IEEE Trans on AES*, vol. 41, no. 4, pp. 1464–1472, 2005.

CHAPTER 11

Polarimetric ISAR

Polarimetric synthetic aperture radar (Pol-SAR) has been widely used for classifying natural and man-made targets. More specifically, Pol-SAR systems are used in land, ice, and ocean remote sensing to obtain extra information about scattering mechanisms that can be exploited for extracting physical parameters of interest [1–3] and used in target classification applications [4–7].

In this chapter, we will (1) outline a framework for polarimetric inverse synthetic aperture radar (Pol-ISAR) imaging by defining the geometry and by introducing suitable signal models, (2) introduce image formation algorithms that use the fully polarimetric information contained in the received signal, and (3) discuss how to interpret and represent fully polarimetric ISAR images. We will be dealing with the subject of defining a framework for Pol-ISAR imaging in Section 11.1, where the Pol-ISAR point spread function (PSF) will be derived in an arbitrary geometry (under suitable assumptions) and with a fully polarimetric received signal. Regarding to the second subject, we will address the issue of target's radial motion compensation, which is one of the most important steps in ISAR image reconstruction. As pointed out in Chapter 5, motion compensation is to remove a phase term of the received signal produced by the target radial motion. Two algorithms, among others from the class of parametric autofocusing techniques, have been proposed in the literature for single polarization ISAR: image contrast based autofocusing (ICBA) and image entropy based autofocusing (IEBA) (see Chapter 5). The image contrast (IC) and the image entropy (IE) represent two different ways of measuring the focus of the ISAR image. In the case of a single polarization ISAR, the success of maximizing the IC or minimizing the IE (which is the way to achieve motion compensation) also depends on the polarization used by the system. Evidently, specific scatterers may produce a more stable signal return in a given polarization rather than in others. Often, in SAR applications, the HH and VV co-polarization channels offer a higher signal-to-noise ratio (SNR) with respect to the cross-polarizations HV and VH. Nevertheless, in ISAR applications, such a-priori knowledge cannot be taken for granted. In any case, the availability of all polarizations provides the basis for optimizing the autofocusing algorithm with respect to the polarimetric space. In the past, polarimetric radars were used for maximizing the SNR with respect to the polarization space in order to improve detection performance [8]. The concept of increasing the performance by finding an optimal polarization can be extended to the ISAR image autofocus. In other words, it can be assumed that the image focus reaches its maximum for a particular polarization. This insight can be justified by considering that the image focus strongly depends on the time invariance of the scatterer's contributions. Moreover, the Doppler components for each scatterer are generally modulated by several causes. These are related to the modulation induced by the scatterer scintillation (superposition of several scatterers in the same resolution cell) and to the effect of noise. Exploiting the full polarization can reduce both these causes. In fact, finding the optimal polarization can jointly reduce both the SNR and the modulation effect induced by a scatterer when illuminated from different aspect angles. The definition of such an optimality criterion will be defined in Section 11.3 after introducing the signal model and calculating the PSF of the Pol-ISAR image in Section 11.2. Section 11.3 will also provide the definition of the polarimetric ICBA (Pol-ICBA), the polarimetric IEBA (Pol-IEBA) and polarimetric PPP (Pol-PPP) techniques. In Section 11.4, we will finally provide some tools for interpreting Pol-ISAR images, which make use of polarimetric decompositions. Since different scatterer's shapes produce different polarimetric responses, some polarimetric decompositions will be introduced that will help in the process of recognizing such shapes.

We will also introduce some ways of visually representing Pol-ISAR images, with particular interest to fusing the multichannel information into a single color-coded image. Such a representation aims at coding polarimetric signatures with colors so that different scattering mechanisms map into different colors, which can be directly interpreted by a visual inspection.

The signal model will be introduced in the next section followed by the analytical derivation of the PSF of the Pol–ISAR image.

11.1 Signal Model

The polarimetric matrix of the received signal, in free-space conditions, can be written in a time-frequency domain by extending the signal model defined in (1.16) and similar to (9.7):

$$S_R(f,t) = W(f,t)\exp\left\{-j\frac{4\pi f}{c}R_0(t)\right\}\int_V \rho(\boldsymbol{x})\exp\left\{-j\frac{4\pi f}{c}[\boldsymbol{x}\cdot\boldsymbol{i}_{\mathrm{LOS}}(t)]\right\}d\boldsymbol{x} \tag{11.1}$$

where, by referring to the two-dimensional (2-D) coordinate system in Figure 1.5, $\boldsymbol{x} = (x, y)$ and also, similar to (8.32), we have $\boldsymbol{x}^T \cdot \boldsymbol{i}_{\mathrm{los}}(t) = x\cos[\theta(t)] + y\sin[\theta(t)]$. It is worth noting that $\boldsymbol{i}_{\mathrm{los}}(t)$ is expressed here in terms of the target's aspect angle: $\boldsymbol{i}_{\mathrm{los}}(t) = [\cos\theta(t)\ \sin\theta(t)]^T$ radar typically stores polarimetric data using a scattering matrix format. This is a result of the way the radar system generally operates with two orthogonally and linearly polarized antennas, which is typically achieved as follows:

1. The signal is received in both H and V polarizations when H is transmitted, so measurements in co-polarization HH and cross-polarization HV are stored;
2. The signal is received in both H and V polarizations when V is transmitted, so measurements in cross-polarization VH and co-polarization VV are stored.

Therefore, the polarimetric received signal matrix can be expressed as follows:

$$\boldsymbol{S}_R(f,t) = \begin{bmatrix} S_R^{HH}(f,t) & S_R^{HV}(f,t) \\ S_R^{VH}(f,t) & S_R^{VV}(f,t) \end{bmatrix}, \tag{11.2}$$

whereas $W(f,t) = \text{rect}\left(\frac{t}{T_{obs}}\right)\text{rect}\left(\frac{f-f_0}{B}\right)$ represents a rectangular window in the time-frequency domain (f, t), f_0 represents the carrier frequency, B is the transmitted signal bandwidth, T_{obs} is the observation time, V is the spatial region where the target scattering matrix, namely, $\rho(\boldsymbol{x}) = \begin{bmatrix} \rho^{HH}(\boldsymbol{x}) & \rho^{HV}(\boldsymbol{x}) \\ \rho^{VH}(\boldsymbol{x}) & \rho^{VV}(\boldsymbol{x}) \end{bmatrix}$, is defined, and again the rect(·) is the rectangular function.

Before proceeding, we will introduce a different polarimetric decomposition, namely, Pauli's decomposition [8], even though a number of others may be used. Most of these decompositions may be found in [4,5,9–16]. The reason for this is that Pol-ISAR processing is independent of the used decomposition method. Nevertheless, Pauli's decomposition can be easily defined with a combination of linear polarizations, which makes it easy to relate physical polarimetric channels with the signal mathematical model. Moreover, we can exploit the symmetry of isotropic media, which implies the equality of the cross-polarization signals: $S_R^{HV}(f,t) = S_R^{VH}(f,t)$. Therefore, the polarimetric data that represents the received signal can be written as follows:

$$\boldsymbol{S}_R(f,t) = \left[S_R^{VV} + S_R^{HH},\quad S_R^{VV} - S_R^{HH},\quad 2S_R^{HV}\right], \tag{11.3}$$

where the dependence on (f, t) is omitted for notation simplicity. The same decomposition applies for the target scattering matrix. Therefore, the scattering vector obtained from the scattering matrix is

$$\rho(\boldsymbol{x}) = \left[\rho^{VV} + \rho^{HH},\quad \rho^{VV} - \rho^{HH},\quad 2\rho^{HV}\right], \tag{11.4}$$

where, in this case, the dependence on the coordinate $\boldsymbol{x}$ has been omitted for the sake of notation simplicity.

As a result, the received signal can be seen as a vector in a complex 3-D polarimetric space. All possible projections can be obtained by means of an internal product between the received signal vector and a generic polarization vector, $\mathbf{p}$:

$$S_R^{(\mathbf{p})} = \boldsymbol{S}_R \cdot \mathbf{p}, \tag{11.5}$$

where $\mathbf{p}$ can be expressed according to the decomposition introduced by Cloude in [10]:

$$\mathbf{p} = \left[p^{VV} + p^{HH},\quad p^{VV} - p^{HH},\quad 2p^{HV}\right] = \begin{bmatrix} \cos\alpha \cdot \exp\{j\varphi\} \\ \sin\alpha \cdot \cos\beta \cdot \exp\{j\delta\} \\ \sin\alpha \cdot \sin\beta \cdot \exp\{j\gamma\} \end{bmatrix}, \tag{11.6}$$

where

1. α is the internal degree of freedom (DOF) of the scatterer, which ranges within [0°, 90°]. The meaning of such an angle is related to the scattering properties of the target; for example, for an ideal dipole the value of α is equal to 45°. A pictorial interpretation of α is given in Figure 11.1.
2. β represents a physical rotation of the scatterer on the plane perpendicular to the direction of the electromagnetic (EM) wave propagation.

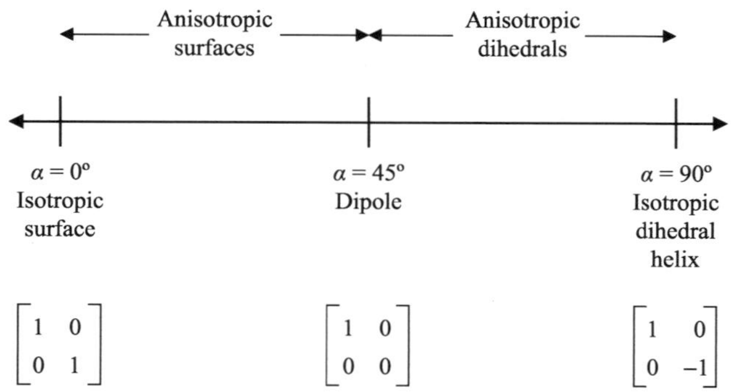

Figure 11.1 Interpretation of the parameter α.

3. φ, δ, and γ are the scatterer's phases of the three Pauli's polarimetric components.

It is important to say that such a representation is meant to highlight the physical properties of the scattering mechanism induced by a given scatterer. Therefore, by defining the unit vector, **p**, it is possible to define a specific polarization that resonates with a scatterer with given physical properties. Moreover, the decomposition proposed in [10] provides, among other polarimetric decompositions, a suitable domain for image contrast (IC) maximization or image entropy (IE) minimization.

11.2 Image Formation and Point Spread Function

The procedure of the image formation will be carried out without including the noise contribution. Therefore, we follow a standard procedure in synthetic aperture radar (SAR) and inverse synthetic aperture radar (ISAR) imaging.

As stated in Chapter 5, radial motion compensation consists of removing the phase term $exp\left\{-j\frac{4\pi f}{c}R_0(t)\right\}$e to the target's radial motion. After motion compensation, the received signal without considering noise can be written as follows:

$$\boldsymbol{S}_R^{(\boldsymbol{p})}(f,t) = W(f,t)\int_V \rho^{(\boldsymbol{p})}(\boldsymbol{x})\exp\left\{-j\frac{4\pi f}{c}[\boldsymbol{x}\cdot\boldsymbol{i}_{\mathrm{LOS}}(t)]\right\}d\boldsymbol{x}, \tag{11.7}$$

where

$$\rho^{(\boldsymbol{p})}(\boldsymbol{x}) = \rho(\boldsymbol{x})\cdot\mathbf{p}. \tag{11.8}$$

To reconstruct the ISAR image, we use the range-Doppler (RD) image formation. As we know, the RD technique uses a 2-D Fourier transform to the motion compensated signal $\boldsymbol{S}_R^{(\boldsymbol{p})}(f,t)$to form a complex image in the time-delay and Doppler shift domain. If we consider the target model as the point scatterer model, that is, the composition of K independent scattering centers, for a given polarization it is possible to define a reflectivity function by $\rho^{(\boldsymbol{p})}(\boldsymbol{x}) = \sum_{k=1}^{k} a_k^{(\boldsymbol{p})}\cdot\boldsymbol{\delta}\left(\boldsymbol{x}-\boldsymbol{x}_k^{(\boldsymbol{p})}\right)$. Each scatterer is characterized by a specific scattering vector, $\boldsymbol{a}_k$, and that $a_k^{(p)} = \boldsymbol{a}_k \cdot \mathbf{p}$ is the complex amplitude of the k-th scatterer when the generic polarization, **p**, is considered. After taking the 2-D Fourier transform (FT) of $\boldsymbol{S}_R^{(p)}(f,t)$, the polarimetric complex ISAR image of the target composed by point scatterers can be written as follows:

$$I^{(p)}(\tau,\nu) = T_{obs}B\sum_{\mathrm{k=1}}^{\mathrm{k}} a_k^{(p)}\mathrm{sinc}\left[T_{obs}\left(\nu-\nu_k^{(p)}\right)\right]\mathrm{sinc}\left[B\left(\tau-\tau_k^{(p)}\right)\right]\exp\left\{-j2\pi f_0\left(\tau-\tau_k^{(p)}\right)\right\} \tag{11.9}$$

where $\nu_k^{(p)}$ and $\tau_k^{(p)}$ represent the coordinates of the scatterers in the time-delay and Doppler shift domain, respectively. The polarization affects both the scatterer amplitude and the scatterer position. This latter point has been taken into account because in the real world the scattering center position is determined by a weighted sum of several contributions. Because such weights depend on the polarization, the scattering center location also depends on the polarization. It is also worth noting that the polarimetric complex ISAR image of a set of semi-ideal scatterers is composed of a sum of sinc-like shaped terms shown in (11.9) as it was in the case of single-polarization ISAR.

11.3 Polarimetric ISAR Image Autofocus

ISAR image autofocus is a task of fine radial motion estimation, which is accomplished by solving a signal phase estimation problem. This may be accomplished using only the received signal via a number of techniques (see Chapter 5). When polarimetric radars become available, more channels and therefore more observations can be exploited to perform such an estimation accurately. In the next subsections, three techniques (Pol-ICBA, Pol-IEBA, and Pol-PPP) will be introduced by extending the corresponding single-polarization (single-channel) techniques (ICBA, IEBA, and PPP).

11.3.1 Polarimetric Image Contrast and Entropy-Based Autofocus

The idea behind Pol-ISAR image autofocus is that of jointly processing all polarimetric channels to obtain well-focused ISAR images. Such an insight relies on the concept of enhancing the image focus by maximizing the IC over the joint space of the focusing parameters $\boldsymbol{\xi}$ and of the polarization, **p**:

$$\left(\hat{\boldsymbol{\xi}}_{IC}, \mathbf{p}\right) = \arg\max_{\boldsymbol{\xi},\mathbf{p}}\{IC(\boldsymbol{\xi},\mathbf{p})\}, \tag{11.10}$$

where the IC is defined by

$$IC(\boldsymbol{\xi},\mathbf{p}) = \frac{\sqrt{A\left\{\left[I^{(\mathbf{p})}(\tau,\nu,\boldsymbol{\xi}) - A\{I^{(\mathbf{p})}(\tau,\nu,\boldsymbol{\xi})\}\right]^2\right\}}}{A\{I^{(\mathbf{p})}(\tau,\nu,\boldsymbol{\xi})\}}, \tag{11.11}$$

and $\boldsymbol{\xi} = [\alpha_1, \alpha_2, \ldots, \alpha_N]$ with α_i the model polynomial coefficients. Equation (11.11) represents the new image contrast function defined in the joint domain and where the operator $A\{\cdot\}$ represents the image spatial mean over the coordinates (τ, ν). In the same way, minimizing the Image Entropy (IE) can enhance the image focus, as follows:

$$\left(\hat{\xi}_{IE}, \mathbf{p}\right) = \arg\max_{\boldsymbol{\xi},\mathbf{p}}\{IE(\boldsymbol{\xi}, \mathbf{p})\}, \tag{11.12}$$

where the IE is defined by

$$IE(\boldsymbol{\xi}, \mathbf{p}) = \iint \bar{I}^{(\mathbf{p})}(\tau, \nu, \boldsymbol{\xi}) \cdot ln\left[\bar{I}^{(\mathbf{p})}(\tau, \nu, \boldsymbol{\xi})\right] d\tau d\nu \tag{11.13}$$

and

$$\bar{I}^{(\mathbf{p})}(\tau, \nu, \boldsymbol{\xi}) = \frac{\left|I^{(\mathbf{p})}(\tau, \nu, \boldsymbol{\xi})\right|^2}{A\left\{\left|I^{(\mathbf{p})}(\tau, \nu, \boldsymbol{\xi})\right|^2\right\}}. \tag{11.14}$$

Initialization. To proceed with the application of the Pol-ICBA and Pol-IEBA to fully polarimetric ISAR data, a solution for the initial polarization vector must also be provided. The problem can be solved by means of the following algorithm:

STEP 1: The polarization vector that provides the maximum SNR is obtained by solving the optimization problem stated by

$$\mathbf{p}_M = \arg\max_{\mathbf{p}}\left\{\frac{\iint |\boldsymbol{S}_R(f,t)|^2 df\,dt}{\iint |\boldsymbol{N}_R(f,t)|^2 df\,dt}\right\} \tag{11.15}$$

where

$$\boldsymbol{N}_R(f,t) = \begin{bmatrix} N_R^{HH}(f,t) & N_R^{HV}(f,t) \\ N_R^{VH}(f,t) & N_R^{VV}(f,t) \end{bmatrix}. \tag{11.16}$$

The SNR can be assumed to reach the maximum value when the signal energy reaches its maximum, provided that the noise level is the same in all the polarization channels (basically when the noise level in the H and V receiving channels are the same). Therefore, (11.15) can be simplified as

$$\mathbf{p}_M = \arg\max_{\mathbf{p}}\left\{\iint |\boldsymbol{S}_R(f,t)|^2 df\,dt\right\}. \tag{11.17}$$

STEP 2: An initial guess for the focusing parameter vector, $\boldsymbol{\xi}$, can be obtained by applying the Radon transform and running a 1-D optimization problem (see Chapter 5 for details). Specifically, the scalar ICBA and IEBA must be applied to the received signal with polarization, $\mathbf{p}_M$, as found at step 1:

$$\hat{\boldsymbol{\xi}}_{IC}^{(\mathbf{p}_M)} = \arg\max_{\boldsymbol{\xi}}\{IC(\boldsymbol{\xi}, \mathbf{p}_M)\} \tag{11.18}$$

and

$$\hat{\boldsymbol{\xi}}_{IE}^{(\mathbf{p}_M)} = \arg\max_{\boldsymbol{\xi}}\{IE(\boldsymbol{\xi}, \mathbf{p}_M)\}, \tag{11.19}$$

where $IC(\boldsymbol{\xi}, \mathbf{p}_M)$ and $IE(\boldsymbol{\xi}, \mathbf{p}_M)$ can be obtained from (11.11) and (11.13). Therefore, the initial guess can be obtained by adjoining the polarization vector, $\mathbf{p}_M$, to the focusing parameter vector $\hat{\boldsymbol{\xi}}_{IC}^{(\mathbf{p}_M)}$ or $\hat{\boldsymbol{\xi}}_{IE}^{(\mathbf{p}_M)}$, which can be expressed as $\left(\hat{\boldsymbol{\xi}}_{IC}^{(\mathbf{p}_M)}, \mathbf{p}_M\right)$ or $\left(\hat{\boldsymbol{\xi}}_{IE}^{(\mathbf{p}_M)}, \mathbf{p}_M\right)$.

Optimization. Once the initial guess is estimated, the optimization problems as stated in (11.18) and (11.19) can be solved iteratively by using numerical methods for a maximum (or minimum) search. Several methods for solving optimization problems have been proposed in existing literature that can be used in our case. The method used here is the simplex method, proposed by Nelder and Mead in [17]. Nevertheless, other solutions may be obtained by using statistical methods, such as genetic algorithms [18–20]. The iterations stop when the difference in the cost function between two consecutive iterations is smaller than a preset value. For clarity, the algorithm flowchart is shown in Figure 11.2.

Most focused ISAR image. The proposed algorithm also provides the most focused ISAR image. The polarization $\mathbf{p}_{IC}$, which maximizes (11.18) or equivalently, the polarization $\mathbf{p}_{IE}$, which minimizes (11.18), are obtained as part of the solution of the optimization problems. Therefore, the ISAR images obtained by processing the received data in the polarizations $\mathbf{p}_{IC}$ and $\mathbf{p}_{IE}$ represent the most focused images according to the IC and IE focus indicators, respectively. It is worth pointing out that the IC and the IE are indicators of image focus. Therefore, their maximization (IC) or minimization (IE) represents the best result in terms of image focus according to one or the other indicator. This should be considered a similar result to the maximization of the SNR with respect to polarimetric space, as introduced by Novak in [7] relatively to the polarimetric whitening filter.

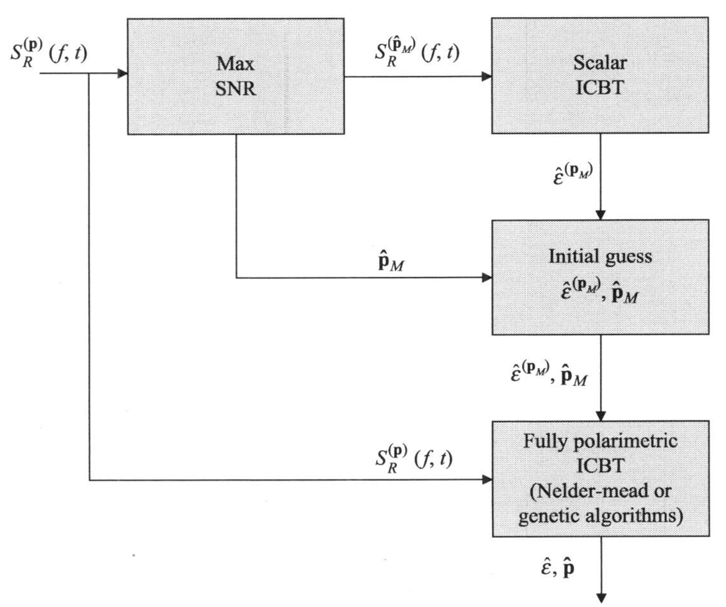

Figure 11.2 Pol-ICBA flowchart.

11.3.2 Polarimetric Prominent Point Processing

In the specific case of PPP processing, the availability of more polarimetric channels provides an extra set of measurements, therefore increasing the probability of finding stable scatterers. To clarify this point, as an example we may assume that a stable scatterer has a strong response in the VV channel but not so in the HH channel (e.g., a vertically oriented dipole-like scatterer). Such a scatterer would be a perfect candidate for PPP processing if VV polarization was available, but it would not help if only HH polarization was available. The same argument can be made if a strong scatterer responds in the HV polarimetric channel. Therefore, fully polarimetric data provides the means for exploiting the presence of strong and stable scatterers in any polarization.

The stability of at least one target scatterer is a key point for the success of the PPP. The idea is to use all available polarimetric channels to find the scatterers that are the most stable. Such an insight is suggested by the fact that scattering centers may have a stronger and more stable return in one polarization rather than in another. Moreover, the presence of several scatterers in a given range cell is often a source of instability. Such scatterers may not be separable in the spatial domain. Nevertheless, they may be separable in the polarimetric domain, and therefore a single predominant scatterer could be selected by choosing a suitable polarimetric channel. To clarify this point, the reader may consider a single polarization system as a reduced set of measurements with respect to a fully polarimetric radar system. As in the case of single polarization PPP (SP-PPP), the proposed method uses a rough range alignment and a phase alignment process, as detailed in Chapter 5.

Rough range alignment. The rough range alignment is performed by cross-correlating adjacent range profiles, as detailed in Chapter 5. Since the nature of the scattering response changes from one polarimetric channel to another, the cross-correlation must be performed using range profiles that belong to the same channel. Nevertheless, to exploit all channels a cross-correlation rough alignment is performed using each polarimetric channel, and then the minimum variance is calculated among the range cells for each channel. The polarimetric channel with the minimum variance is then chosen as the reference channel to perform the rough alignment in the remaining channels.

Phase alignment. As pointed out in Chapter 5, the phase history estimate accuracy is highly related to the presence of a dominant stable scatterer. The main problem, in typical ISAR scenarios, is that it is quite likely to have two or more scattering centers in the same range cell. This typically produces a phase return that is a superposition of all scatterer phase histories. The ability to separate such contributions in a single polarization ISAR system strongly depends on the range cell resolution. Specifically, the finer the range resolution, the higher the likelihood of separating the scattering centers. In a polarimetric ISAR system, such a separation can be achieved in the polarimetric domain. In fact, scatterers that are present in the same range cell may have different polarimetric responses, and therefore they would appear in different polarimetric channels.

In synthesis, the basic idea for increasing the effectiveness of the PPP autofocusing algorithm is to select the range cells that provide the minimum variance among all three Pauli polarimetric channels. In this manner, not only are the scattering centers more likely to be separable, but also the most stable scatterers in the three Pauli components are used to provide a better phase history estimate. This can be expressed mathematically as follows.

Let the signal relative to the m-th selected range cell, r_m, in a given polarization channel, $\mathbf{p}_m$, be represented as

$$s_m(t, r_m, \mathbf{p}_m) = A_{r_m}^{(\mathbf{p}_m)}(t)\exp\left\{j\varphi_{r_m}^{(\mathbf{p}_m)}(t)\right\}, \quad A_{r_m}^{(\mathbf{p}_m)}(t) \simeq A, m = 1, 2, \ldots, M, \tag{11.20}$$

where M is the number of desired reference range cells. Therefore, the phase history estimate can be carried out by averaging the selected phase histories:

$$\varphi_0(t) = \sum_{m=1}^{M} \varphi_{r_m}^{(\mathbf{p}_m)}(t). \tag{11.21}$$

It is important to note that r_m and $\mathbf{p}_m$ are independent indexes. Therefore, it may occur that the same range cell is selected in two or more polarimetric channels simply because the scatterer response is stable in more than one channel or because two or more stable scatterers in the same range cell respond in different polarimetric channels. It may also happen that all the chosen range cells belong to the same polarimetric channel. In such a case, the result obtained by using the multi-polarimetric approach would be identical to the result obtained by using the SP-PPP approach when applied to that particular polarimetric channel. Nevertheless, it must be pointed out that no a priori knowledge about which channel provides the most stable response is generally available.

11.4 Polarimetric ISAR Image Interpretation

Pol-ISAR images may be very useful for target classification and recognition as they carry more information about each target scattering center than a single polarization ISAR image does. In particular, target scattering mechanisms can be estimated from the information available in a Pol-ISAR image. Because such scattering mechanisms are an intrinsic characteristic of each target, they may become important features to be exploited to improve classification performances. The following subsections will introduce some ingredients that will be used to characterize a target by means of its polarimetric signature.

A polarimetric ISAR system is in all senses a multichannel ISAR system. Therefore, the resulting image must be handled as a multichannel image. To be more specific, each pixel of the formed image is a complex valued vector and not simply a complex valued scalar. A natural way of representing Pol-ISAR images would be that of using an image cube, where each layer represents an ISAR image relative to one polarimetric channel.

This concept is depicted in Figure 11.3, where the ISAR image cube is seen as the composition of four complex-valued ISAR images, one for each polarimetric vector component. Moreover, a real data example is provided in Figures 11.4 and 11.5. Specifically, a physical object composed of elementary scatterers is assembled, as shown in Figure 11.4 and used as a target for a fully polarimetric ISAR experiment. The four ISAR intensity images obtained by processing the HH, HV, VH, and VV channels are shown in Figure 11.5. Different shapes produce different polarimetric returns (polarimetric signature). It should also be observed that the HV and VH channel ISAR images are practically identical, which is a result of the reciprocity theorem for reciprocal media.

Although the image cube is a full-information representation of the Pol-ISAR image, some composite Pol-ISAR images may be useful to extract some polarimetric target properties.

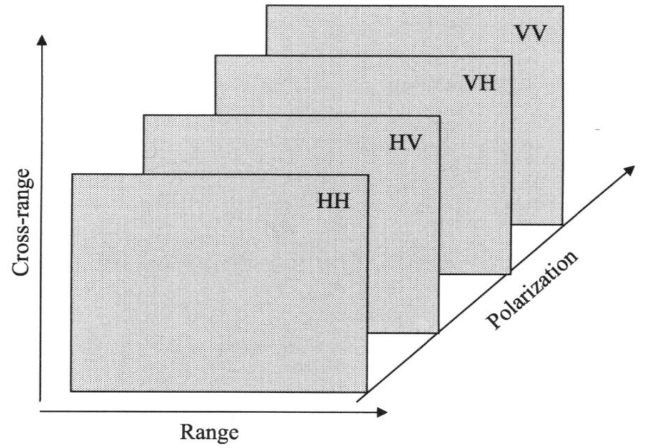

Figure 11.3 Pol-ISAR image representation.

Figure 11.4 Target composed of elementary scatterers.

Composite Pol-ISAR images can be obtained by fusing the polarimetric information. To understand the meaning of this operation, the concept of polarimetric space must first be introduced.

11.4.1 Polarimetric Space

Each pixel of a Pol-ISAR image is represented by a complex-valued vector that belongs to a vector space, namely, the polarimetric space. Conceptually, this means that any linear combination of two or more polarimetric vectors would result in another polarimetric vector. Given a vector space, an infinite number of bases exist. The definition of a polarimetric space basis is important, as the physical meaning of a polarimetric vector representation would strongly depend on it. Some bases are well suited to represent the polarimetric features of elementary scattering mechanisms; for example, trihedral- and

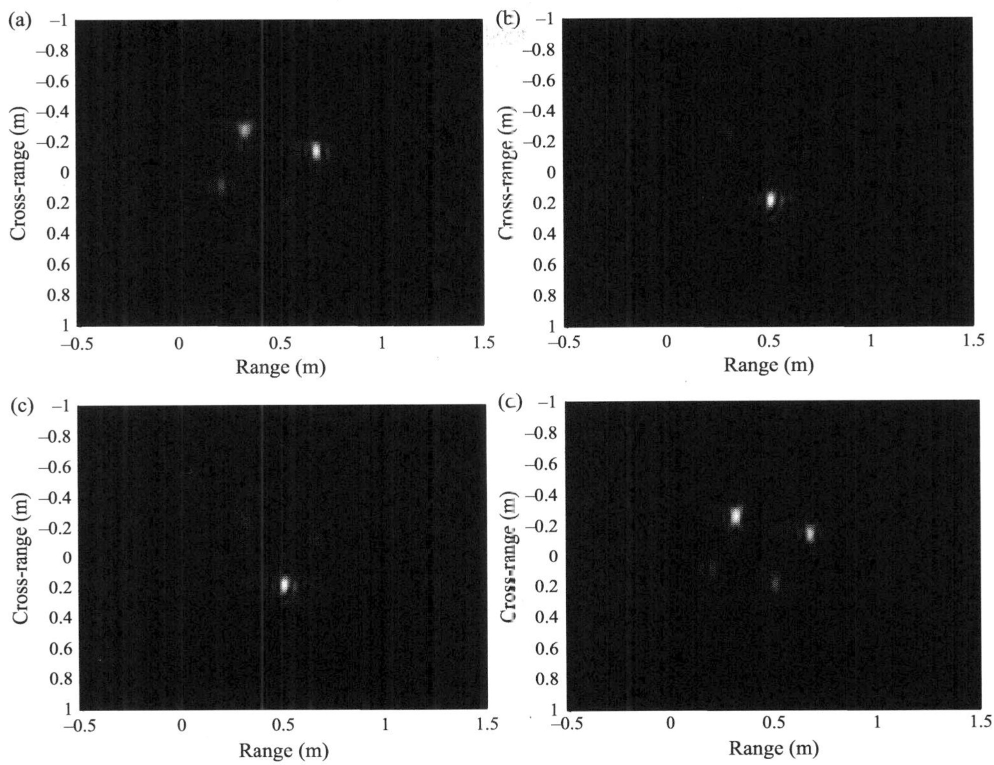

Figure 11.5 Pol-ISAR images of the target in Figure 11.4: (a) HH, (b) HV, (c) VH, and (d) VV.

dihedral-like scattering mechanisms are well represented by the Pauli's basis. As mentioned at the beginning of this chapter, several polarimetric decompositions may be appropriate for the classification of the scattering mechanism type. A comprehensive survey of such decompositions may be found in [21].

11.4.2 Color-Coded POL-ISAR Images

By using Pauli's decomposition, a Pol-ISAR image may be represented by means of three polarimetric channels: HH + VV, VV – HH, and 2HV.

$$I_C(\tau, \nu) = \left[I_C^{VV} + I_C^{HH}, \quad I_C^{VV} - I_C^{HH}, \quad 2I_C^{HV} \right]. \tag{11.22}$$

By assigning a different color to each channel, a color composite image can be obtained. Such a color composite image codes each polarimetric signature with a different color. Therefore, different scattering mechanisms are represented with different colors. In the example in Figure 11.6, the polarimetric channels are coded with the following rule: HH + VV = red, VV – HH = blue, and 2HV = green.

When a polarimetric basis such as that in (11.21) is used, some scattering mechanisms, such as trihedral and dihedral, are clearly distinguishable by looking at the pixel color.

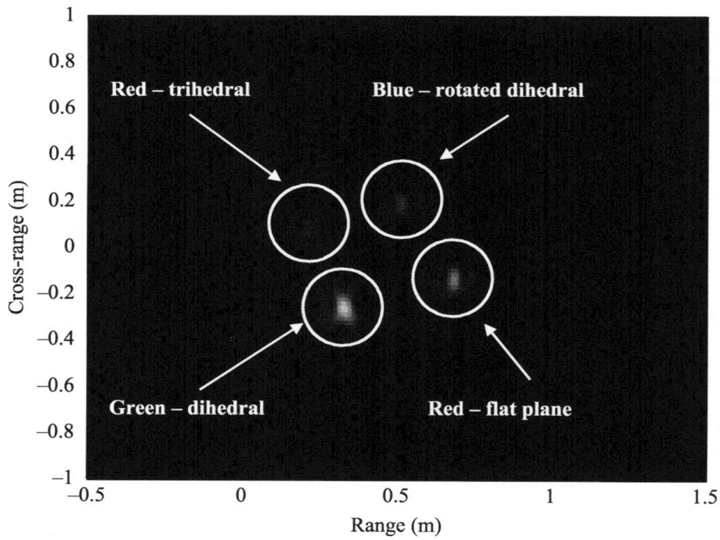

Figure 11.6 RGB Pol-ISAR image of the target in Figure 11.4.

Specifically, red pixels correspond to odd-bounce scattering mechanisms, such as flat surfaces and trihedrals, whereas blue pixels correspond to vertical and horizontally structured even-bounce scattering mechanisms, such as horizontal and vertical dihedrals. A green color would appear in the case of a 45° rotated dihedral, as in the example provided in Figure 11.6.

References

[1] Cloude, S.R., and E. Pottier, "An Entropy Based Classification Scheme for Land Applications of Polarimetric SAR," *IEEE Trans on Geoscience and Remote Sensing*, vol. 35, no.1, pp. 68–78, 1997.

[2] Shi, J., and J. Dozier, "Inferring Snow Wetness Using C-Band Data from SIR-C's Polarimetric Synthetic Aperture Radar," *IEEE Trans on Geoscience and Remote Sensing*, vol. 33, no. 4, pp. 905–914, 1995.

[3] Migliaccio, M., A. Gambardella, and M. Tranfaglia, "SAR Polarimetry to Observe Oil Spills," *IEEE Trans on Geoscience and Remote Sensing*, vol. 45, no. 2, pp. 506–511, 2007.

[4] Cameron, W.L., N.N. Youssef, and L.K. Leung, "Simulated Polarimetric Signatures of Primitive Geometrical Shapes," *IEEE Trans on Geoscience and Remote Sensing*, vol. 34, no. 3, pp. 793–803, 1996.

[5] Touzi, R., and F. Charbonneau, "Characterization of Target Symmetric Scattering Using Polarimetric SARs," *IEEE Trans on Geoscience and Remote Sensing*, vol. 40, no. 11, pp. 2507–2516, 2002.

[6] Cameron, W.L., and H. Rais, "Conservative Polarimetric Scatterers and Their Role in Incorrect Extensions of the Cameron Decomposition," *IEEE Trans on Geoscience and Remote Sensing*, vol. 44, no. 12, pp. 3506–3516, 2006.

[7] Novak, L.M., M.B. Sechtin, and M.J. Cardullo, "Studies of Target Detection Algorithms that Use Polarimetric Radar Data," *IEEE Trans on Aerospace and Electronic Systems*, vol. 25, no. 2, pp.150–165, 1989.

[8] Cloude, S.R., and E. Pottier, "A Review of Target Decomposition Theorems in Radar Polarimetry," *IEEE Trans on Geoscience and Remote Sensing*, vol. 34, no. 2, pp. 498–518, 1996.

[9] Cloude, S.R., *Polarimetry: The Characterization of Polarization Effect in EM Scattering*, Ph.D. dissertation, University of Birmingham, UK, 1986.

[10] Cloude, S.R., "Target Decomposition Theorems in Radar Scattering," *IEEE Electron. Lett.*, vol. 21, no. 1, pp. 22–24, 1985.

[11] Touzi, R., W.M. Boerner, and E. Luneburg, "A Review of Polarimetry in the Context of Synthetic Aperture Radar: Concept and Information Extraction," *Canadian Journal of Remote Sensing*, vol. 30, no. 3, pp. 380–407, 2004.

[12] Cloude S.R., and K.P. Papathanassiou, "Polarimetric SAR Interferometry," *IEEE Trans on Geoscience and Remote Sensing*, vol. 36, no. 5, pp. 1551–1565, 1998.

[13] Giuli, D. "Polarization Diversity in Radars," *Proceedings of the IEEE*, vol. 74, no. 2, pp. 245–269, 1986.

[14] Huynen, J.R., *Phenomenological Theory of Radar Targets*, Ph.D. dissertation, Delft University of Technology, Drukkerij Bronder-Offset N. V., Rotterdam, 1970.

[15] Kennaugh, E.M., *Polarization Properties of Radar Reflection*, Ph.D. dissertation, Department of Electric Engineering, Ohio State University, Columbus, 1952.

[16] Krogager, E., "New Decomposition of the Radar Target Scattering Matrix," *Electronic Letters*, vol. 26, no. 18, pp. 1525–1526, 1990.

[17] Lagarias, J.C., J.A. Reeds, M.H. Wright, and P.E. Wright, "Convergence Properties of the Nelder-Mead Simplex Method in Low Dimensions," *SIAM Journal of Optimization*, vol. 9, pp. 112–147, 1998.

[18] Holland, J., *Adaptation in Natural and Artificial Systems*, University of Michigan Press, 1975.

[19] Michalewicz, Z., *Genetic Algorithms + Data Structures = Evolution Programs*, Springer, 1994.

[20] Martorella, M., F. Berizzi, and S. Bruscoli, "Use of Genetic Algorithms for Contrast Maximization and Entropy Minimization in ISAR Autofocusing," *Journal of Applied Signal Processing*, Special Issue on Inverse Synthetic Aperture Radar, vol. 2006, pp. 1–11, 2006.

[21] Paladini, R., L. Ferro Famil, E. Pottier, M. Martorella, F. Berizzi, and E. Dalle Mese, "Point Target Classification via Fast Lossless and Sufficient Ω-Ψ-Φ Invariant Decomposition of High-Resolution and Fully Polarimetric SAR/ISAR Data," *Proceedings of the IEEE*, vol. 101, no. 3, pp. 798–830, March 2013.

CHAPTER 12

Applications of ISAR Imaging

Inverse synthetic aperture radar (ISAR) imaging is typically useful when there is a need to classify, recognize, or identify a moving target of interest. In fact, an ISAR image highlights two-dimensional (2-D) geometric features of a target, which can provide indications of target's type, size, and other salient information. Such information can be then used for target classification, recognition, and identification.

In this chapter, we will provide five case studies on applications of ISAR imaging. Such case studies are chosen to diversify the type of application and to give additional indications about how to interpret ISAR images. Specifically, in the first case study (Section 12.1), we show an example of ground-based ISAR images of a noncooperative sailing ship. The ISAR techniques used to form the ISAR images are the range-Doppler (Chapter 3) and the ICBA (Chapter 5) techniques. Different time windows are chosen to show the effects of the coherent processing interval (CPI) length on the ISAR image. Since the radar is ground based, this represents a case where the ISAR image is formed by exploiting only the sea surface induced target's motions.

The second case study (Section 12.2) refers to a scenario where the radar is carried by an aircraft and the target is a ship at sea. In this case, both the target and the platform motions concur to the ISAR image formation. Also in this case, the range-Doppler and ICBA are used to form the ISAR image sequence.

The third case study (Section 12.3) concerns dual ground-based/satellite ISAR imaging of a noncooperative sailing ship. In this experiment two radars are used to image a sailing ship at the same time. Two techniques are here used to form the ISAR images: a refocus technique (see Chapter 8) to form the satellite ISAR image; and a range-Doppler with ICBA to form the ground-based ISAR image. In this case study, we show how a dual system is able to form ISAR images of the same target with different target's views and projections. This is a simple case of multiple-perspective ISAR images, which may be looked at as a way to improve a target's classification and recognition [1].

The fourth case study (Section 12.4) shows ISAR images of four aircrafts obtained by processing data acquired with a ground-base radar. This case study may be regarded as the classic ISAR imaging example. Four different airplanes have been selected to show how their characteristics are mapped onto ISAR images. Such characteristics can then be used by classifiers for recognition purposes.

The fifth case study (Section 12.5) concentrates on micro-Doppler feature extraction from ISAR images. The techniques applied to the real data are described in Chapter 7.

The applications of ISAR imaging are not restricted to those shown in this chapter.

12.1 Case Study 1: Ground-Based ISAR Images of a Noncooperative Sailing Ship

In the first case study, we consider a scenario with a ground-based radar looking out at a target on the sea surface (Figure 12.1). This case study aims at showing that ISAR images can be effectively produced when strong sea surface-induced target motions are present. The fast time-varying characteristics of sea surface-induced motions generate effective rotation vectors that vary very rapidly, and therefore the image projection plane (IPP) changes rapidly from one ISAR image frame to the next. In this study, we will describe the radar and target characteristics and then provide the results.

12.1.1 Radar and Target Description

We employ an X-band multipurpose radar, MECORT, designed to accomplish detection, tracking, and imaging and owned by the Council for Scientific and Industrial Research (CSIR) Defence, Peace, Safety and Security of South Africa. When operating in imaging mode, the radar is able to generate stepped frequency waveforms with wide bandwidth. In this case study, the radar generated a stepped-frequency waveform with a central frequency of about 9.6 GHz and a bandwidth of about 624 MHz. More details about the data acquisition are shown in Table 12.1.

The target is a dual mast sailing ship *Esperance*, and a photo is shown in Figure 12.2. During the data acquisition, the target was sailing in a circle, which generated a relatively steady yaw motion. Strong pitch and roll also generated side (profile) and front view components in the image of the sailing ship.

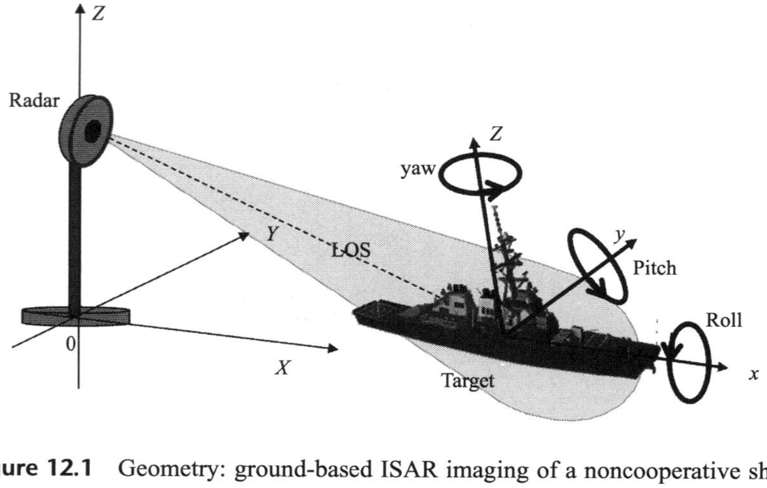

Figure 12.1 Geometry: ground-based ISAR imaging of a noncooperative ship.

Table 12.1 Radar Parameters: First Case Study

No. of sweeps	13416
No. of transmitted frequencies	86
Lowest frequency	9.82 GHz
Frequency step	8 MHz
Range resolution	0.24 m
Radar height	50 m
PRF/sweep rate	20 kHz/123.89 Hz
Data time length	108 sec

Figure 12.2 Sailing ship *Esperance*.

12.1.2 ISAR Images of Sailing Ship

Three sequences of ISAR images are shown in Figures 12.3 through 12.5. As predicted, the IPP changes significantly from frame to frame as the effective rotation vector changes rapidly due to the presence of highly varying pitch, roll, and yaw motions. We should

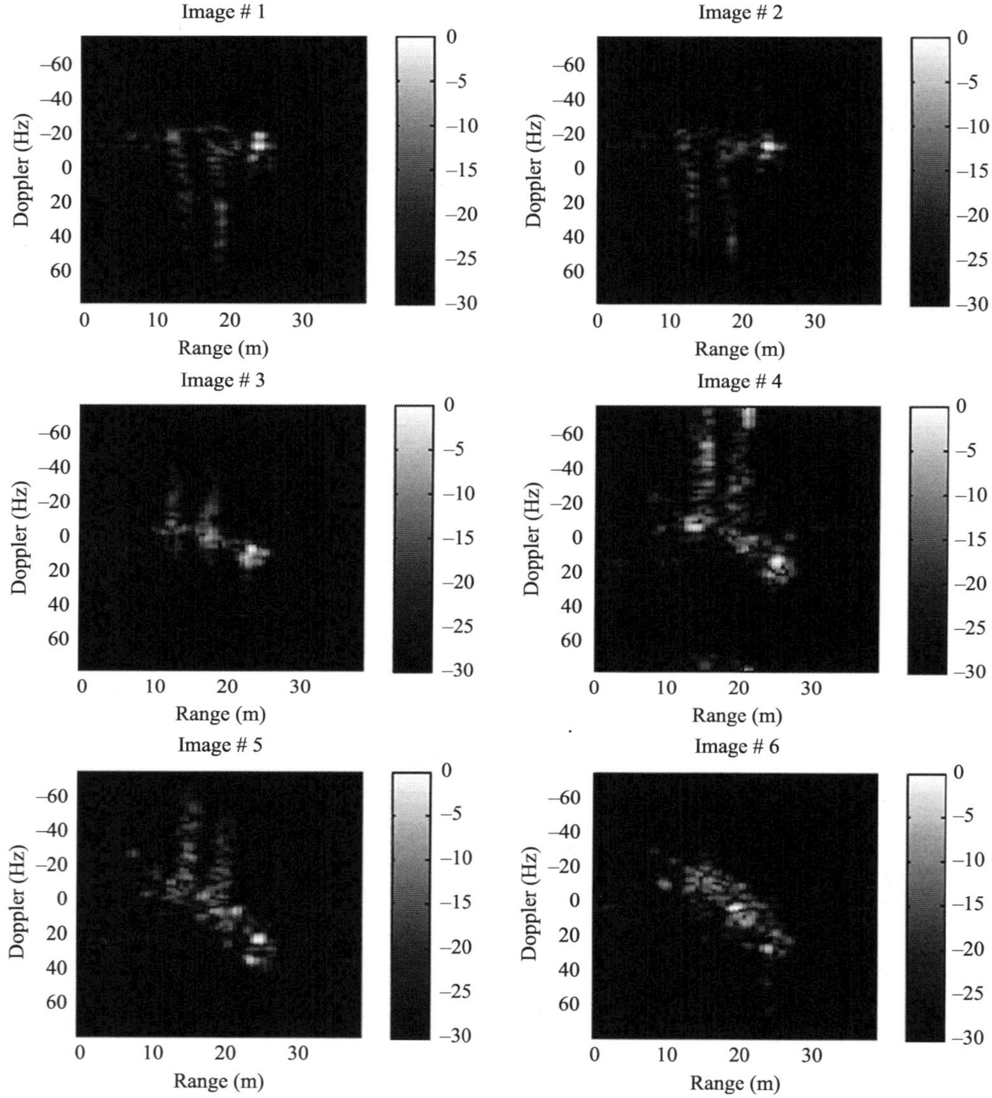

Figure 12.3 Sequence of ISAR images of a sailing ship. Data captured with a ground-based radar. CPI = 0.4 sec, frame-to-frame interval = 1 sec.

highlight here the importance of a correct interpretation of the ISAR image. Such an interpretation strongly depends on the way a target is projected onto the IPP. Straight projections occur when pure side, front, or top views are obtained. Such projections are obtained when the effective rotation vector is aligned with the three main axes of a target (e.g., the axes that define the length, width, and height of a target). All other projections may be regarded as *hybrid* (or composite) views as the effective rotation vector is not aligned to any of the axes.

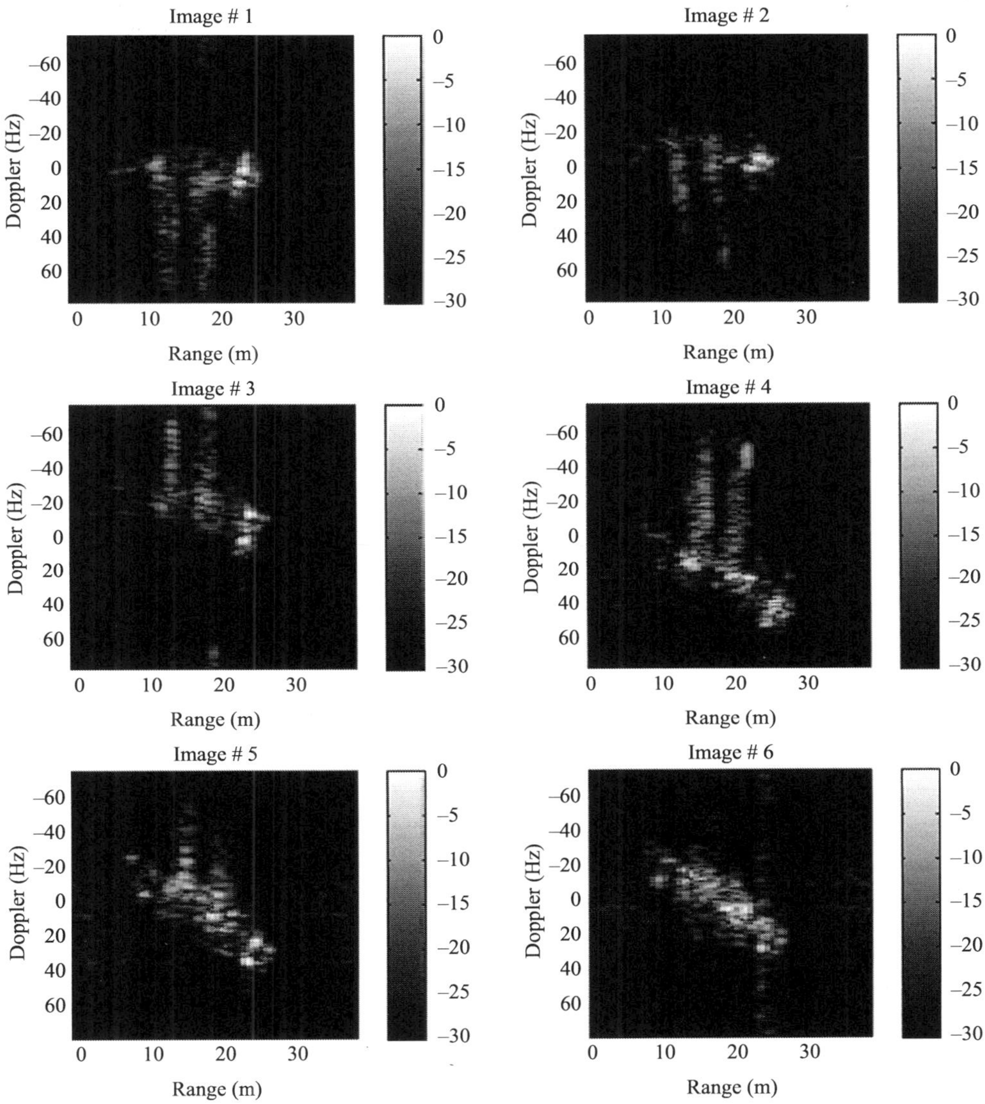

Figure 12.4 Sequence of ISAR images of a sailing ship. Data captured with a ground-based radar. CPI = 0.8 sec, frame-to-frame interval = 1 sec.

It is worth noting that hybrid views are more likely to appear as pure views need one of the three type of motions to be strongly dominant, such as yaw for a top (or plan) view.

Each ISAR image sequence is generated from the same data set, but they differ because of the CPI used to form the image (0.4 sec for the first sequence, 0.8 sec for the second one, and 1.6 sec for the third one). The different values of CPI produce sequences of ISAR images that may be more or less focused. In fact, because of the rapid changes in pitch, roll, and

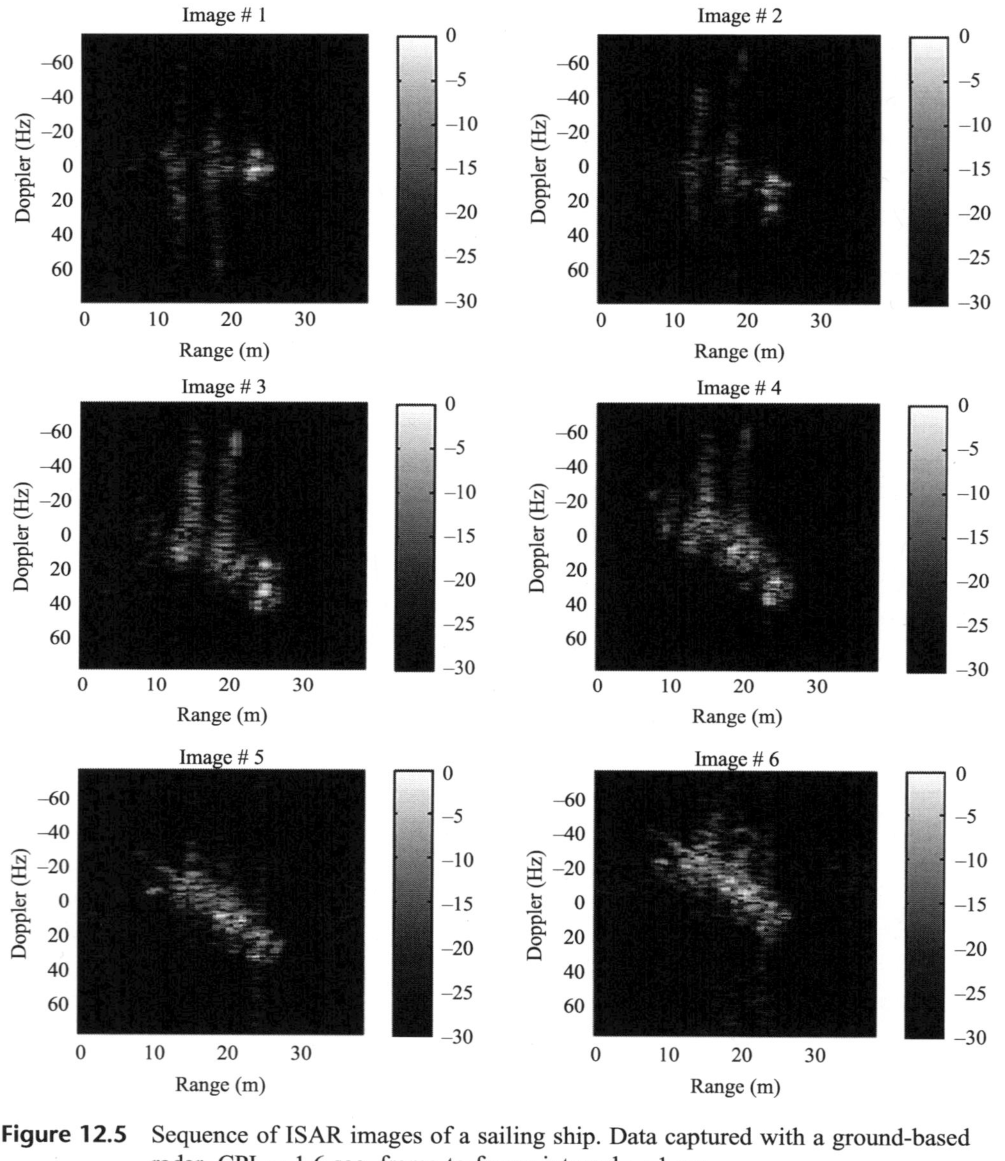

Figure 12.5 Sequence of ISAR images of a sailing ship. Data captured with a ground-based radar. CPI = 1.6 sec, frame-to-frame interval = 1 sec.

yaw, the Doppler components are not constant, originating blurred images. An evident example can be observed in Figure 12.4 at frames 1 and 2, where both positive and negative Doppler frequencies are visible that are produced by the oscillating masts. It is quite clear that within 1.6 seconds of integration the mast oscillation produces a change from positive to negative radial velocity, which may occur when the pitch oscillating period is comparable to or greater than the CPI. It should also be noticed that a short integration time allows for more focused ISAR image to be produced. This is because the likelihood that the effective rotation vector is constant during the integration time is higher than for longer intervals. This effect is clear if we compare the set of ISAR images in Figure 12.3 with those in Figure 12.5.

The application of a time window technique is essential to obtain well-focused ISAR images in cases of oscillating targets. Techniques like the one described in Chapter 6 are strongly recommended specifically when dealing with ISAR imaging of small ships in high sea state.

It is also worth noting that by selecting short CPIs, more focused ISAR images are produced. Nevertheless, they are not necessarily the best images that can be produced since the resolution, in some cases, may be improved by selecting a longer CPI. As pointed out in Chapter 6, the selection of the optimal time window is a delicate task that aims at producing the best resolution combined with the production of a well-focused ISAR image.

12.2 Case Study 2: Airborne ISAR Imaging of a Noncooperative Cargo Ship

The scenario that we are considering in this case study is that of a noncooperative ship imaged by a high-resolution airborne radar. Such a scenario is usually referred to as a hybrid SAR/ISAR system as both the platform and ship motion contribute to the total synthetic aperture formation. As the contribution produced by the ship is unknown and unpredictable, the total synthetic aperture is unknown. This detail makes the use of direct SAR processing nonapplicable to this case. Therefore, ISAR imaging is to be applied to form well-focused radar images. A pictorial view of this scenario is shown in Figure 12.6.

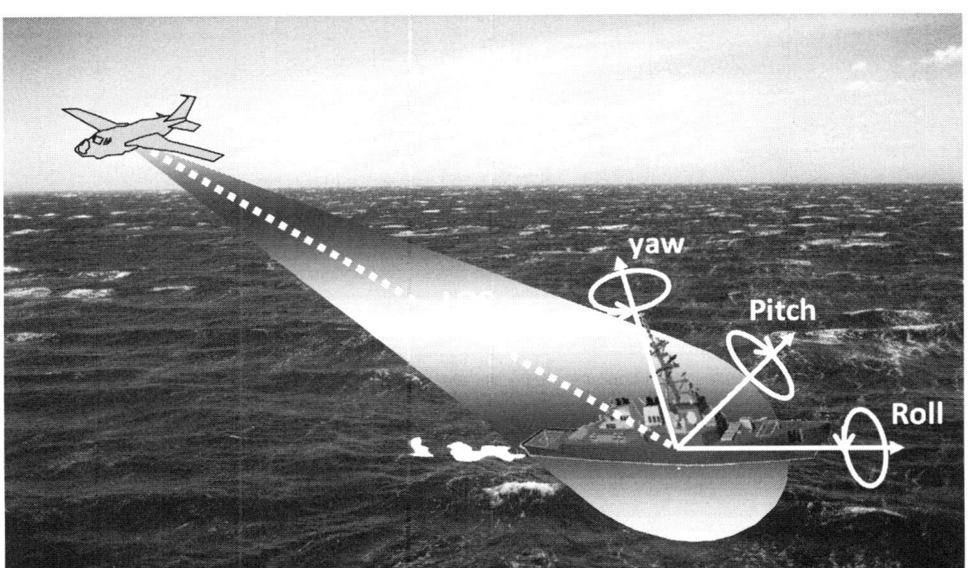

Figure 12.6 Geometry: airborne ISAR imaging of a noncooperative ship.

Table 12.2 Radar Parameters

No. of sweeps	256
No. of transmitted frequencies	256
Lowest frequency	9.16 GHz
Frequency step	0.6 MHz
Range resolution	0.97 m
Radar height	305 m
PRF/sweep rate	20 kHz/78.13 Hz
Data time length	65 sec

Figure 12.7 Target—bulk loader.

The high-resolution radar parameters are shown in Table 12.2, whereas a picture of the target is shown in Figure 12.7.

Since both the radar platform and the ship motion contribute to the formation of the synthetic aperture, the sequence of ISAR images produced in Figure 12.8 shows a change in the IPP, which is due to the ship's motions. Such changes are quite strong, so we may conclude that the ship's motions are dominant with respect to the platform-induced motions. Nevertheless, the rotation vector component generated by the radar platform motion is large and steady enough to maintain the IPP slightly steadier than in the case where only target's motion are present (see Case Study 1).

It can be concluded that the platform motion provides a constant motion that may be driving the synthetic aperture formation when the target's own motions are small. In general, this characteristic increases the probability of obtaining a well-focused and high-resolution

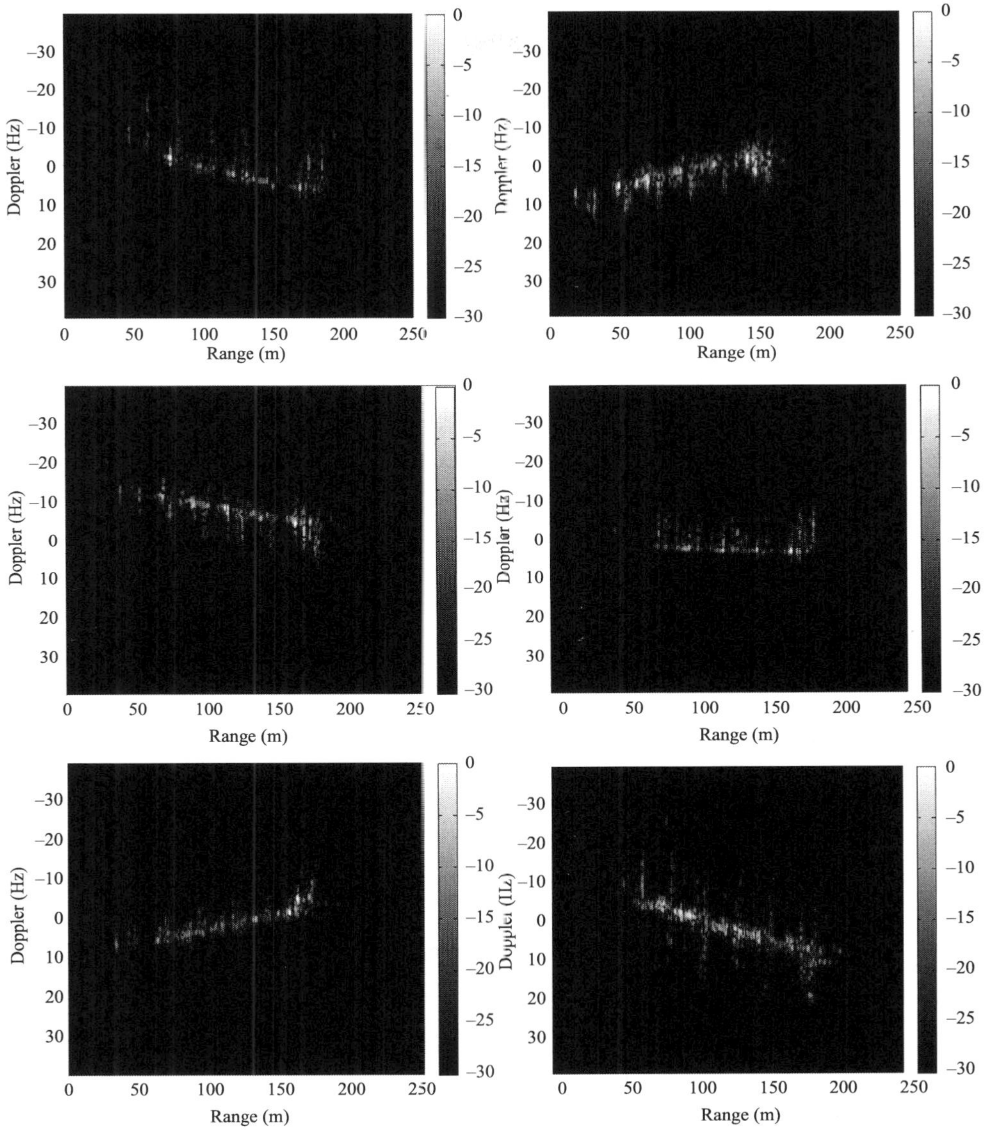

Figure 12.8 ISAR image sequence—bulk loader.

ISAR image also in cases where the ship's motions are small (either in calm sea conditions or in the case of large ships with calm to moderate sea conditions). However, the platform motion cannot be considered the only source of the aspect angle variation and, therefore, the only synthetic aperture driver. As a consequence, ISAR algorithms are to be preferred to SAR algorithms as they are designed to account for target's own motions (translational and rotational) and prove to be effective in those scenarios.

12.3 Case Study 3: Dual Ground-Based/Satellite ISAR Imaging of a Noncooperative Sailing Ship

In this case study, we will focus on the dependence of the image projection plane on the position of the sensor. To appreciate the differences due to the IPP, a scenario with one target and two ISAR sensors is considered. Specifically, the MECORT system is considered as a ground-based radar, and the Cosmo-Skymed (CSK) SAR system is considered as a moving platform. While the first sensor has a very low grazing angle (less than 10°), the second one has a grazing angle of about 49°. The two systems illuminated the target at X-band. The two radars transmitted two frequency separated waveforms, thus guaranteeing the orthogonality of the two transmitted signals. A picture of the sailing ship is shown in Figure 12.9.

The position of the ground-based radar and the target is shown in Figure 12.10. The SAR image formed by the CSK SAR system is shown in Figure 12.11, which also highlights the target position. Although not perfectly superposed, the data were acquired at the same time. Therefore, we can consider the sea surface induced motions to be practically equivalent in both the CSK and MECORT ISAR data acquisitions.

The SAR image of the target shown in Figure 12.12 appears defocused because it underwent a pitch motion during the data acquisition time. This is highlighted by the fact that the ship's mast produces both positive and negative Doppler frequencies in the image.

Figure 12.9 Target—sailing ship.

Figure 12.10 Positions of the MECORT radar and the sailing yacht.

As pointed out in Chapter 8, a refocus technique can be applied to the SAR image of the target to obtain a well-focused image of the moving target. After applying such a technique, an ISAR image of the same target is produced and shown in Figure 12.13. This image was obtained using the time windowing technique detailed in Chapter 6, and the target's bow is not visible since it was probably shaded by the ship's superstructure.

The Mecort ISAR image is shown in Figure 12.14. The ground-based radar image shows a quite different IPP. Also, due to the target's view angle, the bow is visible in the ground-based ISAR image, whereas it is not in the CSK image.

Multiperspective ISAR images of a target can provide more information about a target that can be used for classification and recognition purposes.

12.4 Case Study 4: Ground-Based ISAR Imaging of Airplanes

In this case study we will be showing ISAR images of aircrafts and discuss their characteristics. Also the differences between ISAR imaging of ships and airplanes will be highlighted.

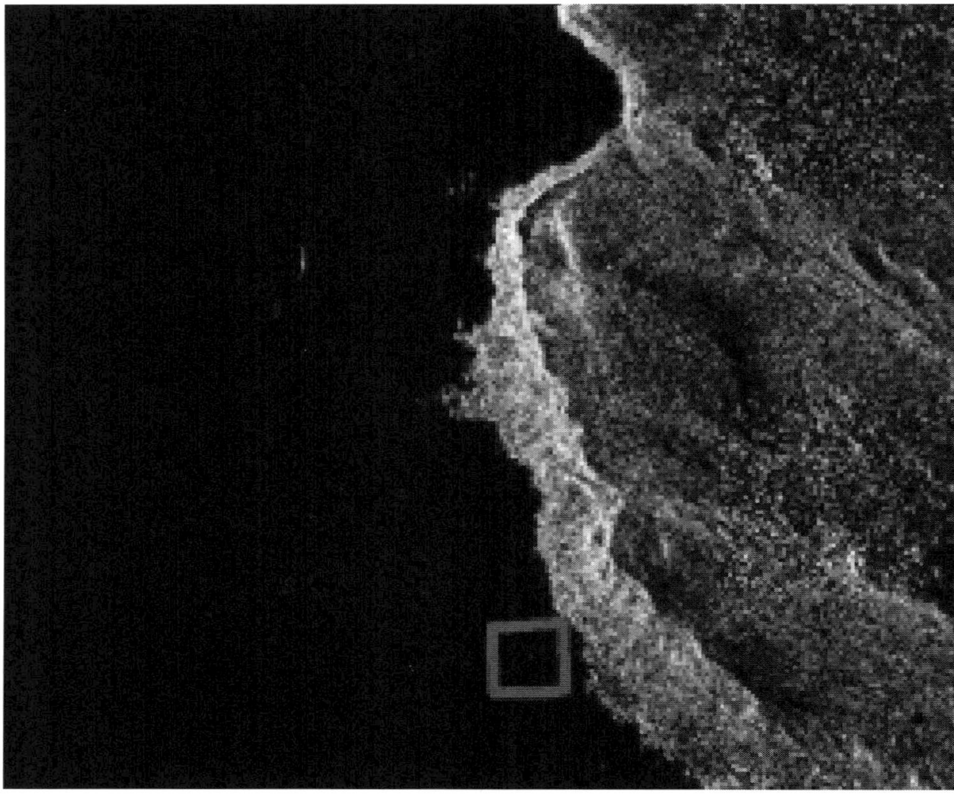

Figure 12.11 SAR image of the area with the sailing yacht position highlighted.

Airplanes generally cruise along rectilinear trajectories. This characteristic provides quite good conditions for ISAR imaging when the path orientation is such that a target aspect angle variation is created during the CPI. Also, rectilinear trajectories produce top-view images of airplanes, therefore enabling relatively easy target classification based on the image geometrical features. When an airplane maneuvers, instead, different IPP may be created that are typically unpredictable.

The case study that we are considering is that of a ground-based radar looking at civilian airplanes either taking off or landing near the Adelaide airport in Australia. The radar was positioned in such a way that the airplanes were flying roughly 90° off the radar line of sight (LOS). Under such conditions, top-view images are produced. In some of the images, masking effects are often present, such as in the case of the fuselage masking one of the wings. Some of the parameters of the radar used to collect the data are shown in Table 12.3.

Four different airplanes are considered in this case study: a Boeing 737, a Boeing 727, a Boeing 747, and a BAE 146. The four ISAR images of the four targets are presented in Figures 12.15 through 12.18, respectively. The ISAR images have been formed by using the ICBA as autofocus algorithm and the RD technique as image formation. Different time

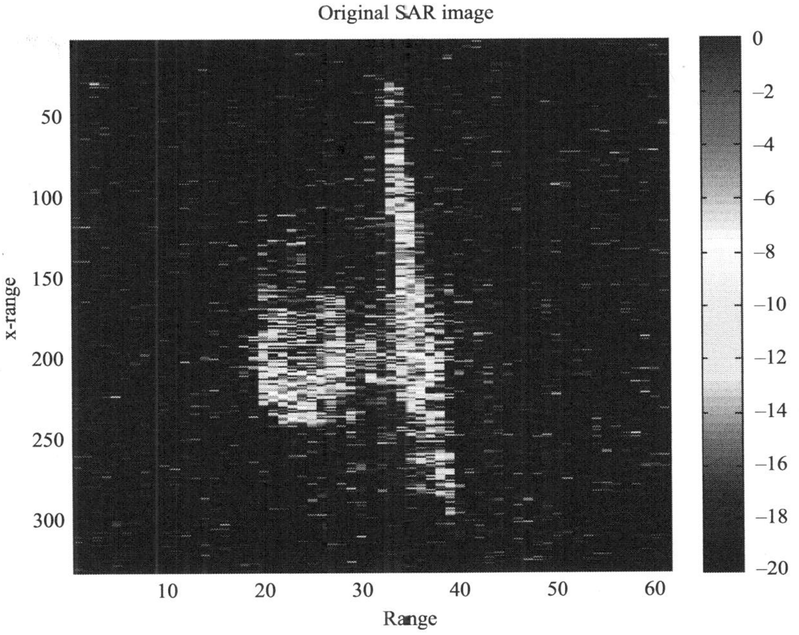

Figure 12.12 Zoomed-in SAR image of the sailing yacht. The image is unfocused due to the ship's pitch motion.

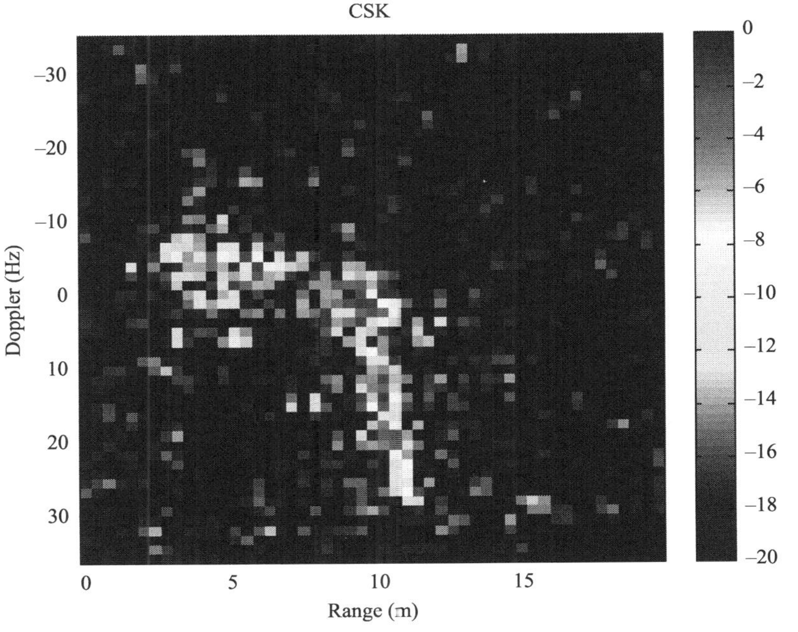

Figure 12.13 CSK refocused ISAR image.

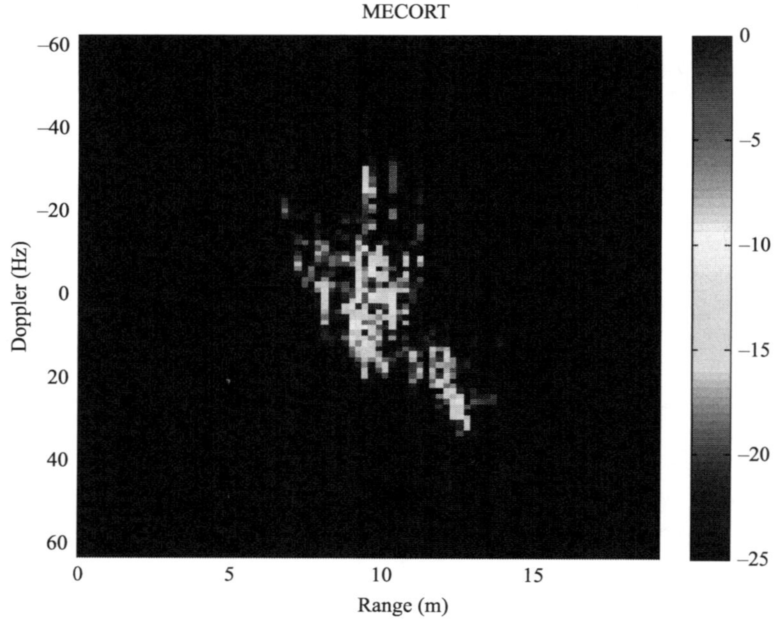

Figure 12.14 MECORT ISAR image.

Table 12.3 Radar Parameters

No. of transmitted frequencies	128
Lowest frequency	9.16 GHz
Frequency step	1.5–2 MHz
Range resolution	0.6–0.8 m
Radar height	0 m
PRF/sweep rate	20 kHz/156.25 Hz
Data time length	3.27 sec

windows have been used for the formation of the four ISAR images. The CPI used for the four images are shown in Table 12.4.

Geometrical features such as the size, the position of the wings with respect to the fuselage, and the position of the motors can be used for classifying or recognizing the target.

12.5 Case Study 5: Extraction of Doppler Features from ISAR Data of Small Vessels in Sea Clutter

In some cases ISAR images of small sea vessels immersed in strong clutter may not be effective. This could be due to the lack of resolution from the presence of a strong sea clutter, especially nonstationary clutter.

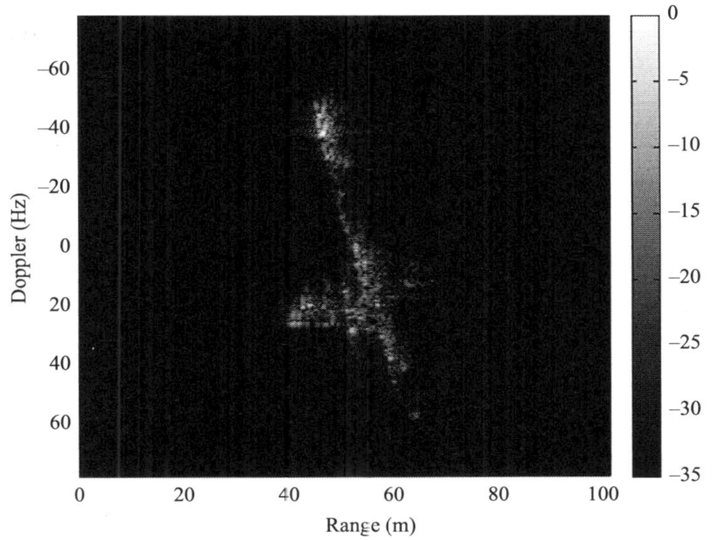

Figure 12.15 ISAR image of the Boeing 737.

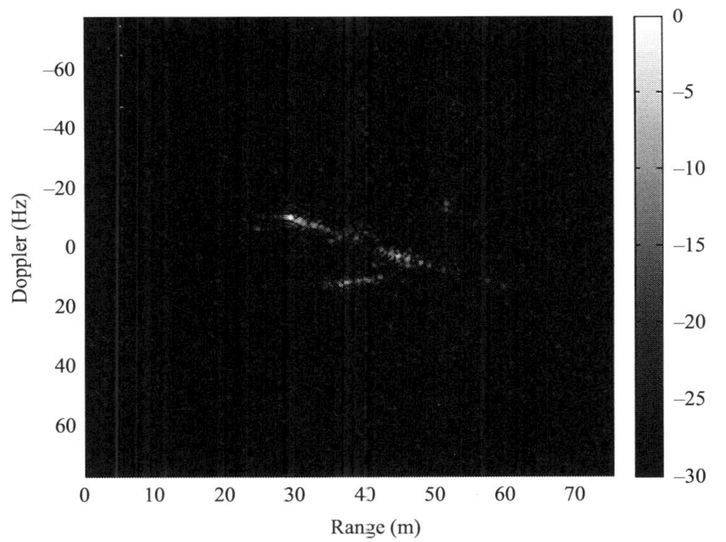

Figure 12.16 ISAR image of the Boeing 727.

In these cases, micro-Doppler analysis may provide a significant aid to extract target's feature that can be used for its classification or recognition. Moreover, micro-Doppler analysis could be used to separate the target from the sea clutter in a more effective way than in the ISAR image domain.

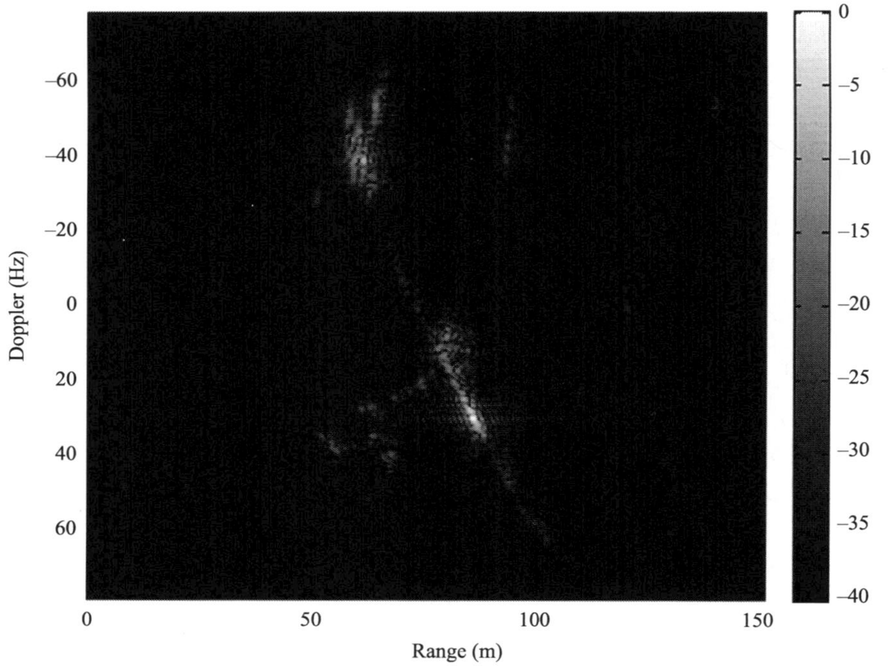

Figure 12.17 ISAR image of the Boeing 747.

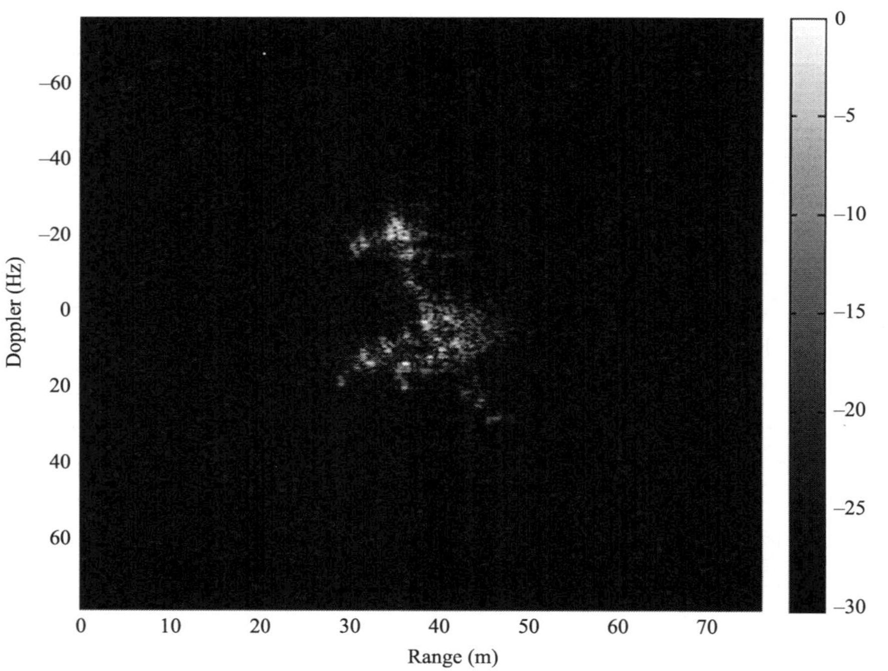

Figure 12.18 ISAR image of the BAE 146.

TABLE 12.4 Coherent Processing Interval

Target	CPI
Boeing 737	1.64
Boeing 727	0.82
Boeing 747	0.82
BAE 146	1.64

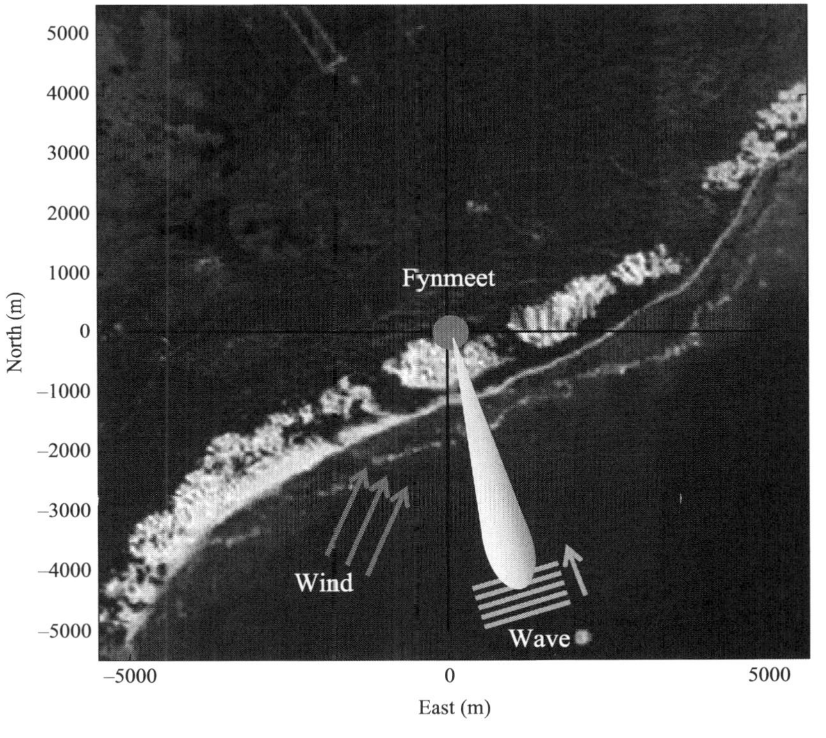

Figure 12.19 X-band radar deployed in CSIR Fynmeet measurement trial. [2]

12.5.1 Rigid Inflatable Boat and Large Birds in Sea Clutter

In this case study, we use CSIR data of small vessels immersed in strong sea clutter [2,3]. The radar system used in these CSIR's measurement trials is an X-band radar called Fynmeet, with lower range resolution (15 m) and located near the coast (Figure 12.19).

The specifications of the Fynmeet system used to collect the data are shown in Table 12.5.

Since the range resolution used in the released database is low (15 m), the vessel, a Class 3–5.5 m long rigid inflatable boat (RIB) (Figure 12.20a), occupies only less than half range cell. The intensity range profiles, represented by dBm^2, are shown in Figure 12.20b, where the higher radar cross section (RCS) of the RIB with respect to the surrounding clutter can be seen during the entire 50 sec time interval. The RIB was moving away from the radar with a speed of 2–3 m/s. The range-Doppler image of the vessel and the sea clutter are shown in Figure 12.21. From the range-Doppler map, it is difficult to distinguish the vessel's features

TABLE 12.5 Fynmeet System Specifications [2]

Frequency range	6.5–17.5 GHz
Peak power	2 kW
PRF range	0–30 kHz
Waveforms	Pulsed CW/pulse-to-pulse agile
Antenna gain	≥30 dB
Antenna beamwidth	≤2°
Capture range	200 m–15 km
Range resolution	15 m or 45 m

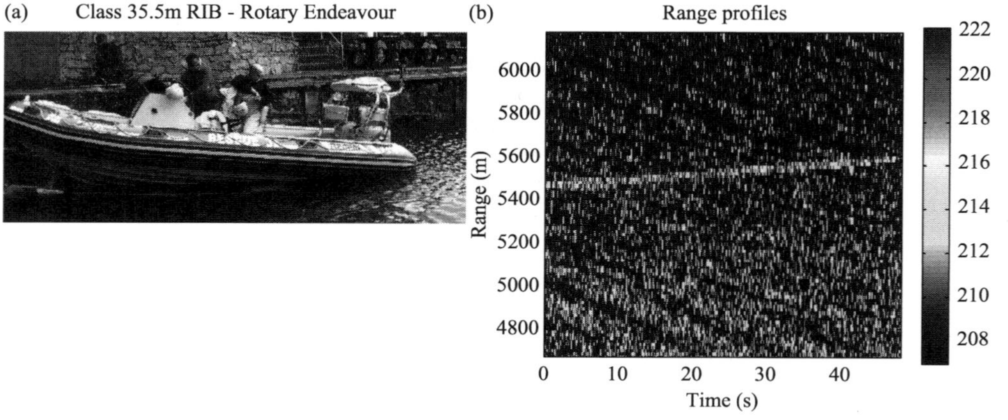

Figure 12.20 (a) *Rotary Endeavour*, Class 35.5 m RIB with two 60 hp Yamaha outboard motors and a single very high frequency (VHF) antenna; (b) range profiles of sea clutter with a moving RIB.

mainly because the sea state is quite high. To extract the vessel's features at low range resolution, a useful analysis method is to analyze Doppler frequency features instead of range domain features. Because longer integration time is possible, high Doppler resolution (1 Hz or less) can be easily reached. Therefore, time-varying Doppler features can be used to detect the small vessel in the presence of strong sea clutter and to track them in coastal areas.

Small targets such as the RIB, sea clutter, and possibly other moving objects like flying birds typically have different spatial or temporal characteristics. This allows for small vessels to be detected and tracked in the joint range and slow-time (pulses) domain, the joint range-Doppler domain, or the joint Doppler and slow-time domain.

In the available data collection, a number of large birds (seagulls) swept through the illuminated area producing radar echoes that were recorded in the same data set.

The time-varying Doppler spectrum of the sea clutter and the RIB are shown in Figure 12.22. The sea clutter Doppler shift is positive, whereas the micro-Doppler signature of the RIB is negative since the clutter and RIB were moving in opposite radial direction with respect to the radar LOS. In the spectrogram, the micro-Doppler signature of flying birds was also captured as indicated in Figure 12.22.

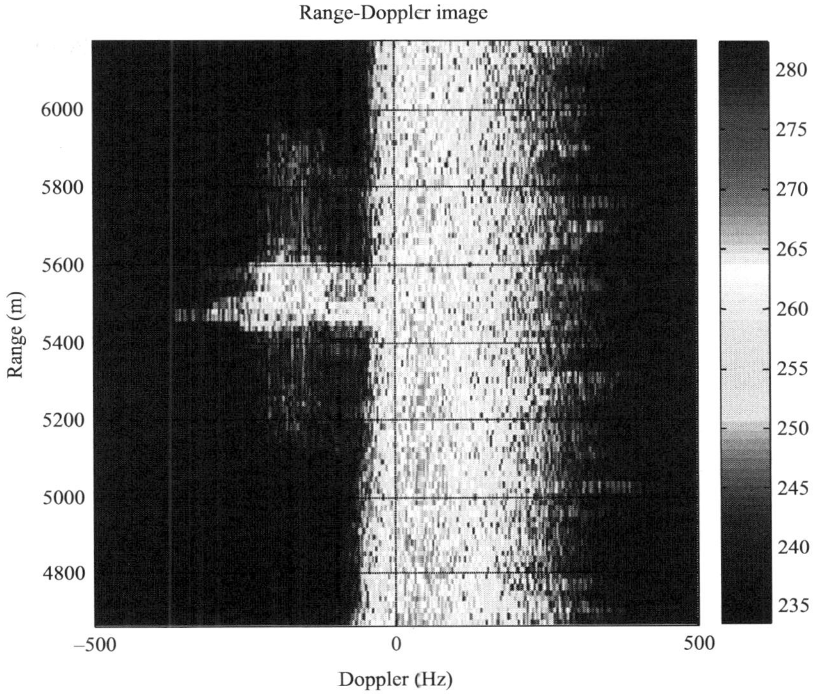

Figure 12.21 Range-Doppler image of the RIB and sea clutter.

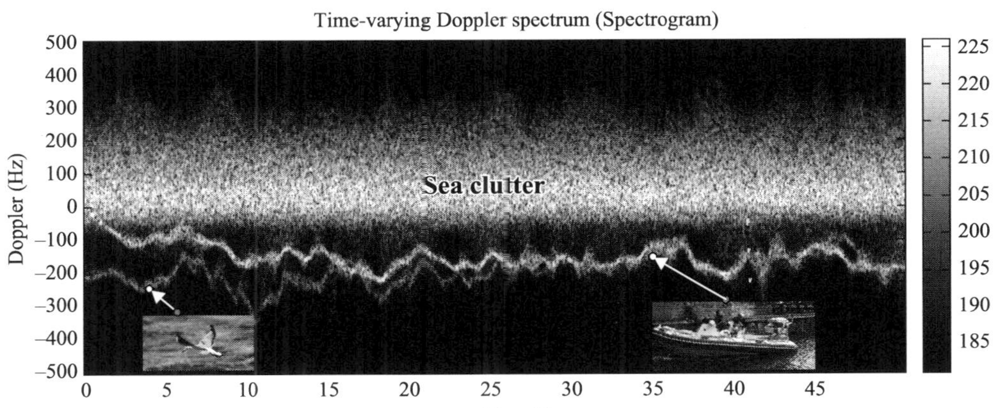

Figure 12.22 The time-varying Doppler spectrum of the RIB, sea clutter, and flying birds can be seen.

As clearly visible in Figure 12.22, time-varying Doppler spectrum proves to be an effective tool to separate the three contributions: the sea clutter, the RIB, and the birds.

12.5.2 A Floating Kayak Boat in Sea Clutter

In this data collection, a kayak floated on the sea surface. The intensity range profiles, represented by dBm^2, are shown in Figure 12.23a, where a total of 11 range bins were recorded for about 40 sec. The kayak was stationed in one to two range bins for the entire data acquisition. The range-Doppler image plot of the sea clutter, and the kayak is shown in Figure 12.23b. The Doppler shifts of the sea clutter and the boat are around zero Doppler. Micro-Doppler signature of the double-bladed paddle stroking is also captured and shown in Figure 12.24.

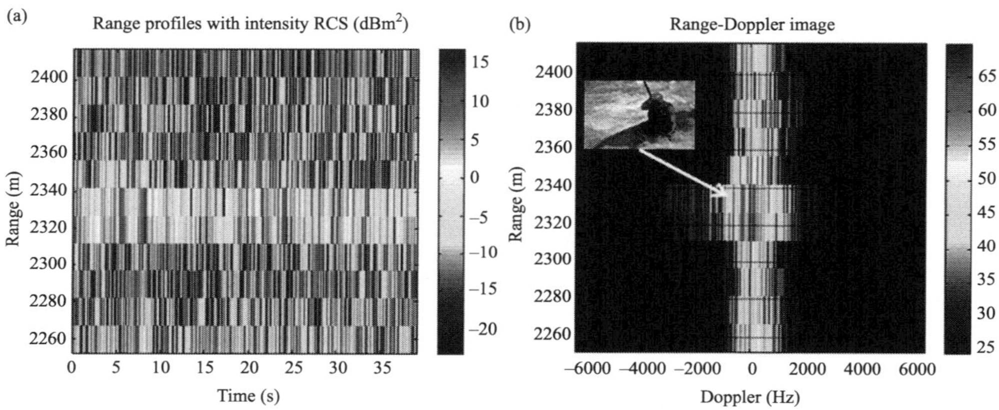

Figure 12.23 Range profiles and range-Doppler plot of the floating kayak in sea clutter collected in the Signal Hill trial 06-029-TTrSA.

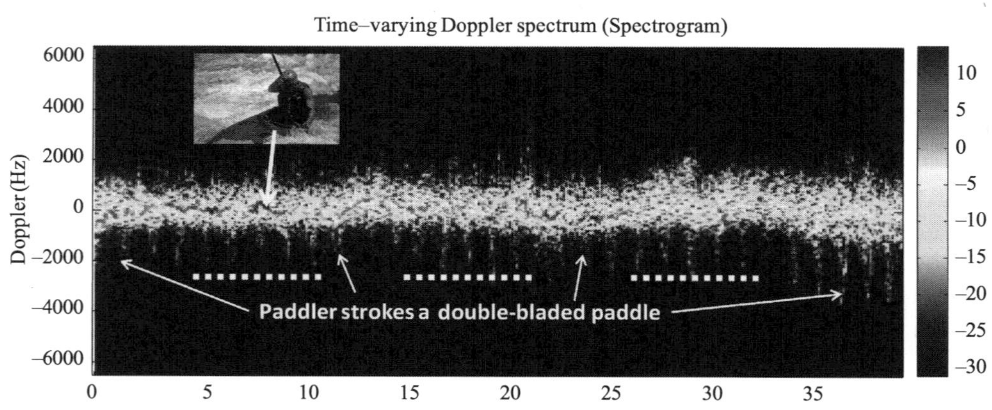

Figure 12.24 Time-varying Doppler spectrum of the Signal Hill trial 06-029-TTrSA, where the micro-Doppler signature of the double-bladed paddle stroking can be seen.

References

[1] Vespe, M., "Multi-Perspective Radar Target Classification," Ph.D. Thesis, Electronic & Electrical Engineering Department, UCL, May 2007.

[2] Herselman, P., C. Baker, and H.J. de Wind, "An Analysis of X-Band Calibrated Sea Clutter and Small Boat Reflectivity at Medium-to-Low Grazing Angles," *International Journal of Navigation and ObservationVolume 2008 (2008), Article ID 347518, 14 pages, http://dx.doi.org/10.1155/2008/347518*

[3] Herselman, P.L. and C.J., Baker, "Analaysis of calibrated sea clutter and boat reflectivity data at C- and X-band in South African coastal waters," *Radar Systems, 2007 IET International Conference on*, pp.1,5, 15-18 Oct. 2007.

Index